Insight Into Images

Insight Into Images

Principles and Practice for Segmentation, Registration, and Image Analysis

Terry S. Yoo, Editor

National Library of Medicine, NIH
Office of High Performance Computing
and Communications

A K Peters
Wellesley, Massachusetts

Editorial, Sales, and Customer Service Office

A K Peters, Ltd.
888 Worcester Street, Suite 230
Wellesey, MA 02482
www.akpeters.com

Library of Congress Cataloging-in-Publication Data

Insight into images : principles and practice for segmentation, registration, and image analysis / Terry S. Yoo, editor.
    p. cm.
  Includes bibliographical references and index.
  ISBN 1-56881-217-5
    1. Diagnostic imaging–Digital techniques. 2. Image processing.
    I. Yoo, Terry S., 1963-

RC78.7.D53I567 2004
616.07'54–dc22

                                                        2003068974

Printed in Canada
08 07 06 05 04   10 9 8 7 6 5 4 3 2 1

To my parents, T.S.Y.

To my wife Geanne, Baby Ada and my parents, D.N.M.

To Konrad, Marcin and Tomasz C.I.

To Stephen Pizer, G.D.S.

To the ITK team at the lab, J.C.G.

To my family and parents, J.K.

To Geoff, L.N.

For Dan and Marlene, R.T.W

To my beloved wife and my parents, Yinpeng Jin

To Daniel and Camilo, L.I.

To the friendships formed while writing ITK, S.R.A.

To my grandmother, P.K.S.

To Rebecca and her big heart, B.B.A.

To lab friends old and new, T.A.S.

To my daughter, Y.Z.

To my parents and friends in the ITK group, T.C.

The purpose of computing is Insight, not numbers.

–Richard Hamming

Contents

Foreword   xi

I Introduction and Basics   1

1 Introduction   3
  1.1   Medical Image Processing 4
  1.2   A Brief Retrospective on 3D Medical Imaging 5
  1.3   Medical Imaging Technology 6
  1.4   Acquisition, Analysis, and Visualization 16
  1.5   Summary 17

2 Basic Image Processing and Linear Operators   19
  2.1   Introduction 19
  2.2   Images 20
  2.3   Point Operators 21
  2.4   Linear Filtering 25
  2.5   The Fourier Transform 37
  2.6   Summary 45

3 Statistics of Pattern Recognition   47
  3.1   Introduction 47
  3.2   Background 49
  3.3   Quantitative Comparison of Classifiers 56
  3.4   Classification Systems 59
  3.5   Summary of Classifiers' Performance 84
  3.6   Goodness-of-Fit 87

3.7   Conclusion   92
3.8   Appendix: Extruded Gaussian Distributions   93

4   Nonlinear Image Filtering with Partial Differential Equations   103
4.1   Introduction 103
4.2   Gaussian Blurring and the Heat Equation 103
4.3   Numerical Implementations 110

II   Segmentation   119

5   Segmentation Basics   121
5.1   Introduction 121
5.2   Statistical Pattern Recognition 123
5.3   Region Growing 124
5.4   Active Surfaces/Front Evolution 126
5.5   Combining Segmentation Techniques 127
5.6   Looking Ahead 128

6   Fuzzy Connectedness   131
6.1   Background 131
6.2   Outline of the Chapter 133
6.3   Basic Notations and Definitions 136
6.4   Theory 138
6.5   Methods and Algorithms 151
6.6   Applications 161
6.7   Concluding Remarks 171

7   Markov Random Field Models   181
7.1   Markov Random Field Models: Introduction and Previous Work . 181
7.2   Gibbs Prior Model Theories 182
7.3   Bayesian Framework and Posterior Energy Function 185
7.4   Energy Minimization 186
7.5   Experiments and Results 186

8   Isosurfaces and Level Sets   193
8.1   Introduction 193
8.2   Deformable Surfaces 195
8.3   Numerical Methods 199
8.4   Applications 208
8.5   Summary 214

9 Deformable Models   219

9.1   Introduction 219
9.2   Previous Work 220
9.3   Deformable Model Theories 222
9.4   Experiments, Results, and Applications in ITK 230

III Registration   237

10 Medical Image Registration: Concepts and Implementation   239

10.1 Introduction 239
10.2 Image Registration Concepts 240
10.3 A Generic Software Framework for Image to Image Registration . 242
10.4 Examples 295

11 Non-Rigid Image Registration   307

11.1 Introduction 307
11.2 Optical Flow: Fast Mono-Modality Non-Rigid Registration . . . 312
11.3 Variational Framework for Computational Anatomy 321
11.4 Review 338

IV Hybrid Methods - Mixed Approaches to Segmentation   349

12 Hybrid Segmentation Methods   351

12.1 Introduction 351
12.2 Review of Segmentation Methods 353
12.3 Hybrid Segmentation Engine 357
12.4 Hybrid Segmentation: Integration of FC, VD, and DM 358
12.5 Evaluation of Segmentation 365
12.6 Results 367
12.7 Conclusions 375

Index   389

Foreword

This book is the product of a extraordinary collaboration among commercial groups, researchers from computer science departments and engineering schools, and researchers and scientists from medical schools and radiology departments. The ultimate result has been the Insight Toolkit (or ITK). ITK is an open source library of software components for data segmentation and registration, provided through federal funding administered by the National Library of Medicine, NIH, and available in the public domain. ITK is designed to complement visualization and user interface systems to provide advanced algorithms for filtering, segmentation, and registration of volumetric data. Created originally to provide preprocessing and analysis tools for medical data, ITK is being used for image processing in a wide range of applications from handwriting recognition to robotic computer vision.

History and Philosophy of ITK

In August of 1991, the National Library of Medicine (NLM) began the acquisition of the Visible Human Project Male and Female (VHP) datasets. The VHP male dataset contains 1,971 digital axial anatomical images (15 Gbytes), and the VHP female dataset contains 5,189 digital images (39 Gbytes). However, researchers have complained that they are drowning in data, due in part to the sheer size of the image information and the available anatomical detail. Also, the imaging community continues to request data with finer resolution, a pressure that will only compound the existing problems posed by large data on analysis and visualization. The NLM and its partner institutes and agencies seek a broadly accepted, lasting response to the issues raised by the segmentation and registration of large 3D medical data.

The Insight Software Research Consortium was formed in 1999 with the mission to create public software tools in 3D segmentation and deformable and rigid registration, capable of analyzing the head-and-neck anatomy of the Visible Human Project data. The result has been the Insight ToolKit (ITK), a publicly available software package for high-dimensional (3D, 4D, and higher) data processing supported by the NLM and its partner institutes and federal agencies. At the time of this writing, the software is soon to be released as ITK 1.4, its third release incorporating software contributions from new contributors including Harvard Medical School, the Georgetown ISIS Center, and the Mayo Clinic. The most aggressive hope for this project has been the creation of a self-sustaining open-source software development community in medical image processing based on this foundation. By creating an open repository built from common principles, we hope to help focus and concentrate research in this field, reduce redundancy of development, and promote cross-institution communication. The long-term impact of this initiative may well reach beyond the initial mission of archiving computer algorithms. The team of contractors and subcontractors includes:

- GE Global Research

- Harvard BWH Surgical Planning Lab

- Kitware, Inc.

- Insightful, Inc.

- The University of Pennsylvania

- The University of North Carolina at Chapel Hill

- The University of Tennessee at Knoxville

- Columbia University

- The University of Pittsburgh

- The University of Utah

- Georgetown University

- The University of Iowa

- The Imperial College/Kings College London

- Cognitica, Inc.

- The Mayo Clinic

- Carnegie Mellon University Robotics Institute

- The University of Utah SCI Institute

ITK has met its primary goals for the creation of an archival vehicle age processing algorithms and an established functioning platform to acce new research efforts. New projects have connected ITK to existing software plications for medical and scientific supercomputing visualization such as "A alyze" from the Mayo Clinic and "SCIRun" from the University of Utah's SCI Institute. These projects were sponsored to demonstrate that the application programmer's interface (API) for ITK was suitable for supporting large applications in established medical and scientific research areas. Also, recent projects have been funded to add algorithms not previously funded to the collection, proving the versatility of the underlying software architecture to accommodate emerging ideas in medical imaging research.

There is growing evidence that ITK is beginning to influence the international research community. In addition to the sponsored growth of the ITK community, the mailing list of users and developers includes over 300 members in over 30 countries. These facts may be indicators that a nascent research community may be forming about the core ITK software. The applications for ITK have begun to diversify including work in liver RF ablation surgical planning, vascular segmentation, live-donor hepatic transplant analysis, intra-operative 3D registration for image-guided intervention, rapid analysis of neurological function in veterans, longitudinal studies of Alzheimer's patients using MRI, and even handwriting analysis. The Insight Program of the Visible Human Project is emerging as a clear success and a significant and permanent contribution to the field of medical imaging.

In 1999, the NLM Office of High Performance Computing and Communications, supported by an alliance of NIH ICs and Federal Funding Agencies, awarded six contracts for the formation of a software development consortium to create and develop an application programmer interface and first implementation of a segmentation and registration toolkit, subsequently named the Insight Toolkit (ITK). The final deliverable product of this group has been a functional collection of software components, compatible for direct insertion into the public domain via Internet access through the NLM or its licensed distributors. Ultimately, NLM hopes to sponsor the creation of a public segmentation and registration software toolkit as a foundation for future medical image understanding research. The intent is to amplify the investment being made through the Visible Human Project and future programs for medical image analysis by reducing the reinvention of basic algorithms. We are also hoping to empower young researchers and small research groups with the kernel of an image analysis system in the public domain.

Scope of This Book

While a complete treatise on medical image processing would be an invaluable contribution to the research community, it would be far too much hubris on our

...rt to attempt to describe it all. The concept of Finite Element Models (FEM) is a book-length treatise in itself, as are each of the topics of Level Set Methods and Statistical Pattern Recognition. Indeed, we reference many valuable texts on these topics in particular.

We therefore limit this material to methods that have been implemented in the Insight ToolKit. All methods described herein are reflected in the source code for ITK. We limit the perspectives in this text to the viewpoints presented in ITK, and though there are many valid and valuable alternate approaches incorporating many of these same ideas, we cannot treat them all, and in fairness, we claim expertise only in those areas that we can deomnstrably express directly through software development.

In keeping with the principles of the Insight project itself, we concentrate only on medical image filtering, segmentation, and registration. These were the primary areas of concentration of the software project, and thus they are the topics described in this book. As such, we do not directly treat the ares of visualization, identification, classification, or computer-aided detection and diagnosis. All of the methods presented in this book are relevant to these topics, and we cannot show our results without some form of visualization, but we do not include an in-depth treatise of these topics or explicitly describe the integration of ITK with visualization environments.

Suggested Uses for This Book

Throughout the development of this text, we have internally described it as the "theory book." We have avoided including source code in an attempt to focus on the principles of the methods employed in ITK. There is a companion text for ITK directed toward implementation and integration details, called the *ITK Software Guide*. These books should be considered complementary to each other and supplementary in support of ITK. If you, the reader, are familiar with the principles of some method or approach in ITK, you only need to read the *ITK Software Guide* to learn about our particular implementation of the method and our structure and intended uses for that method. If, however, you require greater background on the theory behind some method, you should refer to the material in this book. Thus, we perceive this book to be an extended reference for ITK, providing more in-depth presentation of the basic algorithms, structure, and methods in the toolkit than can be gleaned either from the source code or from the base level descriptions in the *ITK Software Guide*.

We also perceive this book to be the basis for reading and curriculum foundations for a graduate course on medical image processing. ITK has been effectively used to create courses on this topic. Its advanced implementations permit the rapid explorations of complex algorithms without requiring the student or the

instructors and their staff to create the software infrastructure neces
high-dimensional image processing software. For such a course, this bo
represent theoretical course material, and the ITK Software Guide would
invaluable resource for laboratory exercises and course projects.

We are extraordinarily proud of ITK and hope that you find our offering in
structive, valuable, and a useful supporting addition to your software projects. We
humbly and gratefully present this material with hope for your future success in
research and medical software application development.

Acknowledgements

Following the consortium-based nature of this project, this book is the collec-
tive effort of many, many people. The contributing authors include in alphabet-
ical order: Brian Avants, Stephen Aylward, Celina Imielinska, Jisung Kim, Bill
Lorensen, Dimitris Metaxas, Lydia Ng, Punam K Saha, George Stetten, Tessa
Sundaram, Jay Udupa, Ross Whitaker, and Terry Yoo. Other people have made
essential contributions, including: Luis Ibanez, Will Schroeder, and others. The
Insight ToolKit would not have been possible without the selfless efforts of Bill
Hoffman, Brad King, Josh Cates, and Peter Ratiu.

We are indebted to the coalition of federal funding partners who have made
this project possible: the National Library of Medicine, the National Institute
for Dental and Craniofacial Research, the National Institute for Mental Health,
the National Eye Institute, the National Institute on Neurological Disorders and
Stroke, the National Science Foundation, the DoD Telemedicine and Advanced
Technologies Research Center, the National Institute on Deafness and other Communication Disorders, and the National Cancer Institute. We would especially
like to recognize the patronage of the Visible Human Project from the Office of
High Performance Computing and Communications of the Lister Hill National
Center for Biomedical Communications of the National Library of Medicine. In
particular, we'd like to thank Drs. Donald Lindberg and Michael Ackerman for
their vision and leadership in sponsoring this project.

Part One

Introduction and Basics

Introduction

Terry S. Yoo
National Library of Medicine, NIH

This book is an introduction to the theory of modern medical image process-ing. Image processing is the manipulation and analysis of image-based digital data to enhance and illuminate information within the data stream. One com-mon approach to this topic developed by many texts is to treat the matter as a high-dimensional form of signal processing, concenterating on filtering and fre-quency analysis as the foundations with later chapters discussing applications and advanced algorithms. This type of organization reflects a bias toward electrical engineering, viewing images as similar to complicated audio or radiofrequency signals. Another approach is to start from the mathematical perspective, treat-ing the issue as a series of exercises in statistics or applied mathematics through equations and theorems. Our approach is to explore the topic through implemen-tations of advanced algorithms, providing the mathematical, statistical, or signal processing as needed for background.

This book takes a software engineer's viewpoint, and describes methods and algorithms available in the Insight Toolkit as open source software. The goal of this book is to explain the background specific to our implementations, enabling the reader or student to use complex tools in medical image processing as rapidly as possible. We do not claim an exhaustive treatment of these ideas, but rather a working and functional introduction to these topics, partnered with working implementations of the methods described.

1.1   Medical Image Processing

The medical mission differs from the other forms of image processing arising
from non-medical data. In satelite surveillance analysis, the purpose is largely a
screening and cartographic task, aligning multiple types of data and corresponding them to a known map and highlighting possible points of interest. In computer
vision, camera views must be analyzed accounting for the perspective geometry
and photogrammetric distortions associated with the optical systems that are the
basis for robotic sensors. In many of these systems, autonomous navigation, target identification and acquisition, and threat avoidance are the primary tasks. For
the most part, the incoming information arrives as 2D images, data arrays that
can be organized using two cartesian dimensions. In addition, the tasks are to
be performed independently by the machine, relying on the development of machine learning and artificial intelligence algorithms to automatically accomplish
the tasks.

In medicine, the problem as well as the input data stream are usually three-
dimensional, and the effort to solve the primary tasks is often a partnership of
human and machine. Medicine is notably a human enterprise, and computers are
merely assistants, not surrogates nor possible replacements for the human expert.
The medical task can often be split into three areas: (1) data operations of filtering,
noise removal, and contrast and feature enhancement, (2) detection of medical
conditions or events, and (3) quantitative analysis of the lesion or detected event.
Of these subtasks, detection of lesions or other pathologies is often a subjective
and qualitative decision, a type of process ill-suited for execution by a computer.
By contrast, the computer is vastly more capable of both quantitative measurent
of the medical condition (such as tumor volume or the length of a bone fracture)
and the preprocessing tasks of filtering, sharpening, and focusing image detail.
The natural partnership of humans and machines in medicine is to provide the
clinician with powerful tools for image analysis and measurement, while relying
on the magnificent capabilities of the human visual system to detect and screen
for the primary findings.

We divide the problems inherent in medical image processing into three basic
categories:

- **Filtering:** These are the basic tasks involved in filtering and preprocessing
  the data before detection and analysis are performed either by the machine
  or the human operator.

- **Segmentation:** This is the task of partitioning an image (2D array or volume) into contiguous regions with cohesive properties.

- **Registration:** This is the task of aligning multiple data streams or images,
  permitting the fusion of different information creating a more powerful diagnostic tool than any single image alone.

Part One

Introduction and Basics

Introduction

Terry S. Yoo
National Library of Medicine, NIH

This book is an introduction to the theory of modern medical image process-ing. Image processing is the manipulation and analysis of image-based digital data to enhance and illuminate information within the data stream. One com-mon approach to this topic developed by many texts is to treat the matter as a high-dimensional form of signal processing, concenterating on filtering and fre-quency analysis as the foundations with later chapters discussing applications and advanced algorithms. This type of organization reflects a bias toward electrical engineering, viewing images as similar to complicated audio or radiofrequency signals. Another approach is to start from the mathematical perspective, treat-ing the issue as a series of exercises in statistics or applied mathematics through equations and theorems. Our approach is to explore the topic through implemen-tations of advanced algorithms, providing the mathematical, statistical, or signal processing as needed for background.

This book takes a software engineer's viewpoint, and describes methods and algorithms available in the Insight Toolkit as open source software. The goal of this book is to explain the background specific to our implementations, enabling the reader or student to use complex tools in medical image processing as rapidly as possible. We do not claim an exhaustive treatment of these ideas, but rather a working and functional introduction to these topics, partnered with working implementations of the methods described.

1.1   Medical Image Processing

The medical mission differs from the other forms of image processing arising from non-medical data. In satelite surveillance analysis, the purpose is largely a screening and cartographic task, aligning multiple types of data and corresponding them to a known map and highlighting possible points of interest. In computer vision, camera views must be analyzed accounting for the perspective geometry and photogrammetric distortions associated with the optical systems that are the basis for robotic sensors. In many of these systems, autonomous navigation, target identification and acquisition, and threat avoidance are the primary tasks. For the most part, the incoming information arrives as 2D images, data arrays that can be organized using two cartesian dimensions. In addition, the tasks are to be performed independently by the machine, relying on the development of machine learning and artificial intelligence algorithms to automatically accomplish the tasks.

In medicine, the problem as well as the input data stream are usually three-dimensional, and the effort to solve the primary tasks is often a partnership of human and machine. Medicine is notably a human enterprise, and computers are merely assistants, not surrogates nor possible replacements for the human expert. The medical task can often be split into three areas: (1) data operations of filtering, noise removal, and contrast and feature enhancement, (2) detection of medical conditions or events, and (3) quantitative analysis of the lesion or detected event. Of these subtasks, detection of lesions or other pathologies is often a subjective and qualitative decision, a type of process ill-suited for execution by a computer. By contrast, the computer is vastly more capable of both quantitative measurent of the medical condition (such as tumor volume or the length of a bone fracture) and the preprocessing tasks of filtering, sharpening, and focusing image detail. The natural partnership of humans and machines in medicine is to provide the clinician with powerful tools for image analysis and measurement, while relying on the magnificent capabilities of the human visual system to detect and screen for the primary findings.

We divide the problems inherent in medical image processing into three basic categories:

- **Filtering:** These are the basic tasks involved in filtering and preprocessing the data before detection and analysis are performed either by the machine or the human operator.

- **Segmentation:** This is the task of partitioning an image (2D array or volume) into contiguous regions with cohesive properties.

- **Registration:** This is the task of aligning multiple data streams or images, permitting the fusion of different information creating a more powerful diagnostic tool than any single image alone.

ITK has met its primary goals for the creation of an archival vehicle for image processing algorithms and an established functioning platform to accelerate new research efforts. New projects have connected ITK to existing software applications for medical and scientific supercomputing visualization such as "Analyze" from the Mayo Clinic and "SCIRun" from the University of Utah's SCI Institute. These projects were sponsored to demonstrate that the application programmer's interface (API) for ITK was suitable for supporting large applications in established medical and scientific research areas. Also, recent projects have been funded to add algorithms not previously funded to the collection, proving the versatility of the underlying software architecture to accommodate emerging ideas in medical imaging research.

There is growing evidence that ITK is beginning to influence the international research community. In addition to the sponsored growth of the ITK community, the mailing list of users and developers includes over 300 members in over 30 countries. These facts may be indicators that a nascent research community may be forming about the core ITK software. The applications for ITK have begun to diversify including work in liver RF ablation surgical planning, vascular segmentation, live-donor hepatic transplant analysis, intra-operative 3D registration for image-guided intervention, rapid analysis of neurological function in veterans, longitudinal studies of Alzheimer's patients using MRI, and even handwriting analysis. The Insight Program of the Visible Human Project is emerging as a clear success and a significant and permanent contribution to the field of medical imaging.

In 1999, the NLM Office of High Performance Computing and Communications, supported by an alliance of NIH ICs and Federal Funding Agencies, awarded six contracts for the formation of a software development consortium to create and develop an application programmer interface and first implementation of a segmentation and registration toolkit, subsequently named the Insight Toolkit (ITK). The final deliverable product of this group has been a functional collection of software components, compatible for direct insertion into the public domain via Internet access through the NLM or its licensed distributors. Ultimately, NLM hopes to sponsor the creation of a public segmentation and registration software toolkit as a foundation for future medical image understanding research. The intent is to amplify the investment being made through the Visible Human Project and future programs for medical image analysis by reducing the reinvention of basic algorithms. We are also hoping to empower young researchers and small research groups with the kernel of an image analysis system in the public domain.

Scope of This Book

While a complete treatise on medical image processing would be an invaluable contribution to the research community, it would be far too much hubris on our part to attempt to describe it all. The concept of Finite Element Models (FEM) is a book-length treatise in itself, as are each of the topics of Level Set Methods and Statistical Pattern Recognition. Indeed, we reference many valuable texts on these topics in particular.

We therefore limit this material to methods that have been implemented in the Insight ToolKit. All methods described herein are reflected in the source code for ITK. We limit the perspectives in this text to the viewpoints presented in ITK, and though there are many valid and valuable alternate approaches incorporating many of these same ideas, we cannot treat them all, and in fairness, we claim expertise only in those areas that we can deomnstrably express directly through software development.

In keeping with the principles of the Insight project itself, we concentrate only on medical image filtering, segmentation, and registration. These were the primary areas of concentration of the software project, and thus they are the topics described in this book. As such, we do not directly treat the ares of visualization, identification, classification, or computer-aided detection and diagnosis. All of the methods presented in this book are relevant to these topics, and we cannot show our results without some form of visualization, but we do not include an in-depth treatise of these topics or explicitly describe the integration of ITK with visualization environments.

Suggested Uses for This Book

Throughout the development of this text, we have internally described it as the "theory book." We have avoided including source code in an attempt to focus on the principles of the methods employed in ITK. There is a companion text for ITK directed toward implementation and integration details, called the *ITK Software Guide*. These books should be considered complementary to each other and supplementary in support of ITK. If you, the reader, are familiar with the principles of some method or approach in ITK, you only need to read the *ITK Software Guide* to learn about our particular implementation of the method and our structure and intended uses for that method. If, however, you require greater background on the theory behind some method, you should refer to the material in this book. Thus, we perceive this book to be an extended reference for ITK, providing more in-depth presentation of the basic algorithms, structure, and methods in the toolkit than can be gleaned either from the source code or from the base level descriptions in the *ITK Software Guide*.

We also perceive this book to be the basis for reading and curriculum foundations for a graduate course on medical image processing. ITK has been effectively used to create courses on this topic. Its advanced implementations permit the rapid explorations of complex algorithms without requiring the student or the instructors and their staff to create the software infrastructure necessary to run high-dimensional image processing software. For such a course, this book would represent theoretical course material, and the ITK Software Guide would be an invaluable resource for laboratory exercises and course projects.

We are extraordinarily proud of ITK and hope that you find our offering instructive, valuable, and a useful supporting addition to your software projects. We humbly and gratefully present this material with hope for your future success in research and medical software application development.

Acknowledgements

Following the consortium-based nature of this project, this book is the collective effort of many, many people. The contributing authors include in alphabetical order: Brian Avants, Stephen Aylward, Celina Imielinska, Jisung Kim, Bill Lorensen, Dimitris Metaxas, Lydia Ng, Punam K Saha, George Stetten, Tessa Sundaram, Jay Udupa, Ross Whitaker, and Terry Yoo. Other people have made essential contributions, including: Luis Ibanez, Will Schroeder, and others. The Insight ToolKit would not have been possible without the selfless efforts of Bill Hoffman, Brad King, Josh Cates, and Peter Ratiu.

We are indebted to the coalition of federal funding partners who have made this project possible: the National Library of Medicine, the National Institute for Dental and Craniofacial Research, the National Institute for Mental Health, the National Eye Institute, the National Institute on Neurological Disorders and Stroke, the National Science Foundation, the DoD Telemedicine and Advanced Technologies Research Center, the National Institute on Deafness and other Communication Disorders, and the National Cancer Institute. We would especially like to recognize the patronage of the Visible Human Project from the Office of High Performance Computing and Communications of the Lister Hill National Center for Biomedical Communications of the National Library of Medicine. In particular, we'd like to thank Drs. Donald Lindberg and Michael Ackerman for their vision and leadership in sponsoring this project.

These three basic divisions represent the organization of both this book and the Insight Toolkit for which this book was written. In order to best understand the approaches and difficulties associated with these tasks, we being with a history of medical imaging and a brief overview of modern imaging modalities.

1.2 A Brief Retrospective on 3D Medical Imaging

In the broad realm of the sciences, medical imaging is a young field. The physics enabling diagnostic imaging are barely one hundred years old. In 1895, Wilhelm Roentgen discovered x-rays while experimenting with a Crookes tube, the precursor to the cathode ray tube common in video applications today. It is worth noting that he immediately recognized the potential of x-ray radiation in diagnostic imaging and that one of his earliest images is of the bones of his wife's hand. Roentgen's discovery was so noteworthy and revolutionary that he received a Nobel prize within six years of announcing his initial work. Thus began a bountiful and tightly-coupled one hundred year partnership between physics and medicine.

The meaning and purpose of medical imaging has been to provide clinicians with the ability to see inside the body, to diagnose the human condition. The primary focus for much of this development has been to improve the quality of the images for humans to evaluate. Only recently has computer technology become sufficiently sophisticated to assist in the process of diagnosis. There is a natural partnership between computer science and radiology.

Early in the twentieth century, a Czech mathematician named Johann Radon derived a transform for reconstructing cross-sectional information from a series of planar projections taken from around an object. While this powerful theory had been known for over fifty years, the ability to compute the transform on real data was not possible until digital computers began to mature in the 1970s.

Imaging in 3D emerged in 1972 when x-ray computed tomography (CT) was developed independently by Godfrey Hounsfield and Alan Cormack. These innovators later shared the 1979 Nobel Prize in Medicine. Their achievement is noteworthy because it is largely based on engineering, the theoretical mathematics and the underlying science had been described decades earlier. The contribution is in the application of mechanical engineering and computer science to complete what had previously only been conceived on paper. Clinical systems were patented in 1975 and began service immediately thereafter.

While techniques for using x-rays in medical imaging were being refined, organic chemists had been exploring the uses of nuclear magnetic resonance (NMR) to analyze chemical samples. Felix Bloch and Edward Purcell were studying NMR in the mid 1940s. Together, they shared the Nobel Prize in Physics in 1952. Paul Lauterbur, Peter Mansfield, and Raymond Damadian were the first to develop imaging applications from NMR phenomena. Lauterbur created tomographic images of a physical phantom constructed of capillary tubes and water in a modified spectrometer in 1972. Damadian later was able to create animal images in 1975. The means of creating medical images using magnetic fields and radio waves was later renamed Magnetic Resonance Imaging (MRI). Lauterbur and Mansfield received the 2003 Nobel Prize in Medicine. Raymond Damadian has not been so honored, but his contribution is significant and should not be overlooked by us.

Radiologists were thus simultaneously presented with the possibilities for slice (or tomographic) images in the axial plane from x-ray CT and from MRI in the early 1970s. Arising from the long development of x-ray technology, CT matured first. The difficulties and expense of generating strong magnetic fields remained an obstacle for clinical MRI scanners until engineers were able to produce practical superconducting magnets. While still expensive, MRI technology is now available worldwide, creating precise images of deep structures within the body, outlining the anatomy and physiology of internal objects other than bones. MRI and CT have joined the established 3D imaging modalities of nuclear imaging and the growing field of volumetric medical ultrasound. The creation of new modalities with differing strengths has led to the need to align or register these multiple data streams.

1.3 Medical Imaging Technology

The majority of medical visualization involves the display of data acquired directly from a patient. The radiologist is trained through long years of education and practice to read relatively simple presentations of the raw data. The key to improving a diagnosis is in the careful crafting of the acquisition, applying the physics of radiology to maximize the contrast among the relevant tissues and suppressing noise, fog, and scatter that may obscure the objects of interest. This is no less true for more complex visualizations that incorporate 3D renderings or volume projections; improving the quality of the acquired data will fundamentally affect the quality of the resulting visualization.

This section will briefly cover some of the more common sources of three or higher-dimensional medical imaging data. This treatment is necessarily superficial, serving mostly as an annotated glossary for many of the terms taken for granted in the radiology community. To create effective medical visualization tools, the computer scientist requires a fundamental understanding of the source of the image data, the technology involved, and the physical principles from which the image values are derived. This cursory introduction will be insufficient for indepth research, but will serve as background for this text; the reader is encouraged to continue the exploration of this topic. This additional command of the basics of medical image acquisition will enhance your practice in medical visualization,

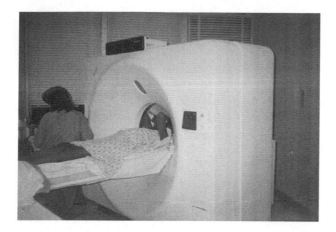

Figure 1.1. The business end of a CT scanner (circa 1998).

ease communication between you and the doctor and the technologist, improve the quality of the source data, and smooth transitions among the many interfaces from acquisition to display.

1.3.1 Computed Tomography

By far the most familiar form of 3D medical imaging is X-ray Computed Tomography or CT (formerly referred to as Computer Assisted Tomography, Computerized Axial Tomography, CAT scanning, Computerized Transaxial Tomography, CTAT, Computerized Reconstruction Tomography, CRT, and Digital Axial Tomography, DAT). The mathematics for tomographic reconstruction from multiple views have been known for most of this century. It took almost fifty years before the components required for x-ray computed tomography were sufficiently developed to make the procedure and the ensemble of instruments economically feasible.

A CT scanner is a room sized x-ray instrument, requiring a shielded environment to protect the technologists and other clinic staff from exposure from routine use. The number of manufacturers of CT scanners has been steadily decreasing with only a handful of vendors providing them today. The cost of these devices ranges from approximately $400,000 to well over $1,000,000. There are some examples of portable CT machines that can be moved into a trauma care center or into operating rooms to aid in the diagnosis and treatment of patients. Emerging technologies in flat-panel digital x-ray devices is enabling 3D imaging using more conventional fluoroscopic tools, but at the time of this writing, the following is the most common configuration for these devices.

An accurately calibrated moving bed to translate the patient through the scanner, an x-ray tube mounted in such a way to allow it to revolve about the patient, and an array of x-ray detectors (gas filled detectors or crystal scintillation detectors) comprise the essential system components of a CT machine. The x-ray tube and detector array are mounted in a gantry that positions the detector assembly directly across from the x-ray source. The x-ray source is collimated by a pair of lead jaws so that the x-rays form a flat fan beam with a thickness determined by the operator. During the acquisition of a "slice" of data, the source-detector ring is rotated around the patient. The raw output from the detector array is backprojected to reconstruct a cross-sectional transaxial image of the patient. By repositioning the patient, a series of slices can be aggregated into a 3D representation of the patient's anatomy. Figure 1.1 is a picture of a CT table and gantry.

Within the last fifteen years there have been significant advances in CT technology, allowing for faster spiral acquisition and reduced dose to the patient as well as multislice detector arrays permitting simultaneous acquisition of several slices at a time. Early CT gantries were constructed with a revolving detector array positioned directly across the patient from the moving x-ray source. Cable lengths connecting the moving detector assembly and the x-ray tube permitted only a single slice to be acquired at one time. The revolving assembly then had to be "unwound", the patient advanced the distance of one slice, and the process repeated.

As small affordable detectors have become available, scanners have been designed with a fixed array of x-ray detectors. The only remaining revolving part of the gantry is the x-ray tube. This has simplified the engineering and cabling of the gantry. The x-ray tube can now be cabled using a slip ring, permitting continuous revolution of the tube about the patient. This type of design is capable of helical (or spiral) CT acquisition. By simultaneously revolving the x-ray source about the patient and continuously moving the patient through the bore of the gantry, the data are acquired via a spiral path. These methods have enabled very fast image acquisition, improving the patient throughput in a CT facility, reducing artifacts from patient motion, and reducing absorbed dose by the patient. Combined with multiple layers of sensors in the detection ring, these new scanners are generating datasets of increasingly larger sizes.

A CT scanner is an x-ray modality and is subject to all of the physics associated with the generation, dispersion, absorption, and scatter associated with all x-ray photons. The process of generating a CT scan is similar to creating a standard x-ray film; however, while a single x-ray exposure generates a complete film-based exam, the CT image is not acquired in a complete form. Rather, it must be reconstructed from multiple views. The advantage of being an x-ray modality is that laymen and clinicians alike have considerable intuition when dealing with x-ray-based images. The concepts of dense objects like bone absorbing more photons relative to less dense tissues like muscle or fat come naturally from our experience and expectations about x-ray imaging. A typical CT scanner can generally acquire the data for a transaxial slice in a matter of seconds (within 1 to 5 seconds). An exam can include several series of slices, in some cases with and without pharmaceutical contrast agents injected into the patient to aid in diagnostic reading. Slices can be spaced such that they are either overlapping or contiguous, though some protocols call for gaps between the slices. A large study can include well over 100 separate 512×512 pixel images. The radiation dose from a CT scan is comparable with that of a series of traditional x-rays.

The concept of "resolution" should be divided in the researcher's mind into spatial resolution (i.e., how much area in each dimension a voxel covers) and sampling resolution (how many voxels in each dimension of the slice). Sampling resolution in current scanners usually creates images that are either 256×256 or 512×512 voxels square. Sampling resolution in the longitudinal direction is limited only by the number of slices acquired. Spatial resolution in the longitudinal direction is bound by physical limitations of collimating the photons into thin planes. The physical lower limit is approximately 1 mm; narrower collimation requires inordinate amounts of x-ray flux to image and also leads to diffraction interference. Sampling resolution in the transaxial dimensions of a voxel is based on the field of view selected by the operator of the CT scanner and the matrix (256×256 or 512×512) yielding pixel dimensions that are generally 0.5 to 2 mm. Attempting to achieve higher spatial resolution will lead to voxels with too little signal to accurately measure x-ray absorption. With today's diagnostic equipment, if pixels smaller than 0.25 mm are attempted, low signal-to-noise ratios become a problem.

Units of measure for the pixel values of CT imaging are standard across the industry. Each pixel ideally represents the absorption characteristics of the small volume within its bounds. By convention, these measurements are normalized relative to the x-ray absorption characteristics of water. This is a unit of measure known as Hounsfield units (HU). The Hounsfield unit scale is calibrated upon the attenuation coefficient for water, with water reading 0 HU. On this scale air is -1000 HU, fat tissue will be in the range of -300 to -100 HU, muscle tissue 10–70 HU and bone above 200 HU.

Every CT image contains artifacts that should be understood and handled when visualizing the data. The process of reconstruction and sampling leads to aliasing artifacts, just as any sampling procedure in computer graphics or signal processing. Another common artifact is partial voluming, the condition where the contents of a pixel are distributed across multiple tissue types, blending the absorption characteristics of different materials. Patient motion while scanning generates a variety of blurring and ring artifacts during reconstruction.

There is another class of artifacts that arises from the fact that CT is an x-ray modality. Embedded dense objects such as dental fixtures and fillings or bullets lead to beam shadows and streak artifacts. More common is the partial filtering

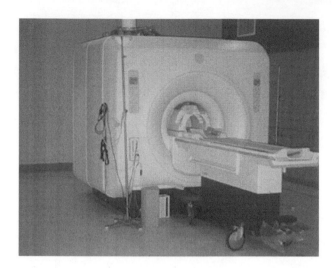

Figure 1.2. The business end of an MRI scanner (circa 1998).

of the x-ray beam by dense tissue such as bone which leads to beam hardening, a condition which causes slight shadows to halo the dense features of the image. When creating visualizations, beam hardening artifacts may cause the researcher to underestimate the volume of dense objects, making them seem smaller than they are.

1.3.2   Magnetic Resonance Imaging

Like CT, Magnetic Resonance Imaging or MRI (formerly referred to as Nuclear Magnetic Resonance imaging) was pioneered in the early 1970s. Introduced under the name "Zeugmatography" (from the Greek word *zeugma*, meaning "that which draws together") it remained an experimental technology for many years. Unlike CT, MRI does not use ionizing radiation to generate cross-sectional images. MRI is considered to be a newer modality since feasible commercial development of diagnostic MRI scanners had to await affordable super-conducting magnets. Today's diagnostic systems ranges from $500,000 to $2 or $3 million.

An MRI scanner is a large magnet, a microwave transmitter, a microwave antenna, and several electronic components that decode the signal and reconstruct cross-sectional images from the data. Generally, the magnet is superconducting and must be operated at cryogenic temperatures (4 degrees Kelvin), necessitating its immersion in a bath of liquid helium. Some MRI scanners with relatively poorer resolution are being constructed using fixed magnets without the need for a liquid helium container, permitting more open designs at the cost of image contrast and resolution. Figure 1.2 is a picture of an MRI magnet and table, showing a head coil, an antenna for imaging the head and neck.

Unlike CT scanners, the bore of an MRI scanner is often up to two meters in length (6–8 feet). Patients are inserted into the middle of the magnetic field, often inducing claustrophobia. The environments for MRI scanners must be shielded for magnetic and radiofrequency interference. Large Faraday cages and substantial masses of iron usually surround the magnet and sometimes the entire room about the scanner.

In MRI scanning, the patient is placed within a high intensity magnetic field. Field strengths vary from 0.35 Tesla to 1.5 Tesla for most diagnostic MRI devices (for reference, 1 Tesla = 10,000 Gauss, and the earth's magnetic field, though variable, is approximate 0.5 Gauss). The induced magnetic field causes the magnetic moments of the hydrogen atoms within the patient to align along the principal direction of the superconducting magnet. Low-level radio waves in the microwave frequencies (approximately 15 to 60 MHz) are then transmitted through the patient, causing the magnetic moments of the hydrogen nuclei to resonate and re-emit microwaves after each pulse. The microwaves emitted by the body are recorded using a radio frequency antenna, filtered, amplified, and reconstructed into tomographic slices. While all of the hydrogen nuclei typically resonate at a frequency fixed by the strength of the induced magnetic field, different tissue types resonate longer than others, allowing the viewer to discriminate among them based on the magnitude of the signal from different points in space over time.

Spatial locations can be determined by varying the magnetic field about the patient in different directions at different times. Linear gradients in the induced magnetic field are produced by magnets that supplement the main superconducting magnet. These gradient coils are activated in a variety of sequences, altering the phase and frequency of the microwave pulses that are received and re-emitted by the hydrogen nuclei. The design and crafting of pulse sequences is a field of research unto itself. The careful selection of pulse sequences can illuminate a variety of clinical conditions including the display of swelling and bleeding, even enhancing the blood vessels deep within the brain. Often, several sequences will be taken of the patient during an exam to capture the variety of information available in the different pulse sequences. The visualization researcher is encouraged to learn about these pulse sequences and their properties in order to select the best ensemble of them to use when crafting visualizations.

The output of an MRI scanner is similar to CT. Slices representing slabs of the object scanned are produced. However, unlike CT which always produces transaxial slices, the slices from MRI can be oriented in any plane. The output values at each image element are not calibrated to any particular scale. Generally they are 10-bit data samples. The values will vary depending upon the scan parameters, and the patient's size and magnetic characteristics. Additionally, the values are not constant over the entire scan space since inhomogeneity in the magnetic field causes pixels that may represent the same tissue, but located some distance apart to give different signals. This lack of an absolute scale for a dataset is a cause of much consternation to the researcher attempting to segment MRI data.

Visualization researchers seldom have good intuition for the meaning of MRI signals. Unlike x-ray-based imaging, there are no physical analogs to what the viewer is seeing. An MRI device measures the radio signals emitted by drops of water over time. Usually skin and fat are much brighter than bone which has virtually no signal at all. Segmentation and classification are therefore significantly harder and are the subject of much research.

As with CT, the concept of resolution must be divided into spatial resolution and sampling resolution. The matrix of the image can typically be selected to be either 256×256 or 512×512, depending on the sampling resolution of the frequency and phase decoders of the receiving equipment. Square pixels in a rectangular matrix are often selected since reducing the number of rows or columns can significantly reduce the time required to complete a scan. Spatial resolution, or field of view, is primarily dependent on the strength of the gradient magnets and their ability to separate the slices along their gradient directions. The radiofrequency receiving equipment must be able to distinguish among frequencies and phases which have only slight variations. Stronger gradients create greater separations and improve spatial resolution.

Remember that the MR scanner is attempting to measure the radio signal resonating from a drop of water that is as small as 1 mm $\times$ 1 mm $\times$ 2 mm. There are significant trades to be made when attempting to increase spatial resolution. Signal to noise will degrade as electronic distortion of the antenna coil and the amplifiers begin to overcome the small signals involved in diagnostic imaging. Imaging with larger voxels, thicker slices, or repeating and averaging multiple acquisitions are solutions. The relative consequences are increased aliasing, partial voluming, and possible patient motion artifact.

MRI data is subject to several artifacts. The issues of partial voluming, patient motion, and aliasing are common with CT. While MRI does not have x-ray-related artifacts, it has its own host of radio and magnetic artifacts of which the visualization expert should be aware. Like any microwave used for cooking, the mass of the patient (or food) will affect how well the material absorbs the radiofrequency energies. This leads to "cold" spots in the imaging volume. In addition, the distribution of the antenna coverage for both transmission and reception and inhomogeneity in the induced magnetic field lead to inconsistent quantitative values and even inaccurate geometry within an image. There is significant research being conducted addressing methods to correct for these distortions.

Most MRI scanners can acquire a set of slices (30 to 50) within five to ten minutes. An entire study of a patient generally represents two to three series of slices, with a study time of 30 to 45 minutes. Each slice generally represents a thickness of 2 to 10 mm and contains 256×256 pixels. As with CT, the pixel dimensions are set by the image matrix and the field of view parameters.

New work in MR imaging is leading to novel capabilities in this modality. Studies in perfusion and diffusion of various agents across the body are being enabled by new capabilities in fast imaging. These studies can illuminate blood flow, creating high-resolution images of vascular structure. Functional MRI has been introduced not only to record the patient's anatomy, but also the physiological functions of the tissues being studied, largely used today to map the cerebral cortex of the brain. The complex pulse sequences of Diffusion Tensor Imaging (or DTI) are beginning to reveal the direction and course of nerve bundles deep within the white matter of the brain, providing new research areas of tractography and patient specific anatomical analysis. The pace of research in MRI is increasing as new capabilities are enabled.

1.3.3 Nuclear Medicine

Imaging using pharmaceuticals tagged with radioactive agents is an older technology than either CT or MRI. In nuclear medicine imaging, a radioactive source is injected into the patient and the distribution of that source is measured using a detector array to catch and quantify the radiation emitted from the body. The detector array is sometimes configured similar to that for CT, while other systems utilize a two-dimensional array of detectors. The major difference between CT and this type of study, is that with CT the radiation source's location is external and known, while in nuclear medicine, the source is internal and its distribution unknown. Figure 1.3 is a picture of a two-headed Gamma Camera (Anger Camera) used for Single Photon Emission Computed Tomography (SPECT). The gantry has two detector arrays for gamma radiation emitted from the patient. The detectors revolve about the patient, and tomographic images are reconstructed.

While the images generated from measuring radioactive decay are often blurry with poor spatial resolution, nuclear medicine allows clinicians to image the physiological activity (the biological activity) of the patient, rather than just the geometry of the patient's anatomy. The selection of the pharmaceutical to which the radionuclei is attached will determine the specificity and performance of the imaging study. For instance, radioactive iodine generates very specific studies of the thyroid and parathyroid glands. Radioactive tracers can be attached to hundreds of different substances.

Nuclear medicine studies produce images with fairly low resolution and a high amount of noise. This is due to our inability to use high radiation doses because of the consequences of high doses of radiation to the patient. Choosing persistent radiopharmaceuticals will cause the agent to reside longer within the body, improving our capacity to resolve interal objects while potentially inflicting additional damage on the patient.

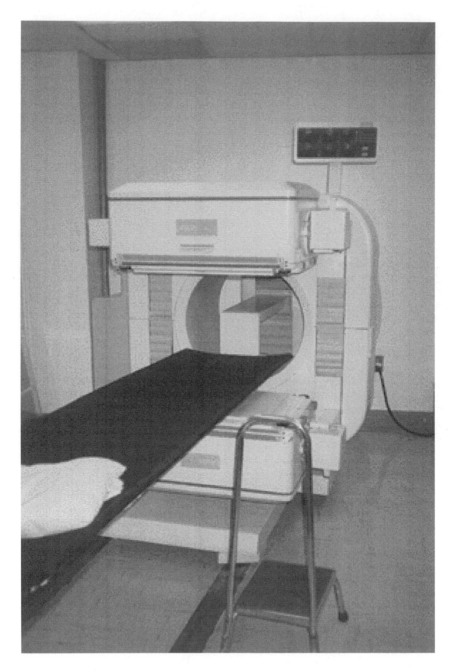

Figure 1.3. A two-headed Gamma Camera used in nuclear medicine (circa 1998).

Most nuclear medicine is still based on the 2D planar representations generated by simply recording the activity of a detector over a period of time. If the detectors are rotated about the patient, reconstruction is possible. The two most common types of 3D nuclear medicine studies are Single Photon Emission Computed Tomography (SPECT) and Positron Emission Tomography (PET). SPECT studies utilize radiotracers that emit photons while decaying. The radioactive agents used in SPECT have half-lives measured in hours and can be easily generated or transported to the clinic. PET studies use radioactive isotopes that decay by positron emission. The resulting positrons annihilate when they encounter electrons generating two high energy photons that conserve momentum by traveling in opposite directions. The half-lives of the isotopes used in PET are often measured in minutes, necessitating their production at the clinic. A particle accelerator such as a cyclotron is required to create most of the radionuclei for PET imaging, increasing the cost of a PET installation.

The output of a SPECT or PET scanner is typically a set of 10 to 30 transaxial slices. Each slice is typically 5 to 10 mm thick and contains pixels 5 to 10 mm in size. There may be gaps between the slices. Often multiple sets of scans are taken five to fifteen minutes apart allowing for time-resolved studies. Nuclear medicine studies produce images which have few anatomic cues. The images mimic physiologic activity which may or may not be easily transferable to the underlying anatomy.

1.3.4   Ultrasound

Unlike the three other modalities described in this chapter, Ultrasonography uses classical physics to perform its imaging rather than the more esoteric phenomena of nuclear decay or x-ray quantum mechanics. No photons are used. A piezoelectric quartz crystal, similar to those used to clock CPUs, wristwatches, or to start electronic ignition lighters for barbeque grills, is used to create high frequency acoustic energy (3 to 10 Megahertz) which is then reflected off surfaces and interfaces between organs deep within the body.

The same device that is used to create the acoustic signal, a transducer, is then used to measure the returning echo information. Partial reflections are created by interfaces between areas of differing acoustical impedance. The result is imaging very similar to the SONAR systems used in maritime and undersea naval imaging. The sonographer places the probe against the patient and moves it to obtain images of various parts of the body. The result is usually a 2D representation of the area under the transducer. Most ultrasound machines consist of a linear array of transducers, and produce an image representing a pie-shaped slice of the body. One of the advantages of ultrasound is that it produces images in real time. Another advantage is the absence of ionizing radiation.

Ultrasound machines are fairly inexpensive compared with the other diagnostic 3D modalities in common use in the health care industry. A high-end diagnostic ultrasound system can be purchased for approximately $250,000, while functional, more moderately priced units can be acquired for less than $100,000 today.

3D images can easily be created by extruding several slices through space. Most commercial ultrasound equipment allows for 3D imaging; however, accurate spatial information is seldom provided. Spatial tracking of the transducer array is imperative for clinical 3D visualization. Some approaches to correcting this defect have been to use physical, optical, and electronic devices for locating the transducer in space. Providing a rigid path for the transducer creates some compelling results. Other techniques involve rotating the transducer assembly, sweeping a volume in a cylindrical section similar to aeronautical RADAR imaging. Finally, recent advances in transducer design have yielded demonstrable 3D transducers that can acquire volume images from ultrasound with no mechanically moving parts.

However, once the position and orientation of the probe are known, the data are still often sampled in irregular intervals, and defy many image processing techniques. Ultrasound images typically contain a large amount of noise termed speckle, that adds to the problem of identifying structures. Object detection in volume ultrasound is the subject of much advanced medical image processing research.

1.4   Acquisition, Analysis, and Visualization

Most acquisition of medical data is not optimized for 3D presentation. Indeed, to improve the signal to noise, thick slices are usually requested. The result is better contrast at the cost of higher partial-voluming artifact and poor spatial resolution in the effective z-direction. In order to reduce the dose to the patient and the time required to perform a scan, slices are not always contiguous. Rather small gaps are introduced between slices. Since most presentation is 2D, these practices seldom affect the clinician; however, they can be fatal to 3D visualization.

Radiologists often must balance the need for image quality with the interests of the patient. Many times the means to improving the diagnostic power of an image will mean increased dose to the patient. The factors affecting the trade-off between image quality and acceptable dose include a wide variety of concerns. For example, since the inadvertent long range biological effects (chromosomal damage) of ionizing radiation are most profound in children, pediatric radiology seldom trades increased dose for improved contrast. Similar trades are made throughout medical imaging (e.g., trading spatial resolution or increased imaging time in MRI for an improved signal-to-noise ratio). The computer scientist interested in medical visualization would be well served to learn where these trades are being made and learn to cope with their consequences.

An even deeper understanding of the clinical findings of a case is required to validate a 3D visualization. Knowledge of what the physician is seeking in the patient will help to direct the acquisition so that the results not only capture the desired pathology in high detail, but also assure the computer scientist that the data are in an appropriate form for advanced rendering.

In his 1993 presentation on 3D Medical Visualization, Derek Ney, an assistant professor for the Johns Hopkins Department of Radiology, wrote:

> A succinct statement of the relationship of visualization and acquisition is that the single easiest method for improving visualization is to use better (higher resolution, higher contrast, higher signal to noise, etc.) acquisition techniques. This implies that the method for acquisition should be as optimal as possible. There is no room in most systems for poor acquisition technique. The researcher in visualization is warned to learn about the potential pitfalls in the acquisition stage, so that they are presented with data that was acquired with the best technique possible.

Indeed, as with all data processing, "if garbage goes in, only garbage comes out." Or more precisely, if one beautifully renders garbage data, then one only has a picture of well-dressed garbage. The result serves no one in particular and wastes valuable resources that would be better used to serve the patient, to serve the community, and to serve future generations of researchers and patients alike.

1.5 Summary

Keeping the basic concepts of medical image acquistion in mind, we turn our attention to the processing and analysis of the generated data. The multiplication of imaging modalities with differing strengths and weaknesses in imaging anatomy v. physiology and hard tissue v. soft tissue requires careful registration and alignment of multimodal data, and differences in size and resolution lead to multiscale methods as well. We refrain from covering the algorithms and methods for generating visualizations of the data, instead concentrating on the analysis and refinement of the data stream in preparation for the ultimate creation of rendered images of the patient anatomy and physiology.

This book is divided into three main parts:

- **Basics:** These chapters describe techniques for preprocessing the image data before complex semantic, knowlege-based operations are performed either by the machine or the human operator. Statistical, linear, and more advanced nonlinear methods are covered, matching implementations from the software.

- **Segmentation:** As stated before, this is the task of partitioning an image (2D array or volume) into contiguous regions with cohesive properties. A variety of approaches are given including hybrid mixtures of statistical and structural methods. We cover the use of finite element models, Voronoi decomposition, fuzzy-connectedness, and other techniques useful for segmentation implemented within ITK.

- **Registration:** This is the task of aligning multiple data streams or images, permitting the fusion of different information creating a more powerful diagnostic tool than any single image alone. We cover basic registration strategies as well as more complex deformable techniques, all implemented under the umbrella of the ITK registration framework.

This book is intended as a companion to the Insight Toolkit covering the practices and principles used to develop the corresponding software. Implementation and integration details are covered elsewhere in the ITK Software Guide. The ultimate arbiter of all of this material is the source code for ITK itself, made available, by intent and design, in the public domain. We hope that this text serves as a summary to guide and explain the many choices made and roads taken in the design and use of the software.

Enjoy ITK.

Basic Image Processing and Linear Operators

Terry S. Yoo
National Library of Medicine, NIH

George D. Stetten
University of Pittsburgh

Bill Lorensen
GE Global Research

2.1 Introduction

The topics discussed throughout this book fall into the realm of what is usually called *image analysis*. Before researchers adopted that term, they often referred to their field as *image processing*, an extension of traditional signal processing to images. An image can be thought of as a signal in space rather than time, arrayed in two or three dimensions rather than one. This leads to fundamental differences between conventional signal processing and image processing. While one generally only moves forward through time, in space, one moves left or right with equal ease, as well as up or down, forwards or backwards. The extension to multiple dimensions raises new issues, such as rotation and other interactions between the axes.

As with conventional signals, images can be either continuous or discrete. Optical images, consisting of light, call for analysis in the continuous domain, as they pass through lenses, for example. In contrast, images found on a computer generally consist of individual samples on a regular grid, and therefore require discrete analysis. Digital sampling theory is intrinsic to image processing. Familiar artifacts such as aliasing can be understood through digital sampling and filtering theory. Likewise, effective reconstruction, the inverse of sampling, requires a clear understanding of digital sampling theory.

Both continuous and discrete image analysis often begin with linear operators adapted from signal processing. These operators range from simple algebra to convolution and the Fourier transform. Linear operators assume that the light or other input adds by superposition, i.e., that individual components of the input do not affect each other within the system. Linear operators preserve the independence of basis functions useful in extracting information from an image. Some linear operators affect each sample in the image, or *pixel*, independently from its neighbors, while others take neighboring pixels, or even the entire image, into consideration at once.

This chapter is a review (or introduction, depending on your background) covering the digital theories that underlie sampling and linear operators. The material will be covered from a practical viewpoint: What should we know about the basic operators of image processing? What do they mean in terms of getting at the underlying information in the image? What are the repercussions of digital sampling theory? The presentation is cursory; we will not explore proofs or rigorous mathematics associated with the presented theorems or tools. For example, the presentation of the Fourier transform will be far from exhaustive, not even listing all of the important mathematical properties of the transform. We also will not be presenting implementations. For more in-depth discussion, we refer the reader to standard textbooks in the field, in particular, *Signals and Systems*, by Oppenheim and Willsky [2], and *Digital Image Processing* by Castleman [1].

2.2   Images

In this book, we often are concerned with 3D, or *volume*, images. A volume image, ϕ, is a mapping from $\mathbb{R}^3$ to $\mathbb{R}^n$ where $n = 1$ for the typical, scalar–valued volume. More precisely:

$$\phi : U \mapsto \mathbb{R} \text{ and } U \subset \mathbb{R}^3, \tag{2.1}$$

where U is the domain of the volume. The image ϕ is often written as a function $\phi(x, y, z)$.

Throughout this chapter, many examples will be presented as either 1D or 2D, where they may be more easily examined (at least in print), and then generalized to higher dimensions. In general, for a specific number of dimensions, we will use the corresponding function notation ($\phi(x)$, $\phi(x, y)$, or $\phi(x, y, z)$).

We denote the sampling of continuous images into discrete, digital images as follows. If F is a discrete sampling of a 2D image $\phi(x,y)$ then we can say that

$$F_{i,j} = \phi(x_i, y_i). \tag{2.2}$$

We likewise use notation to distinguish between discrete and continuous differential operators. To find the partial derivative in the x-direction of a discretely sampled image we can say that

$$\phi_x(x_i, y_i) = \left.\frac{\partial\phi}{\partial x}\right|_{x_i, y_i} \approx \delta_x F_{i,j}, \tag{2.3}$$

using a method such as

$$\delta_x F_{i,j} \equiv \frac{F_{i+1,j} - F_{i-1,j}}{x_{i+1} - x_{i-1}} = \frac{F_{i+1,j} - F_{i-1,j}}{2h}, \tag{2.4}$$

to approximate the partial derivative, where h is the grid spacing (normally assumed to be 1 pixel). The value $\delta_x F_{i,j}$ is an approximation of the instantaneous partial derivative of $\phi(x_i, y_i)$ at (x_i, y_i) in the x-direction. Equation (2.4) is the method of *central differences*, one of a number of commonly used methods to approximate the derivatives of uniformly, discretely sampled datasets.

2.3 Point Operators

The simplest operators on an image are those that treat each pixel independently from its neighbors. Point operators usually ignore information about pixel location, relying only on pixel intensity. A surprising number of useful things can be done with such operators. Some examples follow.

2.3.1 Thresholding

The simplest form of segmentation involves thresholding the image intensity. Thresholding is inherently a binary decision made on each pixel independent of its neighbors. Intensity above the threshold yields one classification, below the threshold another. This simple operation can be surprisingly effective, especially in data such as Computerized Tomography (CT), where the pixel value has real-world significance. The pixel value in a CT image is reported in Hounsfield units, which are calibrated to correspond to the attenuation of x-rays measured within the particular sample of tissue. Since bone is much more radiopaque than other tissues, bone can be segmented effectively in a CT image by setting the threshold between the Hounsfield value for bone and other tissues. An example is shown in Figure 2.1, in which a surface was reconstructed at the threshold between bone and other tissue in a CT scan of a fractured tibia.

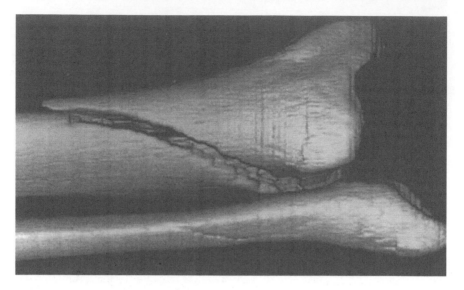

Figure 2.1. A simple threshold segments the bone from other tissue in a CT scan of a fractured tibia.

2.3.2 Adaptive Thresholding

In most applications of thresholding, the optimal threshold is not known before-hand, but depends instead upon the particular imaging modality, tissue-type, pa-tient, and even the individual image acquisition. For example, the intensity of a given pixel in an ultrasound image will depend on the scanner's current gain set-tings and the intervening tissue types. In such cases, an optimal threshold might be chosen manually by looking at the image itself or at a histogram of intensity values from the tissues to be segmented. Ideally, such a histogram will look like Figure 2.2, in which the two tissue types (A and B) have pixel intensities whose ranges hardly overlap. In such a case, where the histogram is bimodal, it is fairly easy to select the optimum threshold. In the real world, this is often not the case, however, and segmentation must take other factors into account, such as neigh-borhood statistics and expected shape.

2.3.3 Windowing

Often it is advantageous to change the range of intensity values for an image to adjust the brightness and contrast of what will be displayed on a screen. Dis-play is an essential part of image analysis, because it is the portal to the human visual system, whose capabilities far surpass any computer system likely to be

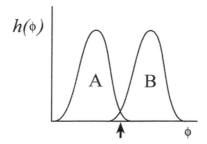

Figure 2.2. Histogram $h(\phi)$ of pixel intensity ϕ for tissue types A and B, with optimum threshold for segmentation (arrow).

built in the near future. The dynamic range of image display systems, between the brightest and darkest pixel the screen can produce, is quite small compared to the dynamic range of the human eye. Many medical imaging modalities are also capable of a greater dynamic range than can be displayed. The display itself therefore represents a bottleneck in the overall system. To minimize loss of information, the brightness and contrast of the image may be adjusted so that the brightest and darkest pixels in the image exactly match the extremes of the display, as shown in Figure 2.3. This is a linear function remapping the intensity of each pixel, with contrast being the slope of the function and brightness being the offset. In some clinical systems, the process is called *windowing*.

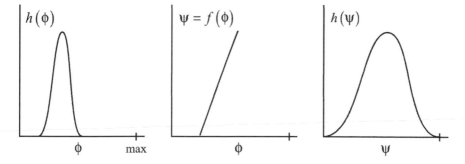

Figure 2.3. Windowing intensity histograms. Left: Histogram $h(\phi)$ of pixel intensity ϕ of original image, on a scale from zero to the maximum value of the display; Center: windowing function $\psi = f(\phi)$ to yield a new intensity ψ; Right: new histogram $h(\psi)$ in which the dynamic range of the image matches that of the display.

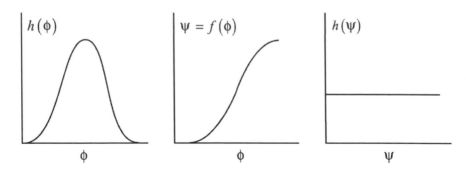

Figure 2.4. Histogram equalization. Left: Histogram $h(\phi)$ of pixel intensity ϕ of original image; Center: equalizing function $\psi = f(\phi)$ to yield a new intensity ψ; Right: new histogram $h(\psi)$ in which each intensity is equally represented.

2.3.4   Histogram Equalization

A further step may be taken to optimize the match between the image and the display system, which ensures that each level of pixel intensity is equally represented in the image. This is called histogram equalization, and although it is non-linear, we include it here as a useful point operator. As shown in Figure 2.4, histogram equalization entails finding a monotonic function $\psi = f(\phi)$ that remaps pixel intensity so that the histogram becomes a constant function of intensity ψ.

2.3.5   Color Maps

Color video images typically have three separate channels to record the red, green, and blue levels from the camera. Color displays are capable of communicating these images to the human visual system, but such displays also can be used to enhance the display of single-channel gray-scale images through the use of colors, artificially assigned to particular pixel intensities. Such color maps are often used to overlay additional information in anatomical images, such as Doppler flow in ultrasound images. The color map usually follows some perceptually consistent ordering scheme, such as the rainbow, or especially bright colors may be assigned to emphasize pixel values above certain levels.

2.3.6   Algebraic Operators

Some point operators combine two images, keeping each pixel separate from its neighbors, while combining it with the corresponding pixel in another image. Algebraic point operators may be used to add images together, for example, in averaging many images acquired from the same sample to reduce noise. Or they may be used to subtract one image from another, such as in Digital Subtraction Angiography where the difference between sequential fluoroscopic images, before and after contrast, leaves just the contrast without the anatomical structures. Another use of algebraic point operators is to mask an image by multiplying it by a second, binary image, containing ones where the first image is to be passed through, and zeroes where it is to be masked.

2.3.7 Location-Dependent Point Operators

All the point operators described so far apply the same operation to every pixel in the image. In some cases it is useful to make the operation depend upon pixel location, for example, to correct for spatial inhomogeneity in the image acquisition. Pixels are kept independent of each other, but in a framework where location within the image makes a difference.

2.4 Linear Filtering

Although many useful things can be done to an image using point operators on individual pixels, most image processing tasks require the simultaneous consideration of multiple pixels. Techniques that combine multiple pixels in a linear and space-invariant manner are known collectively as *linear filtering*. Two standard techniques for linear filtering, *convolution* and the *Fourier transform*, comprise the rest of this chapter.

2.4.1 Convolution

If you have ever adjusted the focus on a camera you have performed convolution in the continuous domain. If you have run a blurring function or performed edge enhancement using a commercial graphics program, it is likely that you were using discrete convolution. In image processing, convolution filters are used to make measurements, detect features, smooth noisy signals, and deconvolve the effects from image acquisition (e.g., deblurring the known optical artifacts from a telescope). The effect of the particular filter depends on the nature of the input filter, or *kernel*.

Convolution, denoted with the operator $\otimes$ can be described in 1D as a continuous operation applying the filter kernel $h(x)$ (itself an image) to an image $\phi(x)$ using the following integral form:

$$\phi(x) \otimes h(x) = \int_{-\infty}^{\infty} \phi(x - \tau) h(\tau) \, d\tau. \tag{2.5}$$

Note that the expression $\phi(x) \otimes h(x)$ itself describes an image mapping. Thus the following equality holds:

$$\phi(x) \otimes h(x) = (\phi \otimes h)(x). \tag{2.6}$$

The expressions $\phi(x) \otimes h(x)$ and $(\phi \otimes h)(x)$ are used interchangeably, with the position index x often omitted to help streamline and clarify the notation especially when higher dimensional (e.g., 3D) images are involved. In the practice of digital image processing, $h(x)$ typically does not have infinite extent, nor is it infinitely dense; however, when working in the continuous domain of linear filtering theory, considering kernels that are infinitely wide (thus avoiding the complications of truncation) is often more convenient. We will elaborate later on discrete functions with finite extent in the discussion on sampling.

We can generalize the 1D convolution operation to 2D and 3D images. Thus, in 2D, convolution becomes

$$\phi(x,y) \otimes h(x,y) = \int_{-\infty}^{\infty} \int_{-\infty}^{\infty} \phi(x-\tau, y-\nu) h(\tau, \nu) \, d\nu \, d\tau. \tag{2.7}$$

Similarly, convolution in 3D is expressed as

$$\phi(x,y,z) \otimes h(x,y,z) =$$
$$\int_{-\infty}^{\infty} \int_{-\infty}^{\infty} \int_{-\infty}^{\infty} \phi(x-\tau, y-\nu, z-\omega) h(\tau, \nu, \omega) \, d\omega \, d\nu \, d\tau. \tag{2.8}$$

2.4.2 Example—Convolution as Noise Reduction (1D)

One of the most common uses of convolution is as a filter to suppress noise in an image. Consider the Gaussian function as a smoothing filter kernel. If convolved with an image containing high-frequency noise, the result is an output image that locally averages the image intensities, reducing noise at the cost of some "sharpness" of the original image. That is, the resulting image has less high-frequency noise, but also has blurred edges.

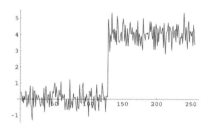

Figure 2.5. A 1D noisy step edge: an example input signal $\phi(x)$.

Figure 2.5 shows a noisy 1D step discontinuity, or edge. The function $\phi(x)$ is shown from $x = 0$ to $x = 255$ with a step function at $x = 128$. The signal to noise ratio is approximately 4:1.

Figures 2.6, 2.7, and 2.8, show a progression of filtered versions of the input from Figure 2.5 with Gaussian filter kernels of increasing aperture or width σ. The Gaussian function $g(x)$ is defined as

$$g(x,\sigma) = \frac{1}{\sqrt{2\pi}\sigma} e^{-\frac{x^2}{2\sigma^2}}. \tag{2.9}$$

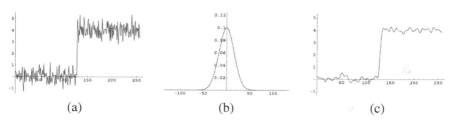

(a)   (b)   (c)

Figure 2.6. Gaussian filtering of input $\phi(x)$ where $\sigma = 16$. (a) Input $\phi(x)$; (b) a 1D Gaussian kernel, $g(x,\sigma)$ where $\sigma = 16$; (c) output of $\phi(x) \otimes g(x,16)$.

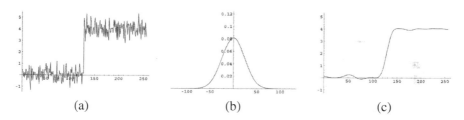

(a)   (b)   (c)

Figure 2.7. Gaussian filtering of input $\phi(x)$ where $\sigma = 24$. (a) Input $\phi(x)$; (b) a 1D Gaussian kernel, $g(x,\sigma)$ where $\sigma = 24$; (c) output of $\phi(x) \otimes g(x,24)$.

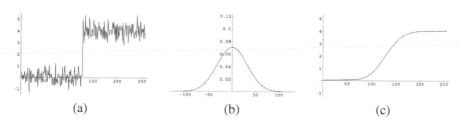

(a)   (b)   (c)

Figure 2.8. Gaussian filtering of input $\phi(x)$ where $\sigma = 32$. (a) Input $\phi(x)$; (b) a 1D Gaussian kernel, $g(x,\sigma)$ where $\sigma = 32$; (c) output of $\phi(x) \otimes g(x,32)$.

Notice how increasing the aperture of the Gaussian decreases the noise but also blurs the edge. This trade-off of resolution for noise reduction is one of the many considerations in the design of linear filter systems. An entire approach to image analysis, called "scale space," depends upon the Gaussian to produce a broad range of blurred versions of a given image.

2.4.3   Properties of the Convolution Operation

The nature of a linear filtering operation depends on the properties of the filter kernel. For instance, the "shape" of a 2D or 3D kernel will determine whether the operation remains invariant with respect to rotation. Independent of kernel shape, however, the convolution operation has many useful properties, including the following:

Convolution is linear:

$$(Ap + Bq) \otimes r = A(p \otimes r) + B(q \otimes r). \tag{2.10}$$

Convolution is commutative:

$$p \otimes q = q \otimes p. \tag{2.11}$$

Convolution is associative:

$$(p \otimes q) \otimes r = p \otimes (q \otimes r). \tag{2.12}$$

Convolution is distributive over addition:

$$p \otimes (q + r) = p \otimes q + p \otimes r. \tag{2.13}$$

These combined properties create the justification for using convolution as the principal operation in linear filtering.

2.4.4   Differentiation by Convolution

Differentiation may be accomplished by using convolution. We will see how this works in the discrete domain a little later, but first let us consider differentiation in the continuous domain. We can explicitly denote differentiation as convolution using the $\otimes$ operator:

$$\frac{\partial}{\partial x}\phi = \frac{\partial}{\partial x} \otimes \phi. \tag{2.14}$$

Equation (2.14) depicts the differential operator as a kernel by which convolution accomplishes differentiation. This is hard to illustrate using the ideal differential

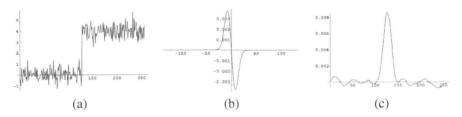

Figure 2.9. Taking derivatives of a noisy input by convolution with the derivative of a Gaussian kernel. (a) 1-D input $\phi(x)$; (b) $\frac{\partial}{\partial x}g(x,\sigma)$ where $\sigma = 3$; (c) $\phi(x) \otimes \frac{\partial}{\partial x}g(x,\sigma)$.

kernel, because such a kernel has infinitesimal width and infinite height. In the real world such a kernel is impossible. In any event, because differentiation enhances high-frequency noise, it is often necessary to regularize (smooth) a noisy image before computing its derivative, by first convolving the function with a smoothing kernel $h(x)$. It follows from the associative and commutative properties of convolution that

$$h(x) \otimes \frac{\partial}{\partial x}\phi(x) = \frac{\partial}{\partial x}h(x) \otimes \phi(x). \tag{2.15}$$

In other words, the derivative of a function $\phi(x)$ convolved with a filter kernel $h(x)$ is equivalent to convolving $\phi(x)$ with the derivative of $h(x)$. This suggests that one of the easiest ways to compute the derivative of a function is through convolution with the derivative of some smoothing kernel. We have already demonstrated the use of the Gaussian $g(x)$ as a smoothing kernel. The Gaussian's infinitely differentiable properties make it attractive for taking derivatives as well as smoothing.

Figure 2.9 depicts a noisy 1D input signal for which derivative information is desired. When convolved with the derivative of a Gaussian, the resulting output reports the *derivative* of a *smoothed version* of the input. By the commutative and associative properties of convolution, it can just as well be described as a *smoothed version* of the *derivative* of the input. This technique for differentiating functions can be extended to higher order derivatives, and to higher numbers of dimensions.

2.4.5 Convolution of Discretely Sampled Data

Convolution of discretely sampled data is similar to the continuous form, except that discrete summation is substituted for integration. Also, since discrete filter kernels cannot be implemented with infinite extent, the limits of the summations do not range from $-\infty$ to ∞ but are rather constrained to the size of the filter kernel. In 2D, a discrete convolution of image P with finite kernel Q looks like

$$P_{x,y} \otimes Q_{x,y} = \sum_{i}^{domain_x[Q]} \sum_{j}^{domain_y[Q]} P_{x-i,y-j} Q_{i,j}. \qquad (2.16)$$

The discrete version of convolution shares all the above-mentioned attributes of its continuous cousin. Convolution of discretely sampled data is linear, commutative, associative, and distributive over addition. As in the continuous domain, convolution in the discrete domain can be used to smooth noisy data or detect boundaries, depending on the choice of kernel. An example of discrete convolution for differentiation in 2D has already been seen in the *central differences operation* (Equation (2.4)). The kernel in that case would be

$$\begin{bmatrix} 0 & 0 & 0 \\ -0.5 & 0 & 0.5 \\ 0 & 0 & 0 \end{bmatrix}, \qquad (2.17)$$

which has the effect of subtracting the pixel to the left from the pixel to the right, just as Equation (2.4) specifies, to represent the x-component of the gradient. The following section shows more examples using convolution for smoothing and taking derivatives in 2D.

2.4.6   The Binomial Kernel

One particularly useful kernel for discrete convolution is the binomial kernel, generated by repeated convolutions with a simple box of identical values, as shown here:

$$[1], \begin{bmatrix} 1 & 1 \\ 1 & 1 \end{bmatrix}, \begin{bmatrix} 1 & 2 & 1 \\ 2 & 4 & 2 \\ 1 & 2 & 1 \end{bmatrix}. \qquad (2.18)$$

With successive iterations, the binomial kernel approaches a Gaussian shape (by the Central Limit Theorem). The scale of the Gaussian is determined by the number of iterations. (The kernels shown in Equation (2.18) would be normalized by 1/4 with each iteration.)

2.4.7   Difference of Offset Gaussians

Boundaries in an image represent areas of high gradient magnitude, i.e., the image intensity increases or decreases rapidly across the edge. The gradient vector,

$$\begin{bmatrix} \frac{\partial \phi}{\partial x}, \frac{\partial \phi}{\partial y} \end{bmatrix}, \qquad (2.19)$$

is oriented in the direction of the steepest change in image intensity, normal to the implied boundary. The gradient magnitude,

$$\left(\frac{\partial \phi}{\partial x}\right)^2 + \left(\frac{\partial \phi}{\partial y}\right)^2, \tag{2.20}$$

represents an orientation-independent measure of boundary strength. How can we determine the individual components of the gradient vector?

We have seen the *central difference operator* already. Another way to represent a gradient component is to use a kernel known as the *difference of offset gaussians* (DoOG).

$$\left[\begin{array}{ccc} -1 & 0 & 1 \end{array}\right], \left[\begin{array}{ccccc} -1 & -2 & 0 & 2 & 1 \\ -2 & -4 & 0 & 4 & 2 \\ -1 & -2 & 0 & 2 & 1 \end{array}\right]. \tag{2.21}$$

The two DoOG kernels shown in Equation (2.21) correspond to iterations 1 and 3 in Equation (2.18). Each kernel is the difference between two copies of the corresponding binomial kernels displaced along the x-axis. Convolution with these DoOG kernels measures the gradient component in the x-direction. Similar kernels can be constructed by displacing copies of the binomial kernel in the y-direction to measure the y-component of the gradient. DoOG kernels can be constructed for images in 3D or higher dimensions. The number of iterations of the underlying binomial kernel determines the scale of the DoOG.

Figure 2.10 shows the results of detecting the individual gradient components as well as the gradient magnitude of a simple image containing a rectangular object. Notice the orientations of the edges detected by the individual gradient components, and the orientation-independence of the gradient magnitude.

2.4.8   2D Example—Higher-Order Differentiation

We have seen how the gradient vector represents the strength and orientation of the boundary. But what of higher order differentials? Where the first derivative (gradient) is represented as a vector, the second derivative is a matrix, known as the Jacobian

$$\left[\begin{array}{cc} \frac{\partial^2 \phi}{\partial x^2} & \frac{\partial^2 \phi}{\partial x \partial y} \\ \frac{\partial^2 \phi}{\partial y \partial x} & \frac{\partial^2 \phi}{\partial y^2} \end{array}\right]. \tag{2.22}$$

The elements of this matrix are useful in a number of ways. For example, the Laplacian,

$$\frac{\partial^2 \phi}{\partial x^2} + \frac{\partial^2 \phi}{\partial y^2}, \tag{2.23}$$

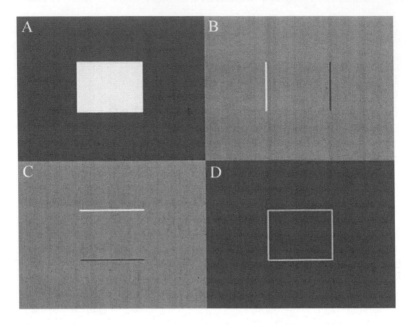

Figure 2.10. Gradient components and gradient magnitude: (A) original image; (B) x-component of gradient; (C) y-component of gradient; (D) gradient magnitude.

sums the diagonal terms of the Jacobian to yield a rotationally invariant representation of the second derivative of image intensity. A common 3×3 kernel representing the Laplacian in 2D is

$$\begin{bmatrix} -1 & -1 & -1 \\ -1 & 8 & -1 \\ -1 & -1 & -1 \end{bmatrix}. \tag{2.24}$$

Convolution with this matrix yields zero along a boundary, no matter what orientation the boundary has. For example, convolution with the following 3×3 patch of an image containing a diagonal boundary (between regions with intensities of 1 and 3 respectively),

$$\begin{bmatrix} 2 & 3 & 3 \\ 1 & 2 & 3 \\ 1 & 1 & 2 \end{bmatrix}, \tag{2.25}$$

yields zero for the center pixel (which is directly on the boundary).

For an intuitive understanding of why the second derivative should be zero on the boundary let us examine the 1D case.

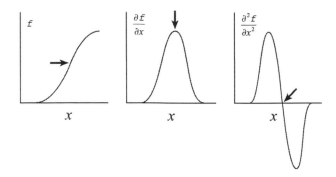

Figure 2.11. 1D case of the intensity function across a boundary, its first derivative, and its second derivative showing a zero crossing at the boundary location.

As shown in Figure 2.11, the boundary is located at an inflection point (arrow) in the intensity curve

$$\phi(x)$$

corresponding to a maximum in the first derivative

$$\frac{\partial \phi(x)}{\partial x},$$

and a zero crossing in the second derivative

$$\frac{\partial^2 \phi(x)}{\partial x^2}.$$

In two or more dimensions, we use the Laplacian to capture the "total" second derivative at a pixel. Mathematically, the Laplacian is the divergence of the gradient. Divergence is a common concept in fluid dynamics, describing how much more fluid enters a region than leaves it. In our case, it describes how much more gradient is "entering" a pixel than "leaving" it from any direction (making it rotationally invariant). A boundary exists where there is no net change in gradient (i.e., the gradient is at a maximum on the boundary), making the Laplacian zero, just as in the 1D case. The difference of Gaussian (DOG) kernel, which we will discuss in Section 2.4.9, yields results similar to the Laplacian, because the DOG kernel has a similar shape to the Laplacian.

Let us now consider the terms *off* the diagonal of the Jacobian matrix (Equation (2.22)). These can be used to measure the curvature of boundaries. The individual components of the gradient at a curved boundary change as one moves orthogonally to that particular component. This type of change shows up as partial second derivatives of intensity off the diagonal of the Jacobian. For illustration,

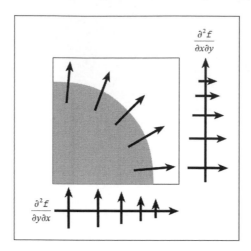

Figure 2.12. The curvature of a boundary shows up in the terms of the Jacobian off the diagonal.

consider the curved boundary segment shown in Figure 2.12. The component of the Jacobian,

$$\frac{\partial^2 \phi}{\partial y \partial x},$$

indicates a decrease in the y-component of the gradient as one moves in the positive x-direction, whereas

$$\frac{\partial^2 \phi}{\partial x \partial y}$$

shows a decrease in the x-component of the gradient as one moves in the positive y-direction.

It is possible to construct features using even higher order derivatives. In general, approaches to image analysis based on partial differential equations (PDEs) have been quite popular and well researched.

2.4.9   Difference of Gaussians (DOG)

Results similar to those just described for the Laplacian can be achieved using the difference between two concentric Gaussian functions with different apertures. This function is called a difference of Gaussians, or DOG, kernel (not to be confused with the difference of *offset* Gaussians, or DoOG, kernel already described; see Equation (2.21)). As with the Laplacian, the DOG kernel can detect edges independent of orientation, but not the orientation itself. The process is also called "unsharp masking." The result is an *edge-enhanced* image.

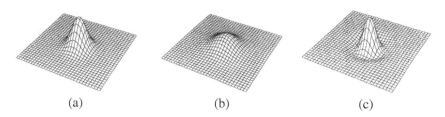

(a)   (b)   (c)

Figure 2.13. A difference of Gaussians filter kernel (depicted in 3D as a height field): (a) 2D Gaussian kernel $g(x,y,\sigma_1 = 16)$; (b) 2D Gaussian kernel $g(x,y,\sigma_2 = 32)$; (c) $g(x,y,\sigma_1) - g(x,y,\sigma_2)$.

Specifically, given a 2D input image $\phi(x,y)$ and two Gaussian filter kernels of differing aperture, $g(x,y,\sigma_1)$ and $g(x,y,\sigma_2)$ where $\sigma_1 < \sigma_2$, an edge enhanced image $\phi'(x,y)$ can be formed by a linear combination of the two filters. That is,

$$\phi'(x,y) = \phi(x,y) \otimes (g(x,y,\sigma_1) - g(x,y,\sigma_2)). \qquad (2.26)$$

From Equation (2.13) it follows that $\phi'(x,y)$ is also the subtraction of two filtered images

$$\phi'(x,y) = (\phi(x,y) \otimes g(x,y,\sigma_1)) - (\phi(x,y) \otimes g(x,y,\sigma_2)). \qquad (2.27)$$

In other words, the difference of two "blurred" versions of the same image yields a version of the original image where edge information is accentuated, and where regions of continuous intensity have a zero value. Areas near boundaries are strongly negative or strongly positive (depending on which side of a boundary you are on) and the boundary itself is denoted by the closed curve of *zero-crossings* (as with the Laplacian), where the image intensity crosses from positive to negative values.

Figure 2.13 shows a particular 2D difference of Gaussians kernel

$$g(x,y,\sigma_1 = 16) - g(x,y,\sigma_2 = 32) \qquad (2.28)$$

as a 3D function where intensity is plotted as a height field above an x-y plane. Figure 2.14 is the same kernel show in Figure 2.13 applied to a circular pulse function (white circle on a gray background), also represented as a height field. Figure 2.15 shows the same functions as Figure 2.14, but depicted as a 2D intensity field. The areas in the original image with constant intensity, inside and outside the circle, result in a flat signal value of approximately zero (gray). Moving outward from the center of the circle, at the boundary there is a rise in the output intensity followed by a negative dip. The zero crossing between them represents the boundary itself.

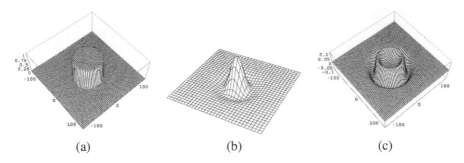

(a)   (b)   (c)

Figure 2.14. 2D image filtered by a difference of Gaussians (depicted in 3D as a height field): (a) 2D input $\phi(x,y)$; (b) $g(x,y,\sigma_1 = 16) - g(x,y,\sigma_2 = 32)$; (c) $\phi(x,y) \otimes (g(x,y,\sigma_1) - g(x,y,\sigma_2))$.

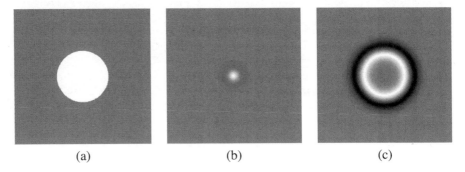

(a)   (b)   (c)

Figure 2.15. 2D image filtered by a difference of Gaussians (depicted as a 2D grayscale density field). (a) 2D input $\phi(x,y)$; (b) $g(x,y,\sigma_1 = 16) - g(x,y,\sigma_2 = 32)$; (c) $\phi(x,y) \otimes (g(x,y,\sigma_1) - g(x,y,\sigma_2))$.

Another version of the DOG kernel deserves mention. By doubling the amplitude of the Gaussian with the smaller aperture in Equation (2.28) to yield a new kernel,

$$2g(x,y,\sigma_1 = 16) - g(x,y,\sigma_2 = 32), \qquad (2.29)$$

we modify its behavior from *edge detection* to *contrast enhancement*. The distinction is somewhat subtle. The area under the kernel is now 1 instead of 0. Thus, in image regions of constant pixel value, convolution passes through the pixel value unchanged rather than setting it to 0. Figure 2.16 shows the results. This DOG kernel returns unchanged the non-zero pixel value inside the circle, while exaggerating the discontinuity in a band around the boundary. Compare Figure 2.15(c), in which the interior of the circle is set to 0 (gray), to Figure 2.16(c), in which the interior of the circle is passed through unchanged (white).

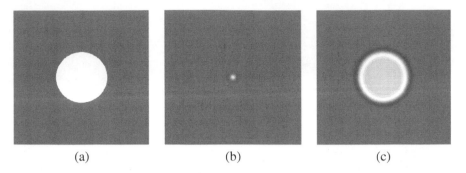

(a)   (b)   (c)

Figure 2.16. A difference of Gausssians filter used for contrast enhancement (depicted as a 2D greyscale density field). This method is also known as *unsharp masking*: (a) 2D input $\phi(x,y)$; (b) $2g(x,y,\sigma_1 = 16) - g(x,y,\sigma_2 = 32)$; (c) $\phi(x,y) \otimes (2g(x,y,\sigma_1) - g(x,y,\sigma_2))$.

2.4.10   Beware of Aliasing

The primary caveat regarding discrete convolution is that all such discrete operations are subject to error induced by the effects of sampling, a class of artifacts known as *aliasing*. Before we can appreciate the problems associated with sampling, we must first understand the nature of the sampling operation and its effects on the information embedded within the image. For that, we need the Fourier transform.

2.5   The Fourier Transform

One of the most important tools for understanding images (1D, 2D, 3D or even higher) is the Fourier transform. The Fourier transform of an image decomposes it into component sinusoidal spatial functions. The representation of an image as its constituent frequencies is a computationally useful means of viewing and manipulating image data. The relationship between an input function and its Fourier transform is governed by the following equation:

$$\mathcal{F}(\phi(x)) = \int_{-\infty}^{\infty} \phi(x)e^{-i2\pi\omega x}dx = \Phi(\omega). \tag{2.30}$$

This relationship maps the spatial domain x to the frequency domain ω. These are often referred to, respectively, as "image space" and "frequency space." The Fourier transform $\mathcal{F}$ of $\phi(x)$ is a complex number $\Phi(\omega)$, encompassing magnitude and phase at a particular frequency ω. As a transform, $\mathcal{F}$ is reversible. The inverse Fourier transform $\mathcal{F}^{-1}$ is

$$\mathcal{F}^{-1}(\Phi(\omega)) = \int_{-\infty}^{\infty} \Phi(\omega)e^{i2\pi\omega x}d\omega = \phi(x), \tag{2.31}$$

which reconstructs $\phi(x)$ from its spectrum $\Phi(\omega)$. It should be emphasized that, in the case of images, these are *spatial*, not temporal, frequencies: periodic variations of intensity in space. Fourier transforms can also be performed in 2D and 3D, as will be described in Section 2.5.3.

2.5.1   Key Properties of the Fourier Transforms

There are some key properties of frequency space and the Fourier transform that creates it, which we will discuss for both the continuous and discrete domains. We will emphasize two particular characteristics of the Fourier transform, which have to do with the effects on frequency space of scaling and convolution in image space.

Scaling in Image Space ↔ Inverse Scaling in Frequency Space

Compression of a function in the spatial domain leads to an inversely proportional broadening of the spectrum in the frequency domain, i.e., reducing the size of an image increases its spatial frequencies. Mathematically speaking, for $a \geq 1$,

$$\mathcal{F}(\phi(ax)) = \frac{1}{a}\Phi\left(\frac{\omega}{a}\right). \tag{2.32}$$

The opposite is also true. Increasing the size of an image decreases its spatial frequencies, which is to say, Equation (2.32) with $0 < a \leq 1$.

Convolution in Image Space ↔ Multiplication in Frequency Space

Convolution between an image and a kernel in image space translates to multiplication of their spectra in frequency space. That is, given a kernel $h(x)$, and its Fourier transform, $H(\omega)$, the following relation holds:

$$\mathcal{F}(\phi(x) \otimes h(x)) = \Phi(\omega)H(\omega). \tag{2.33}$$

This is known as the *convolution theorem*, and it is perhaps the most important theorem in linear system analysis. Given the convolution theorem, it is easy to conceive of why convolution is distributive over addition, symmetric, commutative, etc. It shares most of the properties of multiplication.

Linear filtering under these circumstances can now be considered in a different light. It can be changed from an integral with infinite extent in image space (a difficult continuous operation to implement with a digital computer) to a multiplication in frequency space. Moving to and from frequency space (simple multiplication replaces convolution) requires that only two forward Fourier transforms and one inverse Fourier transform be calculated

$$\phi(x) * h(x) = \mathcal{F}^{-1}(\mathcal{F}(\phi(x))\mathcal{F}(h(x))). \tag{2.34}$$

Depending on the size of both the filter kernel and the input signal, casting the convolution problem in the frequency domain often makes difficult iterative problems more tractable.

Due to the symmetry of image space and frequency space (note the similarity between Equations (2.30) and (2.31)), the converse of the convolution theorem is also true: Multiplication in image space corresponds to convolution in frequency space. This leads to the sampling artifact, *aliasing*, as will be discussed with regards to the Comb function in Section 2.5.2.

There are many other properties of the Fourier transform associated with its symmetry and its use of complex exponentials, which we cannot hope to cover here. However, the two properties just discussed, of spatial/frequency scaling and convolution/multiplication, allow us to make some important points in the following sections.

2.5.2 Four Important Transform Pairs

Four functions commonly used in manipulating images are the comb function (used to sample data discretely in the image domain), the box filter (a square pulse, used as a nearest neighbor interpolant in reconstruction), the pyramid filter (a triangular shaped filter used for linear interpolation in reconstruction), and the Gaussian filter (a good all purpose filter, except for its infinite extent).

Figures 2.17, 2.18, 2.19, and 2.20 show these functions represented in 1D image space, along with their Fourier transforms in frequency space. By studying these four transform pairs, a number of important characteristics of the relationships between image space and frequency space become clear.

Comb Function

A comb function is a series of *impulses*. Impulses, or *delta functions*, play a central role in systems theory as the identity element of convolution. In the continuous domain, the impulse is an infinitely narrow, infinitely high spike with finite area. Convolution of any function $f(x)$ with an impulse yields a perfect copy of $f(x)$, as if taking a picture with an infinitely high-resolution camera. This is because multiplication of $f(x)$ by an impulse isolates a perfect sample of $f(x)$, suitable for integration, at the location of the impulse. Multiplying $f(x)$ by the comb function captures a set of such samples, evenly spaced, to produce a digital image.

It is essential to understand what happens in frequency space, as the price for such sampling. As shown in Figure 2.17, the spectrum of a comb function is itself a comb function in frequency space. Each impulse in the spectrum represents a single frequency of infinite energy but finite power, i.e., a sinusoid. Multiplying (sampling) $f(x)$ with the comb function in image space leads to convolving their

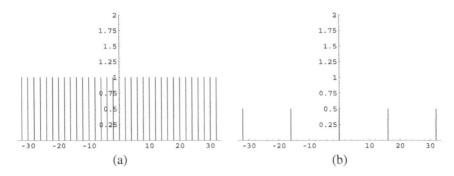

Figure 2.17. The comb filter and its Fourier transform. This function is used for sampling continuous functions into discrete grids: (a) sampling comb; (b) the Fourier transform of the sampling comb.

Fourier transforms, as already discussed in Section 2.5.1. So whatever spectrum is produced by the Fourier transform $F(x) = \mathcal{F}(f(x))$ will be repeated by the comb function in frequency space at each impulse. As long as $F(x)$ is narrow enough not to overlap with its neighboring copy, this is not a problem. But if $F(x)$ is broader than the distance between two adjacent impulses (in the spectrum of the comb function), aliasing will occur. This restriction on the width of the spectrum $F(x)$ is known as the *Nyquist criterion*. (It is also known as the *Shannon sampling theorem*). The maximum allowable frequency to avoid aliasing is called the *Nyquist frequency*, which is half the sampling frequency.

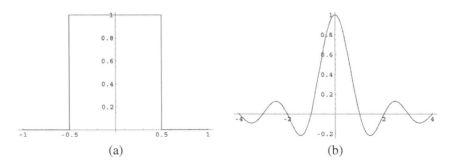

Figure 2.18. The box filter and its Fourier transform, the sinc function. This function is used as a nearest-neighbor interpolant to quickly reconstruct sampled functions: (a) the box function (or the nearest-neighbor interpolant); (b) the Fourier transform of the box function (also known as the sinc function).

Box Filter

The box filter, shown in Figure 2.18(a), is used as a reconstruction function in nearest-neighbor interpolation. Interpolation is a common task in image analysis, to re-sample an already discrete image to a different lattice of sample locations. Nearest-neighbor interpolation, as its name suggests, entails copying the value of the nearest pixel, a brute-force but rapid procedure. The Fourier transform of the box filter in Figure 2.18(b) demonstrates the potential for aliasing. The spectrum takes the form of a *sinc function*. Whereas the primary lobe of the sinc function is fairly well contained, side-lobes continue at regular intervals quite high up in frequency. If significant lobes extend beyond the Nyquist frequency, aliasing will occur. In a graphical display of the image, this artifact expresses itself as "jaggies."

Pyramid Filter

While the box filter just discussed is used for nearest-neighbor interpolation, the pyramid filter, shown in Figure 2.19(a), is used for linear interpolation. This is a computationally more expensive, but more accurate, operation than nearest-neighbor interpolation, with a correspondingly better behavior in terms of aliasing. The spectrum of the pyramid filter, shown in Figure 2.19(b), has smaller side-lobes than those of the box filter in Figure 2.18(b), and therefore has less tendency to produce aliasing.

Gaussian Filter

The ultimate filter in terms of avoiding side lobes is the Gaussian filter. Since the Fourier transform of a Gaussian is itself a Gaussian, as shown in Figure 2.20,

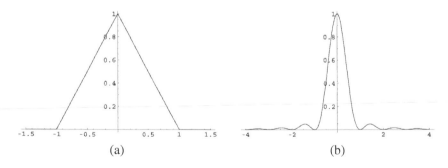

(a)   (b)

Figure 2.19. The pyramid filter and its Fourier transform. This function is used for linear interpolation and reconstruction of sampled functions. (a) The pyramid function (or the linear interpolant); (b) The Fourier transform of the pyramid function.

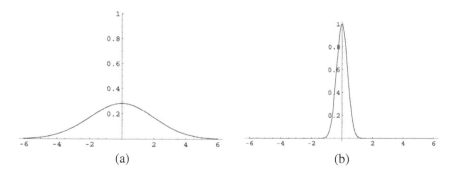

Figure 2.20. The Gaussian filter and its Fourier transform. The Gaussian is a smooth, continuous filter kernel in both image space as well as frequency space. The Fourier transform of a Gaussian function is also a Gaussian: (a) image space Gaussian; (b) the Fourier transform of a Gaussain (a Gaussian).

there are no side-lobes at all. Gaussian filters are widely used to blur images, and form the cornerstone of an entire approach to image analysis known as *scale space*.

2.5.3   The Fourier Transform in 2-D and 3-D

Thus far we have discussed Fourier transforms in only one dimension. Images, of course, tend to have greater numbers of dimensions. The Fourier transform, as shown in Equation (2.30), can be extended to 2D as follows:

$$\mathcal{F}(\phi(x,y)) = \int_{-\infty}^{\infty}\int_{-\infty}^{\infty} \phi(x,y)e^{-i2\pi(\omega x+\nu y)}\,dxdy = \Phi(\omega,\nu). \qquad (2.35)$$

The inverse Fourier transform (Equation (2.31)) becomes

$$\mathcal{F}^{-1}(\Phi(\omega,\nu)) = \int_{-\infty}^{\infty}\int_{-\infty}^{\infty} \Phi(\omega,\nu)e^{i2\pi(\omega x+\nu y)}\,d\omega d\nu = \phi(x,y). \qquad (2.36)$$

This process can be extended to 3D and beyond, in a similar manner. A number of important new properties apply to the multi-dimensional Fourier transform, of which we will mention three.

Separability

Among its most important attributes, the multi-dimensional Fourier transform is separable in each of the orthonormal basis dimensions. Thus Equation (2.35) can be rewritten as

$$\mathcal{F}(\phi(x,y)) = \int_{-\infty}^{\infty}\left[\int_{-\infty}^{\infty}\phi(x,y)e^{-i2\pi\omega x}dx\right]e^{-i2\pi vy}dy = \Phi(\omega,v). \qquad (2.37)$$

If the function $\phi(x,y)$ can be separated into component functions

$$\phi(x,y) = \phi_1(x)\phi_2(y), \qquad (2.38)$$

then the Fourier transform can also be separated into components

$$\Phi(\omega,v) = \Phi_1(\omega)\Phi_2(v), \qquad (2.39)$$

where

$$\Phi_1(\omega) = \mathcal{F}(\phi_1(x)) \text{ and } \Phi_2(v) = \mathcal{F}(\phi_2(y)). \qquad (2.40)$$

Separability is applicable, for example, to the 2D Gaussian function, which is itself separable:

$$g(x,y) = g_1(x)g_2(y). \qquad (2.41)$$

Therefore, the Fourier transform of $g(x,y)$ is

$$\mathcal{F}(g(x,y)) = G(\omega,v) = G_1(\omega)G_2(v), \qquad (2.42)$$

where

$$G_1(\omega) = \mathcal{F}(g_1(x)) \text{ and } G_2(v) = \mathcal{F}(g_2(y)). \qquad (2.43)$$

Rotational Invariance

Consider rotation of the coordinate system of a 2D image about the origin of the (x,y) plane by an angle θ through which each location (x,y) becomes (x',y'):

$$\begin{bmatrix} x' \\ y' \end{bmatrix} = \begin{bmatrix} \cos\theta & \sin\theta \\ -\sin\theta & \cos\theta \end{bmatrix}\begin{bmatrix} x \\ y \end{bmatrix}. \qquad (2.44)$$

The 2D Fourier transform will likewise be rotated about the origin of the (ω,v) plane by an angle θ with each location (ω,v) becoming (ω',v'), where

$$\begin{bmatrix} \omega' \\ v' \end{bmatrix} = \begin{bmatrix} \cos\theta & \sin\theta \\ -\sin\theta & \cos\theta \end{bmatrix}\begin{bmatrix} \omega \\ v \end{bmatrix}. \qquad (2.45)$$

The image and its transform rotate together about their origins.

Projection

Consider the projection $\mathcal{P}_x$ of $\phi(x,y)$ onto the x-axis:

$$\mathcal{P}_x(\phi(x,y)) = \int_{-\infty}^{\infty} \phi(x,y)dy. \tag{2.46}$$

Here, seperability comes into play. By setting $v = 0$, Equation (2.37), reduces to

$$\int_{-\infty}^{\infty}\int_{-\infty}^{\infty} \phi(x,y)e^{-i2\pi\omega x}dxdy = \Phi(\omega,0) = \mathcal{F}(\mathcal{P}_x(\phi(x,y))). \tag{2.47}$$

In other words, the Fourier transform of the projection of $\phi(x,y)$ onto the x-axis is nothing but the 2D Fourier transform $\Phi(\omega,v)$ evaluated along the v-axis. Combining this result with rotational invariance leads to the somewhat surprising result that projecting a 2D image onto any line passing through the origin in image space transforms into a 1D spectrum taken from the same line along the Fourier transform in frequency space. This leads to the Radon transform, central to the process of filtered back projection, by which images are formed in computed tomography (CT), and other useful applications.

2.5.4   The Image as a Periodic Signal

The Fourier transform as shown in Equation (2.30) is an integral with infinite extent. A discrete implementation that can run on a digital computer is required to translate it from a useful mathematical abstraction to a practical tool. This raises several issues. In Section 2.5.2, while discussing the comb function, we addressed the issue of aliasing, which occurs when the spectrum contains frequencies above the Nyquist frequency. Now we take another look at sampling as it applies to Fourier analysis of images. The adaptation of the Fourier transform to discretely sampled data results in the *discrete Fourier transform*. Like its continuous cousin, the discrete Fourier transform can be applied to any signal, whether it is periodic or not. However, it is usually advantageous to treat an image as periodic. A digital image does not have infinite extent, nor can it be sampled at infinitesimal intervals. Computationally, it makes sense to use the Fourier series, instead of the transform. The Fourier series decomposes any periodic signal into discrete harmonics of the fundamental frequency.

To create a periodic signal, an image with n uniformly spaced samples is "wrapped" so that samples 0 through $n - 1$ index the data array exactly, while sample n "wraps" to 0. In fact, if n is the number of uniform samples, the effective index x_{eff} of a location x is, $x_{eff} = x \bmod n$. This repetition is applied to each of the dimensions (x, y, and z).

The interpretation of a discrete image as a periodic signal can also be viewed as an infinite concatenation of the image with itself. The wavelength corresponding to the fundamental frequency is the width of the image. This allows us to represent the image as a sum of a finite number of sinusoidal functions: the discrete Fourier series, which is a linear transformation between n samples in image space and n samples in frequency space. The frequency space representation is capable of reproducing the original sampled image exactly without loss of information.

Computationally, the algorithm of choice for doing this is the fast Fourier transform (FFT), which is efficient even for very large data sets. One restriction of the FFT is that the number of samples in each dimension be a power of 2 (e.g., 128, 256, 512,...), but luckily many images already satisfy this requirement. When it is not the case, images may be "padded" with extra zeros.

2.6 Summary

In this chapter we have reviewed basic image processing and linear operators. These operators may affect pixels individually, or combine pixels using convolution and Fourier analysis. Filtering and frequency space are essential to the understanding of digital sampling theory. As we have seen in the convolution theorem, linear filtering and Fourier analysis are tightly linked, with profound implications in the understanding and manipulation of digital images. Aliasing and other sampling errors are directly related to the properties of the Fourier transform and to the behavior of filters and input signals when discretized. These concepts are borrowed from classical signal processing, and are widely used within the scanners that form the images to begin with. Adapted to the multidimensionality of image space, linear operators form the foundation upon which the higher levels of image analysis discussed in the remainder of this book are built.

References

[1] Castleman, K. *Digital Image Processing*. Englewood Cliffs, NJ, Prentice Hall, 1996.

[2] Oppenheim, A., and A. Willsky. *Signals and Systems*, (2nd Edition). Englewood Cliffs, NJ, Prentice Hall, 1996.

Statistics of Pattern Recognition

Stephen R. Aylward
University of North Carolina at Chapel Hill

Jisung Kim
University of North Carolina at Chapel Hill

> Normality is a myth. There never was, and never will be, a normal distribution.

– Geary, 1947 [21]

> The art of being wise is the art of knowing what to overlook.

– James, 1890

3.1 Introduction

Most scientists encounter problems which involve statistical analysis. What populations are present in my data? How do these populations differ? Have I collected enough data? From which population did this instance originate? In pattern recognition, collections of instances, *samples*, are used to form models of populations, and those models are used to answer questions such as the ones listed above. This chapter is an illustrative introduction to the subset of statistical pattern recognition concerned with the classification/labeling of unknown instances given collections of labeled instances. For an excellent and more complete overview of statistical pattern recognition methods, the reader is encouraged to consult the synopsis provided by Duda, *et al.* [18].

In this chapter we define an *instance* as n measurements obtained from a single object. An instance captures the *features* of an object and maps that object to

47

a point in an *n*-dimensional *feature space*. A sample is a collection of instances from a population. Presumably, every object from a population will share a set of common traits that are captured by the measurements used to define the instances in a sample. Variations among instances from the same population are indicative of correlations, noise, and/or a lack of correspondence between the measurements and the common traits of the population. This variance determines how a population's samples will be distributed in feature space.

Classification methods attempt to assign class labels to unlabeled instances based on information derived from labeled instances from different classes. Typically classifiers model, via implicitly or explicitly estimated *density functions*, the feature-space distribution of the different classes' training instances. A density function for a class is used to estimate the probability that an instance originated from that class. Density functions are distinguished by the assumptions they employ and their parameter specification ("training") process. A frequently used density function is the multivariate normal, i.e., Gaussian, distribution. Gaussian density functions are completely parameterized by a mean vector and a covariance matrix. While Geary [21] and others may claim that Gaussian assumptions are rarely correct, they have been shown to be applicable to many real-world populations [15, 2, 3, 39, 60]. In general, when a particular density function shape, such as Gaussian, is assumed by a classifier and that assumption is correct, the classifier produces consistent and optimally accurate labelings with respect to the features being used. Additionally, even when the assumed density functions are not correct, the populations' separation may be large enough that suboptimal density functions will still provide sufficient accuracy.

In this chapter we primarily focus on the continuum of classifiers that implicitly or explicitly utilize Gaussian components: *k*-nearest neighbor, Parzen windowing, *k*-means, linear, Gaussian, and finite Gaussian mixture modeling. We also discuss multi-layered perceptrons trained via backpropagation. Note that *k*-means and finite Gaussian mixture modeling are often applied as methods for clustering data; in this chapter, clusterers are treated as another set of methods for density estimation for maximum likelihood classification. Given this common view, accuracy and consistency given different problems and samples can be quantified and compared.

Three methods are commonly used to estimate the accuracy and consistency of classifiers for particular problems: Monte Carlo simulation, goodness-of-fit measures (for classifiers that form explicit density estimates), and statistical theorems (for select, well understood, classifiers). In this chapter we utilize Monte Carlo simulations to compare the accuracy and consistency of a wide range of classification methods on two medically motivated simulated data problems. We also introduce goodness-of-fit measures. Again, the reader is encouraged to consult other texts [18, 52] for additional details.

This chapter is organized as follows. We begin with a discussion of density estimation, i.e., density functions, feature spaces, and scattergrams. We also detail the two simulated problems that are used to compare and illustrate the operation of various statistical pattern recognition methods. We conclude with a summary of goodness-of-fit measures that must be applied to ultimately assess any density estimation method's validity for a given problem.

3.2   Background

This section explains the concepts necessary for understanding the operating characteristics, strengths, and weaknesses of several common classification systems. First we introduce the two related, two-class, two-feature classification problems that are used for illustration throughout the remainder of the chapter. Second we discuss how samples, density functions, decision bounds, and decision regions can be visualized in feature space.

3.2.1   Two Related Two-Feature, Two-Class Problems

For the visualization and analysis of the operation and performance of the pattern recognition systems discussed in this chapter, two, two-feature ($n = 2$), two-class ($n_c = 2$) problems are presented. For both problems, the two features are designated f_0 and f_1. These features are discrete; they can attain any integer value from 0 to 255. The distributions in these problems are motivated by the intensity distributions of tissues in inhomogeneous magnetic resonance (MR) images.

Consider the proton density MR image shown in Figure 3.1. It contains an inhomogeneity which is revealed by a dimming in the inferior cerebellum (lower

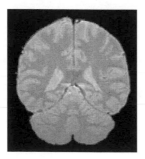

Figure 3.1. Proton density (PD) MR image. [The medical data used to generate this and all of the other images in this chapter were acquired in and provided by the Department of Radiology at The University of North Carolina at Chapel Hill]

portion of the brain). Locally within this image, the tissues have Gaussian distributions, but the means and covariances of those distributions change across the image because of the intensity inhomogeneity. Such spatially varying distributions produce *extruded Gaussian distributions*.

Problem 1: Description and Justification

For Problem 1, the two populations are designated Class A and Class B. Class B is an extruded Gaussian distribution—see this chapter's appendix for additional details. Class A, on the other hand, does not suffer from the inhomogeneity. Class A is represented by a multivariate Gaussian distribution with a large isotropic variance, i.e., its variance is circularly symmetric. Its parameters are given in Table 3.1.

Class A

μ_{f_0}	μ_{f_1}	σ^2
128	128	1296

Table 3.1.   The parameters of the Gaussian that is Class A.

Problem 2: Description and Justification

The second two-feature, two-class problem that we will use to evaluate the statistical pattern recognition methods of this chapter arises when the intensity inhomogeneities are effectively eliminated from Problem 1's data. Image processing techniques, e.g., mean field correction, can remove intensity inhomogeneities from many images [28, 60]. We simulate the application of such methods to Problem 1's data, and thereby define Class A' and Class B'. Class A and Class A' have identical parameterizations. Class B, however, is "transformed" to an elliptical Gaussian distribution, Class B'. The parameters of Problem 2's classes are given in Table 3.2.

Problem 2's training and testing sample sizes are the same as those in Problem 1. Specifically: $|S(\text{train}|A)| = 900$, $|S(\text{test}|A)| = 2700$, $|S(\text{train}|B)| = 900$, and $|S(\text{test}|B)| = 2700$.

Problems 1 and 2: Sample Size and Data Normalization

The random effects inherent in finite empirical sampling cause estimated parameters to differ from a population's true parameters. As previously stated, the accuracy of the assumptions made during parameter estimation drives the accuracy and consistency of the estimated parameters compared with the underlying population parameters. Furthermore, large sample sizes will produce more accurate and consistent parameter estimates.

Class A'

		f_0	f_1
Mean		128	128
Covariance	f_0	1296	0
	f_1	0	1296

Class B'

		f_0	f_1
Mean		128	56
Covariance	f_0	576	0
	f_1	0	324

Table 3.2.   The parameters of the Gaussians for Class A' and Class B'.

Using high-gradient MR and multi-detector CT systems, small tumors, multiple sclerosis lesions, and other pathologies can encompass 900 voxels within an image. Therefore, for Problems 1 and 2, we set the training sample sizes for class B and B' to 900 instances and set the testing sample sizes to 2700 instances:

- $|S(\text{train}|B)| = 900$

- $|S(\text{test}|B)| = 2700$

- $|S(\text{train}|B')| = 900$

- $|S(\text{test}|B')| = 2700$

To indicate equal class *a priori* probabilities, the competing populations, Class A and A', are represented by an equal number of training and testing instances:

- $|S(\text{train}|A)| = 900$

- $|S(\text{test}|A)| = 2700$

- $|S(\text{train}|A')| = 900$

- $|S(\text{test}|A')| = 2700$

The effect of empirical sampling is illustrated by comparing the estimated parameters in Table 3.3 with the population parameters in Tables 3.1–3.2. When an assumed distribution matches the underlying population's distribution, the estimated parameters closely match the population's actual parameters, e.g., Class A, A', and B'.

Class A

		f_0	f_1
Mean		128.345	128.288
Covariance	f_0	1247.353	0.055
	f_1	0.055	1335.568

Class B

		f_0	f_1
Mean		131.113	84.201
Covariance	f_0	1480.644	0.237
	f_1	0.237	477.967

Class A$'$

		f_0	f_1
Mean		128.345	128.288
Covariance	f_0	1247.353	0.055
	f_1	0.055	1335.568

Class B$'$

		f_0	f_1
Mean		128.230	56.144
Covariance	f_0	554.379	0.018
	f_1	0.018	333.920

Table 3.3. Estimated Gaussian parameters of Class A, B, A$'$, and B$'$.

One key concept of the data of these problems is that all features are considered to be of commensurate units in a Euclidean space. For most pattern recognition problems this is an acceptable assumption. When it does not hold, techniques exist for rescaling the feature values so that their marginal/individual variances are normalized and thus their units are made commensurate, e.g., data whitening and factor analysis of correlation and covariance [17, 26, 27, 36]. Neural network researchers have demonstrated numerous applications in which such transforms have proven to be beneficial to the parameter selection processes [36]. However, this normalization should be applied with care since it is possible to eliminate exactly the relationship sought in the data.

Another key concept, given the digital domain in which the algorithms of this book are being implemented, is that only discrete feature spaces are being considered. That is, feature values have a fixed level of precision, or, equivalently, they can only attain a finite set of values. This constraint is not limiting in most pattern recognition problems. Digital imaging techniques already exhibit these constraints. Binning is the process by which continuous data are mapped to a discrete format, i.e., a collection of bins/cells/buckets. A number of binning techniques exist which minimize the possibility that this transform will have a significant effect on the representation of the continuous distribution. Binning is briefly discussed in the section on goodness-of-fit measures.

The remainder of this chapter introduces a variety of classification methods and quantifies their accuracy and consistency for random samplings of Class A, B, A′, and B′. To introduce the classifiers, we use visualizations of the training data, testing data, and the labels generated by each classifier. To quantify accuracy and consistency, the repeated evaluation of each classification method using random samplings from each problem allow Monte Carlo analyses to be applied. Methods for such visual and quantitative comparisons of classifiers are discussed next.

3.2.2 Visual Comparison of Classifiers

In the analysis of a single set of training and testing data, the visualization of those data, as well as the labelings produced by a specific classifier using those data, can provide insight into the strengths and weaknesses of the selected features and classifiers.

Scattergrams: Viewing Data

Scattergrams are n-dimensional plots in which the intensity at each point corresponds to the relative frequency of occurrence of a combination of n feature values in a sample. The shape of the distribution of the sample is better revealed in this manner; see Figure 3.2. Groups of instances having similar values produce clouds of high intensity.

If $n > 2$, it is possible to develop scattergrams whose axes correspond to the projection of the samples onto, for example, a linear combination of features, e.g., via principal component analysis and eigenplots [17, 27]. Furthermore, scattergrams are not limited to rectilinear axes; certain applications benefit from visualizing the data via scattergrams having polar axes or axes that follow the mean track of a cluster [3].

Density Functions: Viewing Populations

Given an instance, a population's density function approximates the probability of that instance having originated from the population. A density function's domain is feature space, and it has a unit integral. Density functions can be visualized in a scattergram to provide a qualitative understanding of their shape and overlap, i.e., consider Figure 3.3 in which brighter intensities correspond to higher probabilities.

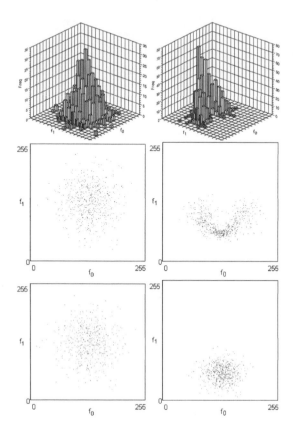

Figure 3.2. Histogram representations for Class A and Class B as well as training data. Top two plots are 2D histograms of the training data for Class A and Class B. Middle two plots are the corresponding scattergrams; intensity depicts frequency. Bottom two plots are the scattergrams for the training data for Class A' and B'.

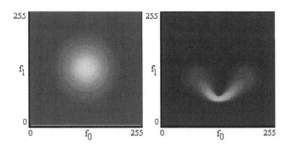

Figure 3.3. Population densities, i.e., known density functions, of Class A and Class B.

Labeling: Viewing Decision Bounds

Decision bounds are hypersurfaces in feature space that delineate the regions, "decision regions," within which every test instance will be assigned the same label. Decision bound specification is the deciding factor in the labeling accuracy of any classification system. As with population density functions, decision bounds can be either explicitly or implicitly represented by a pattern recognition system.

The explicit representation of the decision bounds eliminates the need for the explicit representation of a density function. It follows that assumptions concerning the shapes of the decision bounds imply assumptions regarding the shapes of the distributions and vice versa. Consider the well known k-means clustering system; it makes. the implicit assumption that the different clusters are well separated by piecewise linear decision bounds.

When decision bounds are implicit, explicit density functions are used by the corresponding pattern recognition system. Labels are typically assigned using Bayes' rule. The probability that an instance, $\vec{x}$, came from class C is computed based on the class conditional probability of that instance, $p(\vec{x}|C)$, the class' *a priori* probability, $P(C)$, and that instance's probability of being generated by any class, $P(\vec{x})$:

$$p(C|\vec{x}) = \frac{P(C)p(\vec{x}|C)}{p(\vec{x})}. \qquad (3.1)$$

The value of $p(\vec{x}|C)$ is provided by the density function, and the value of $P(C)$ is usually equal to the portion of training instances from Class C. An instance is assigned a population's label based on which class is the most likely to have generated that sample. When comparing $P(C|\vec{x})$ across classes, the instance's prior probability, $P(\vec{x})$, can be factored out of Equation (3.1). When the classes have equal priors, $P(C)$, that value can also be eliminated. As a result, an instance, $\vec{x}$, can be assigned a label, $C = 1...n_c$, via Equation (3.2):

$$\text{Class of } \vec{x} = \text{argmax}_{C=1...n_c}(p(\vec{x}|C)). \qquad (3.2)$$

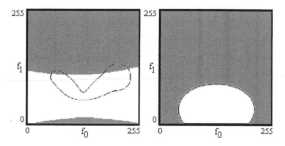

Figure 3.4. Class A/B and Class A'/B' decision regions produced by Gaussian classification with optimal decision bounds overlaid.

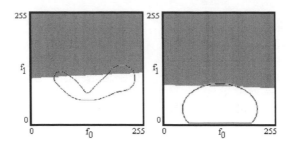

Figure 3.5. Class A / B and Class A' / B' decision regions produced by linear decision bounds with optimal decision bounds overlaid.

By evaluating every point in feature space and mapping each class label to a unique intensity, an image of the decision regions of feature space can be created. The feature space labelings in Figures 3.4 and 3.5 result from the application of Gaussian and linear classifiers, defined from the samples $S(\text{train}|A)$ versus $S(\text{train}|B)$ and $S(\text{train}|A')$ versus $S(\text{train}|B')$, to every point in feature space. The optimal decision bounds for each problem are shown as black curves.

3.3   Quantitative Comparison of Classifiers

The performance of a classifier can be quantified using several criteria: development/training memory requirements, development time, operating memory requirements, labeling speed, labeling accuracy, labeling consistency, and ease of qualitative and quantitative analysis.

For the types of problems being addressed by this chapter, classifier development memory requirements, development time, operating memory requirements, and labeling speed are not considered. For situations in which these factors are important, a different set of pattern recognition systems would perhaps need to be considered.

3.3.1   Labeling Accuracy

The accuracy of a classification technique must be judged by its performance on a specific application. Classifier performance is commonly quantified using true-positive and false-positive rates.

Because different classifiers make different distribution shape assumptions, the specification of a problem is important. If a classifier's assumptions are correct for the chosen problem, that classifier will provide optimal accuracy limited only by the quality of the features. However, if for a different problem its assumptions are incorrect, extremely poor labelings can result.

| | | Actual class | |
		Negative	Positive
Predicted	Negative	True Negative (TN)	False Negative (FN)
class	Positive	False Positive (FP)	True Positive (TP)

Table 3.4. Two class confusion matrix.

One way of summarizing classification results (labeling) is creating a confusion matrix for each class. As illustrated in Table 3.4, a confusion matrix is composed of four cells. Each cell is the number of instances that meet the criteria of it. For instance, "True Negative (TN)" cell shows the number of instances that do not belong to the class and predicted (by a classifier) as not belonging to the class. There are several standard quality terms derived from this confusion matrix. Among such quality terms, we use two terms, true positive rate (TPR) and false positive rate (FPR):

$$TPR = \frac{TP}{FN + TP}, \tag{3.3}$$

$$FPR = \frac{FP}{TN + FP}. \tag{3.4}$$

TPR indicates the portion of correctly classified instances as instances of the class among all the instances belonging to the class. FPR is the portion of incorrectly classified instances as instances of the class among all the instances not belonging to the class. As an accuracy measure, larger TPR value and smaller FPR values indicate better accuracy.

3.3.2 Labeling Consistency

Classifier consistency is quantified by the Monte Carlo oneSigma range of the true positive and false positive rates [54]. A oneSigma range for a measure is related to the standard error range of that measure, and smaller oneSigma ranges correspond to reduced measure variability. That is, classifiers with smaller true positive and false positive oneSigma ranges given different training and testing samples from the same populations are said to be more consistent.

Given a sufficient number of training instances and accurate distribution shape assumptions, the TPRs and FPRs associated with a pattern recognition system for a particular problem will be consistent despite the specific training sample used. However, when the classifier's assumptions are false or when the training sample size is not sufficient, then the performance of a classification system will vary when the training sample is changed.

Classifier performance can also vary because of multiple, non-optimal local extrema in the measure being optimized during the development/training of a classifier. This is usually the case when a classifier's parameter values are determined using an iterative technique or require the prior specification of a hyperparameter, e.g., the number of components. The initial values of the parameters may influence the final accuracy of the classifier as much as the training sample used. Multilayered perceptrons trained via backpropagation and numerous other pattern recognition systems require such additional considerations. These extrema can result in large variations in classifier accuracy, dependent on the training sample used and the starting point in the parameter space.

In the analyses presented in this chapter, labeling consistency is measured via Monte Carlo simulation in which a classifier's TPRs and FPRs are recorded for different, yet constant in size, training and testing samples.

Monte Carlo oneSigma values are proportional to standard error estimates. They specify the 67% confidence intervals for the values of interest, e.g., for the average true positive and false positive rates. If n_r Monte Carlo runs record an average TPR value of μ_{TPR} and a TPR standard deviation of σ_{TPR}, the Monte Carlo true positive oneSigma value is defined as

$$\text{oneSigma}_{TPR} = \frac{\sigma_{TPR}}{\sqrt{n_r}}, \tag{3.5}$$

and the 67% confidence interval for the true positive rate is thus

$$\mu_{TPR} - \frac{\sigma_{TPR}}{\sqrt{n_r}} < TPR < \mu_{TPR} + \frac{\sigma_{TPR}}{\sqrt{n_r}}. \tag{3.6}$$

Knowing that a method is accurate, however, may not always be sufficient; it is often equally important to understand how it is operating, how the features are being combined to make labeling decision, which instances are going to be mislabeled, how confident is the classifier of any particular labeling, etc. These are questions that require qualitative and quantitative analysis of the operation of a classifier.

3.3.3   Ease of Qualitative and Quantitative Operational Analysis

The insight gained through the analysis of the decision process of a pattern recognition system may be as important as the labelings they produce. Such analyses allow the questions listed at the beginning of the chapter and at the end of the last subsection to be answered. For example, if a population is known to have a Gaussian distribution, various methods exist for:

1. identifying outlying samples,

2. specifying confidence intervals for the estimated parameters based on the number of samples used,

3. producing receiver-operator characteristic curves which define the progression of TPR-versus-FPR given different error costs.

Additionally, the concepts of mean and covariance are simple enough to facilitate qualitative interpretation. These values can provide significant insight into the source of the populations and the nature/difficulty of the pattern recognition problem at hand.

3.4   Classification Systems

This section compares seven different classification techniques: linear, Gaussian, k-nearest neighbor (KNN), Parzen windows (PW), multilayered perceptron (MLP), k-means (KM), and finite Gaussian mixture modeling via maximum likelihood expectation maximization (FGMM via MLEM). The presentation of each classifier is organized into three sections: operation, labeling accuracy and consistency, and ease of analysis.

Operation.   These sections provide a high level description of each classification method. No effort is made to provide specific implementation details. A variety of books contain such information [17, 27, 47].

These sections also contain labeled scattergrams resulting from the application of the classification techniques to the training data for Problems 1 and 2. Figure 3.6 is an example of one such labeling generated by a FGMM via MLEM classifier for Problem 1. Dark gray regions are associated with Class A/A' labelings. Regions labeled in light gray correspond to Class B/B' samples. When multiple components are used to model a distribution, different shades of gray are used to

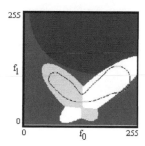

Figure 3.6. Example labeling of the Class A/B (Problem 1) scattergram provided by a FGMM via MLEM classifier with $k = 2$ per class (FGMM02).

distinguish their subregions. In Figure 3.6, there are two Gaussian components per class model as indicated. Overlaid onto these scattergrams are outlines of each problem's optimal decision bound.

Labeling Accuracy and Consistency.   These sections provide the summary statistics from a Monte Carlo study involving each classifier's true-positive and false-positive rates on $n_r = 1000$ different collections of training and testing data from the classification problems presented in Section 3.2.1. The average and Monte Carlo oneSigma values of the Class B and Class B$'$ TPRs and FPRs are given in table form as in Table 3.5. Plots of Class B and Class B$'$ FPR-versus-TPR for all of the classifiers are given at the end of this chapter.

Method	TPR	FPR	oneSigma$_{TPR}$	oneSigma$_{FPR}$
Gauss	0.86057	0.29396	0.01756	0.01737

Table 3.5. Example measures of accuracy and consistency for Gaussian classifier for Class B.

When hyperparameters exist for a classification technique, e.g., the parameter k in k-means, the classifier's performance for a variety of hyperparameter values is explored as well as graphs of a classifier's performance (e.g., TFP and FPR)

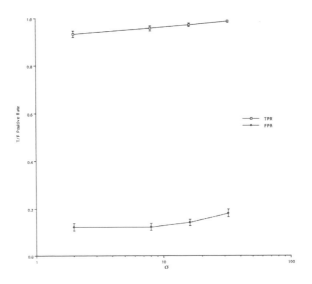

Figure 3.7. Example plot of FPR/TPR versus parameter σ for Parzen windows for Class B$'$; includes Monte Carlo oneSigma range for each rate.

versus hyperparameter. The Monte Carlo oneSigma ranges for the TRP and FPR values for each hyperparameter value are also indicated on the graphs using high and low markers. Figure 3.7 is an example of such a graph.

Ease of Analysis.   These sections discuss the qualitative and quantitative analysis of the representations formed by each classification system. The goal is to provide a rough assessment of how easy it is to interpret the representations provided by each method. References will be provided, and a few of the key strengths and weakness will be mentioned briefly.

Using these categories, we begin with the analysis of linear classification—a constrained Gaussian classifier.

3.4.1   Linear

Linear classifiers operate using linear, i.e., hyperplane, decision bounds. These classifiers make the assumption that the populations are well represented by a threshold (or univariate Gaussians with equal variance) applied to a weighted linear combination of the features.

Operation.   In a popular implementation of this classifier, the weighted linear combination of features is made to maximize the between class scatter while minimizing the within class scatter. Such a vector is defined by the maximum eigenvalued eigenvector, i.e., Fisher's linear discriminate, of the Hotelling matrix of the training samples. The instances are then projected onto this vector and labels are assigned using a threshold that is determined using least squared error minimization or by fitting univariate, equal variance Gaussians to the projections. This combination of features is said to define a *hyperfeature* or *latent feature* that best distinguishes the populations.

The Hotelling matrix is calculated using the class priors, $P(A)$ and $P(B)$, the class mean vectors, $\vec{\mu_A}$ and $\vec{\mu_B}$, and the class covariance matrices, Σ_A and Σ_B. It is a multivariate signal-to-noise measure.

The global mean is calculated as

$$\vec{\mu} = \sum_{c \in A,B} P(c)\vec{\mu_c}. \tag{3.7}$$

The signal matrix captures the spread of the means of the classes about the global mean:

$$S(i,j) = \sum_{c \in A,B} P(c)(\vec{\mu_c}(i) - \vec{\mu}(i)) * (\vec{\mu_c}(j) - \vec{\mu}(j)). \tag{3.8}$$

The noise matrix is the weighted sum of the spread of each class' data about their means:

$$N(i,j) = \sum_{c \in A,B} P(c)\Sigma_c(i,j). \tag{3.9}$$

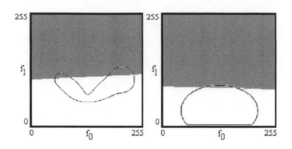

Figure 3.8. Class A/B (left) and Class A'/B' (right) decision regions produced by linear classification.

The Hotelling matrix, H, is the ratio of these two matrices.

$$H = N^{-1}S. \qquad (3.10)$$

Figure 3.8 shows the decision regions developed via linear classification given the data represented by the scattergrams shown in Figure 3.8.

Labeling Accuracy and Consistency. When a linear classifier's assumptions are correct, its results are optimally accurate and consistent given the quality of the features being used. Linear classifiers are also popular when the classes are well separated and/or in a high-dimensional feature space. Neither of these situations apply for the two problems at hand. Tables 3.6 and 3.7 summarize the accuracy and consistency of the linear classifier for these problems.

Method	TPR	FPR	oneSigma$_{TPR}$	oneSigma$_{FPR}$
Linear	0.86497	0.31235	0.06153	0.06942

Table 3.6. TPRs, FPRs, and consistency for Class B (Problem 1) from linear classification.

Method	TPR	FPR	oneSigma$_{TPR}$	oneSigma$_{FPR}$
Linear'	0.95270	0.12630	0.02965	0.03227

Table 3.7. TPRs, FPRs, and consistency for Class B' (Problem 2) from linear classification.

Ease of Analysis. The hyperfeature can provide significant insight into a problem as well as help speed feature collection. For example, when the features are generated via image filters, such as Gabor, texture, or Gaussian derivative filters, the features' weights can be used to specify a weighted linear combination of the image filters to specify a single filter, a *hyperfilter*, which provides an optimal, linear differentiation of the populations in the problem at hand; only that hyperfilter needs to be applied to future images to collect the hyperfeature and distinguish the populations. By visualizing that filter, an understanding of the visual differences between the populations may be attained.

The study of the linear separation of data also leads to principle component analysis and other eigen-analysis and dimensionality reduction methods. The reader is refered to [18] for additional information on such feature weighting and feature selection methods.

3.4.2 Gaussian

Gaussian classifiers operate under the assumption that the populations are well represented by Gaussian-shaped distributions. Gaussian classifiers form explicit Gaussian representations of the distributions and are one of the most popular techniques used in classification. Linear, elliptical, and hyperbolic decision bounds may be implicitly formed between two competing Gaussian distributions.

Operation. The mean and covariance matrix of a population specify its Gaussian representation. The Gaussian probability at an instance $\vec{x} \in \mathbb{R}^n$ is defined by

$$p(\vec{x}|\vec{\mu},\Sigma) = \frac{1}{(2\pi)^{n/2}|\Sigma|^{1/2}} e^{-\frac{1}{2}(\vec{x}-\vec{\mu})^T \Sigma^{-1}(\vec{x}-\vec{\mu})}, \qquad (3.11)$$

and test instances are assigned the label of the class, c, to which they are "closest":

$$\text{Class of } \vec{x} = \text{argmax}_{c=1...n_c} P(c)p(\vec{x}|\vec{\mu_c},\Sigma_c). \qquad (3.12)$$

Figure 3.9 shows the decision regions formed for the two pattern recognition problems using Gaussian classification.

Labeling Accuracy and Consistency. While the Gaussian assumptions are incorrect for the first problem, they are correct for the second; note the drastic change in absolute and relative accuracy and consistency (Tables 3.8 and 3.9).

Method	TPR	FPR	oneSigma$_{TPR}$	oneSigma$_{FPR}$
Gauss	0.86057	0.29396	0.01756	0.01737

Table 3.8. TPRs, FPRs, and consistency for Class B (Problem 1) from Gaussian classification.

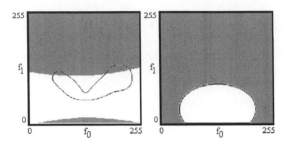

Figure 3.9. Class A/B (left) and Class A′/B′ (right) decision regions produced by Gaussian classification.

Method	TPR	FPR	oneSigma$_{TPR}$	oneSigma$_{FPR}$
Gauss′	0.95211	0.11352	0.00966	0.01185

Table 3.9. TPRs, FPRs, and consistency for Class B′ (Problem 2) from Gaussian classification.

Ease of Analysis.   As a result of being one of the most popular classification techniques and its relatively few and clearly defined parameters, Gaussian classification is probably one of the best understood and most intuitively informative representation methods. Gaussian classification is the standard against which all other techniques are compared in regard to ease of qualitative and quantitative analysis.

3.4.3 k-Nearest Neighbor

If we allow the assignment of labels to be completely data driven, i.e., not attempt to parameterize or limit the shape of the underlying density function, k-nearest neighbor (KNN) classification results.

Operation.   KNN classification assumes that the most common label among the K closest samples is the most probable label. Closeness is commonly judged using the Euclidean distance measure.

    Results are given in Figure 3.10 for both of the pattern recognition problems for a variety of K values.

Labeling Accuracy and Consistency.   it has been proven that the asymptotic total error rate, i.e., (1-TPR) + FPR, of K=1 nearest neighbor classification is at most twice the optimal Bayesian total error rate [17, 51].

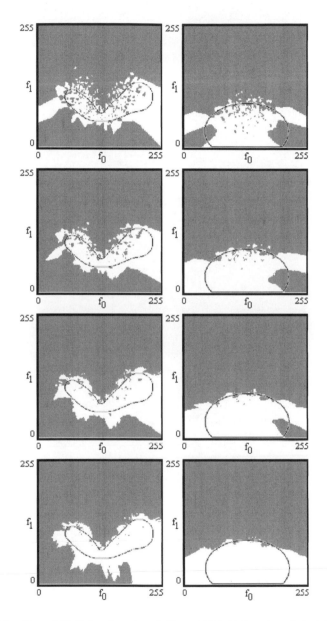

Figure 3.10. Class A/B (left column) and Class A′/B′ (right column) decision regions produced by *k*-nearest neighbor classification for *K* = 1 (row 1), 3 (row 2), 7 (row 3), and 11 (row 4).

Method	K	TPR	FPR	oneSigma$_{TPR}$	oneSigma$_{FPR}$
KNN	1	0.76094	0.24433	0.02126	0.01861
	3	0.82346	0.23834	0.02152	0.01895
	7	0.86584	0.24230	0.02137	0.02049
	11	0.88107	0.24680	0.02085	0.02146

Table 3.10. TPRs, FPRs, and consistency for Class B (Problem 1) from k-nearest-neighbor classification.

Method	K	TPR	FPR	oneSigma$_{TPR}$	oneSigma$_{FPR}$
KNN$'$	1	0.87729	0.12305	0.01832	0.01420
	3	0.92383	0.11823	0.01567	0.01395
	7	0.94232	0.11722	0.01426	0.01441
	11	0.94764	0.11751	0.01397	0.01481

Table 3.11. TPRs, FPRs, and consistency for Class B$'$ (Problem 2) from k-nearest neighbor classification.

The most significant consideration for KNN classification is the value of K. The localized voting process of KNN makes it an approximate smoothing technique. As K increases, the region in feature space over which the labels are averaged increases. Too small of a K value results in undersmoothing the distribution of the labels. Too large of a K value produces oversmoothing. The optimal K value varies for each problem. Parzen windows research indicates that the optimal neighborhood size/scale and thus the optimal K value may even vary throughout feature space for a single problem [43, 52, 55, 29, 34, 4].

3.4.4   Parzen Windows

Parzen windows (PW) is a kernel density estimation technique. Like KNN classification, it is considered a data driven technique. It is also considered a density interpolation technique because of its close relationship to convolution.

Its hyperparameters are the shape of the kernel and the size/scale/bandwidth of that kernel. It has been demonstrated that kernel scale, not kernel shape, is key to the accuracy and consistency of the representations formed by this technique [53]. As a result, the major area of research in regard to kernel density estimation concerns optimal scale/bandwidth estimation. Specifically, Parzen windowing has been extended to include functions which specify variations in kernel size based on the local distribution of the samples. Dependent on the error measure, e.g., mean integrated squared error, used to quantify the difference between the estimated and ideal density, automated, principled methods for variable kernel scale specification have been developed [52, 29].

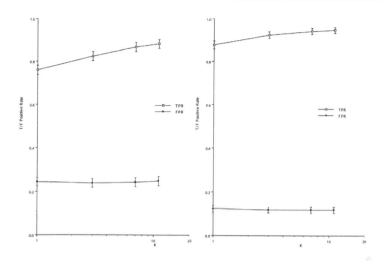

Figure 3.11. K versus TPR/FPR for Class A/B (left) and Class A′/B′ (right).

Operation. Instead of using a triangle or step function as Parzen did [43], we promote the use of a Gaussian-shaped, scale σ kernel.

$$G(\vec{z},\sigma) = \frac{1}{(2\pi\sigma^2)^{n/2}} e^{-\frac{\vec{z}^T \vec{z}}{2\sigma^2}}. \tag{3.13}$$

Given a set of samples S, the Parzen window density function using a window width of σ is

$$p(\vec{x},\sigma) = \frac{1}{|S|} \sum_{j=1}^{|S|} G(||\vec{x} - \overrightarrow{S(j)}||_2,\sigma), \tag{3.14}$$

where $\overrightarrow{S(j)}$ refers to the j-th instance in the set S.

The results are given in Figure 3.12 for a variety of kernel scales. The σ = 2 feature space labelings also contain regions labeled with black to indicate that no label could be assigned to those samples. The implementation used in this study only evaluated training samples within a fixed distance, 3σ, of each testing sample. When no training samples were within that distance, no density was estimated, and a black class label was assigned. These regions are a failure of the implementation and not the technique.

Labeling Accuracy and Consistency. Kernel density estimation techniques generally provide excellent accuracy and consistency for a range of kernel shapes and sizes. Significant research has gone into deriving functions for optimal kernel size specification [52, 55, 29, 34, 4]. As with KNN classification, using too small

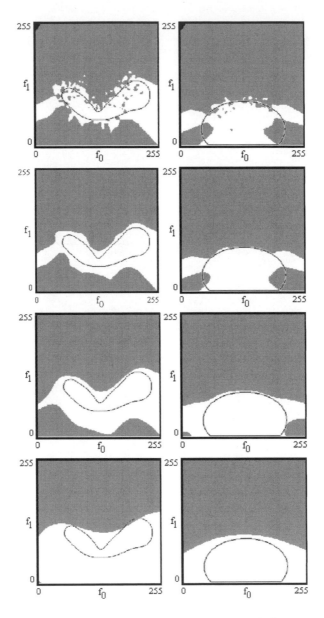

Figure 3.12.   TPR/FPR for Class A/B (left colum) and Class A′/B′ (right column) using Parzen window sizes $\sigma = 2$ (row 1), 8 (row 2), 16 (row 3), and 32 (row 4).

Method	Window σ	TPR	FPR	oneSigma$_{TPR}$	oneSigma$_{FPR}$
PW	2	0.84960	0.24390	0.02069	0.01960
	8	0.90534	0.25473	0.01700	0.01962
	16	0.91394	0.28018	0.01458	0.01801
	32	0.89889	0.30073	0.01420	0.01750

Table 3.12. TPRs, FPRs, and consistency for Class B (Problem 1) from Parzen window classification.

Method	Window σ	TPR	FPR	oneSigma$_{TPR}$	oneSigma$_{FPR}$
PW$'$	2	0.93420	0.12209	0.01428	0.01476
	8	0.95858	0.12271	0.01032	0.01375
	16	0.97261	0.14125	0.00728	0.01416
	32	0.98743	0.18084	0.00448	0.01625

Table 3.13. TPRs, FPRs, and consistency for Class B$'$ (Problem 2) from Parzen window classification.

of a neighborhood/scale/bandwidth results in undersmoothing, and too large of a neighborhood produces oversmoothing.

Ease of Analysis. In isolation, kernel density estimation techniques provide little qualitative or quantitative insight into the problems at hand. However, the resulting interpolated/smoothed density surfaces enable the statistical analysis of the data. For example, mode identification, ridge traversal, valley traver-

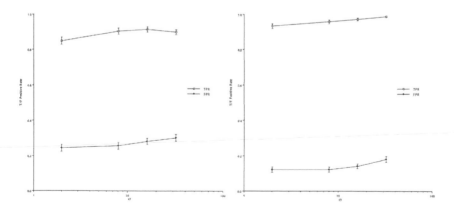

Figure 3.13. TPR/FPR/Consistency versus σ for Class B (left column) and for Class B$'$ (right column) for Parzen window classification.

sal, and numerous other techniques can be applied in a straightforward manner
[52, 58, 32, 8, 2, 3].

3.4.5   Multilayered Perceptron

Multilayered perceptron neural networks (MLPs) are commonly called backpropagation networks. Backpropagation refers to the average root mean squared error
(ARMSE) gradient descent parameter estimation technique, which is often applied to this style of feedforward network [33, 5].

MLPs develop decision bound representations. The complexity of the representations they form are dependent on the network architecture, the gradient
descent step size, and the training time. Each of these considerations has been the
subject of various gradient, genetic algorithm, simulated annealing, and heuristic
strategies in an effort to automate the application of MLPs [51, 5, 45]. Of special
interest is the modification to the network algorithm which results in the definition of radial basis function networks. These networks are finite Gaussian mixture
models that will be discussed later in this chapter [63].

Here we discuss MLPs having two hidden layers and full feedforward connectivity between the layers and sigmoidal activation functions (Figures 3.14 and
3.15). A network having 6 nodes in the first hidden layer and 3 nodes in the
second hidden layer is referenced as MLP6x3.

Operation.   Each input node corresponds to a different feature. Each output node
corresponds to a different class. A test sample is assigned the label associated with
the output node which has the largest output after the input is propagated through
the network.

For Problems 1 and 2 training consists of 3,600,000 iterations, i.e., 667 passes
through the training data. The weights are updated after each sample is presented,
a strategy called "iterative" training. Using a root mean squared error measure,
the weights are updated by taking a step in the gradient direction a distance of 1%
of the gradient magnitude.

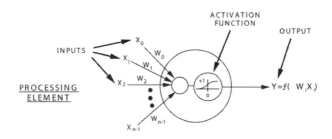

Figure 3.14. A single node of a multilayered perceptron.

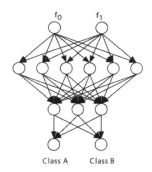

Figure 3.15. Perceptron with two hidden layered (MLP6x3).

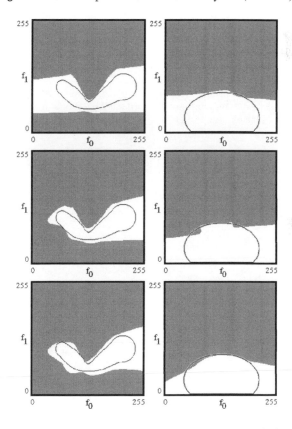

Figure 3.16. Class A/B decision regions (left column) and Class A′/B′ decision regions (right column) produced by MLP classification using architectures 6×3 (row 1), 12×6 (row 2), and 24×12 (row 3).

While not a criterion for the comparison of the classifiers in this Chapter, it is important to note that the training times associated with traditional backpropagation can be excessive. The time required was such that only $n_r = 100$ runs of the Monte Carlo simulation could be performed in a reasonable amount of time. The Monte Carlo oneSigma values have been adjusted accordingly. The results for a variety of different sizes of two hidden layered MLPs are shown in Figure 3.16.

Labeling Accuracy and Consistency.   The Stone-Weierstrauss theorem has been used to prove that MLPs can represent any function to an arbitrary degree of accuracy using one hidden layer, and by using two hidden layers, any decision bound can be represented [24, 46, 42, 7]. The Monte Carlo results for Problems 1 and 2 are given in Tables 3.14 and 3.15. These results are plotted in Figure 3.17.

Method	Net	TPR	FPR	oneSigma$_{TPR}$	oneSigma$_{FPR}$
MLP	6x3	0.90327	0.26495	0.05007	0.05858
	12x6	0.88089	0.24267	0.06650	0.05939
	24x12	0.88044	0.24143	0.06509	0.06155

Table 3.14. TPRs, FPRs, and consistency for Class B (Problem 1) from multilayered perceptron classification.

Method	Net	TPR	FPR	oneSigma$_{TPR}$	oneSigma$_{FPR}$
MLP$'$	6x3	0.94569	0.11549	0.03472	0.03211
	12x6	0.93519	0.10756	0.04273	0.03246
	24x12	0.93376	0.10504	0.03667	0.02764

Table 3.15. TPRs, FPRs, and consistency for Class B$'$ (Problem 2) from multilayered perceptron classification.

One of the problems with MLPs is overtraining. Consider the function

$$F(x) = 4.26(e^{-x} - 4e^{-2x} + 3e^{-3x}) + G(0, \sigma). \qquad (3.15)$$

$G(0, \sigma)$ represents Gaussian additive noise with a zero mean. The $\sigma = 0$ function is shown as the dotted line in Figure 3.18. The thin solid line in Figure 3.18 corresponds to the $\sigma = 0.1$ function. When the latter is given to a MLP with one hidden layer of 30 nodes and trained for 4×10^6 iterations, overtraining occurs. The function approximation developed by the MLP is shown as the thick line in Figure 3.18. The network has begun to fit to the noise of the samples. Figure 3.19 provides a plot of ARMSE for the training and testing data as the training of the network progresses. Overtraining is indicated by a rise in testing ARMSE despite the continual decrease in training ARMSE. Overtraining can also occur

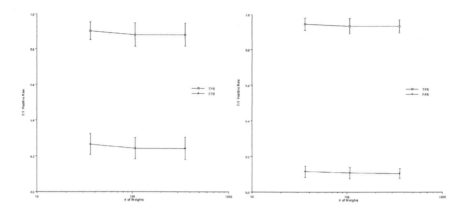

Figure 3.17. Class A/B (left) and Class A′/B′ (right) TPR, FPR, and consistency versus network size for MLP Classification.

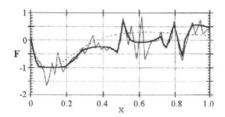

Figure 3.18. Underlying function (dotted), noisy sampling (thin line), and MLP generated model (thick line). Overtraining fits noise.

when a FGMM is given too many components to represent a distribution or when a polynomial of too high of a degree is used to represent a sampled function. With MLPs, however, controlling such error is not straightforward [45].

Ease of Analysis. One of the most common complaints concerning MLPs is that they are often viewed as black box solutions to problems. They provide little intuitive insight into the criterion with which they are making their decisions. Significant research has gone into developing algorithms for the quantitative analysis of their performance as well as providing a more intuitive representation of their decision process.

One of the most common techniques relies on the conversion of the network to a decision tree or table lookup process. Further qualitative and quantitative analysis can then be performed on that structure. These approaches, however, do not completely address the validation and verification process required of black box methods.

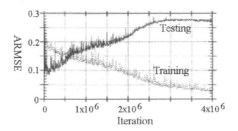

Figure 3.19. Plot of training data's ARMSE and testing data's ARMSE versus training time. Overtraining increases testing ARMSE.

3.4.6 k-Means

k-means (KM) operates under the assumption that the distribution of a collection of samples can be well represented by multiple circularly symmetric Gaussians having equal variance. As a result, k-means representations can be considered finite Gaussian mixture models having constrained Gaussian components. They implicitly form piecewise linear decision bounds.

k-means can be applied as part of a classification scheme so as to model the distribution of a single population, or it can be applied to the training samples of multiple populations as a clusterer with the goal of automatically and efficiently distributing the means among the populations.

In this chapter k-means, like Parzen windowing, is being used for classification. That is, it is being applied to each class independently to model its distribution and in turn provide a class conditional probability for Bayesian classification. Therefore, $K = 1$ refers to modeling each class using one component and $K = 2$ corresponds with two components per class and so forth.

Operation.   The application of k-means to a class C involves the following steps:

1. Choose the number of components, k.

2. Choose initial values for the component means μ_i for i = 1...k.

3. Label each $\vec{x} \in$ S(train|C) to the nearest component mean using minimum Euclidean distance criterion.

4. IF every instance's component label is unchanged THEN end.

5. ELSE Recompute μ_i for i = 1...k using labeled training data and goto Step 3.

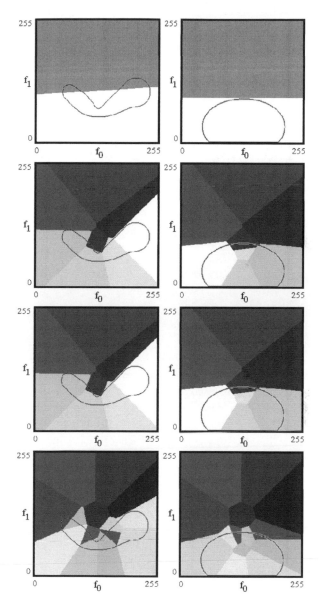

Figure 3.20. Class A/B (left column) and Class A'/B' (right column) decision regions produced by k-means clustering per class using k=1 (row 1), 2 (row 2), 4 (row 3), and 7 (row 4).

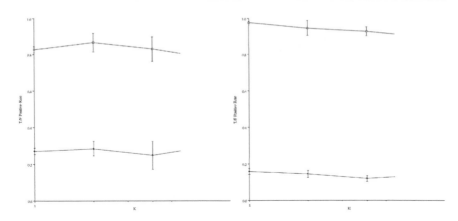

Figure 3.21. TPR/FPR/Consistency versus k for Class B (left column) and for Class B' (right column) for k-means clustering for classification.

A variety of heuristics exist for choosing the initial means for the components (the μ_i in step 2). For the work presented in this chapter, the μ_i were chosen using k random selections from the training set.

Labeling Accuracy and Consistency.   The principal problems with k-means are its dependence on the initial mean values, its dependence on the order in which the training data are presented, its dependence on the specific set of training data being used, and the reliance on the user to specify K. That is to say, k-means is subject to local maxima and therefore provides poor labeling consistency. This is well illustrated by the oneSigma ranges in the graphs in Figure 3.21.

For Problem 1, the drastic change in the oneSigma ranges for the TPR and FPR values as K increases indicates a severe decrease in labeling consistency.

For Problem 1, ideally $K = 1$ for Class A, but since Class B is an extruded Gaussian, it is difficult to determine an appropriate finite K value *a priori*. For Problem 2, K is known to be 1 for both classes. However, for Problem 2 the Class B Gaussian is not circularly symmetric, i.e., isotropic. Nevertheless, for both problems, the additional resources provided by larger K values appear to confound the solution. The problems with parameter initialization, user specification of K, and iterative parameter optimization techniques are studied more closely for FGMMs via MLEM in the next section.

Ease of Analysis.   The use of K means to represent a population's distribution can provide significant qualitative and quantitative insight. The theorems of statistical analysis for Gaussian, i.e., $K = 1$, and FGMM, i.e., $K > 1$, also apply to K means.   However, the existence of local maxima degrades the

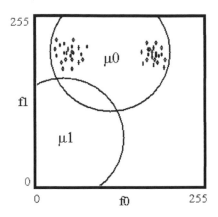

Figure 3.22.   Simulated local maximum for $K = 2$ means.

possibility of significant qualitative insight unless the quantitative analysis is also performed.

Consider the hypothetical training data and the $(K = 2)$-means description of its distribution as shown via the means and isoprobability curves in Figure 3.22. Although a local maximum has been achieved, the performance is far from that of the ideal representation. Poor accuracy results, and any qualitative analysis of the means without a quantitative analysis would be misleading.

Method	K	TPR	FPR	oneSigma$_{TPR}$	oneSigma$_{FPR}$
KM	1	0.82714	0.27073	0.01705	0.01739
	2	0.86754	0.28483	0.05093	0.03924
	4	0.83296	0.24890	0.06654	0.07657
	7	0.79113	0.29009	0.05728	0.12592

Table 3.16. TPRs, FPRs, and consistency for Class B (Problem 1) from k-means classification.

Method	K	TPR	FPR	oneSigma$_{TPR}$	oneSigma$_{FPR}$
KM$'$	1	0.97706	0.15863	0.00650	0.01508
	2	0.94669	0.14549	0.04098	0.01911
	4	0.92937	0.11924	0.02515	0.01614
	7	0.90416	0.13444	0.03270	0.02753

Table 3.17. TPRs, FPRs, and consistency for Class B$'$ (Problem 2) from k-means classification.

3.4.7   Finite Gaussian Mixture Modeling

Finite Gaussian mixture modeling develops representations of complex distributions through the weighted linear combination of multiple Gaussian component distributions. The probability of an instance $\vec{x}$ arising from a population which is represented by the FGMM parameterized by Ψ is given by

$$p(\vec{x}|\Psi) = \sum_{i=1}^{K} \omega_i G(\vec{x}|\mu_i,\Sigma_i) \text{ where } \Psi = \{\{\omega,\mu,\Sigma\}_i | i = 1...K\}. \qquad (3.16)$$

In 1886, Newcomb wrote the seminal paper on finite mixture modeling [41]. He used a mixture of two univariate normals to model a distribution and its outliers. Pearson in 1894 presented a method of moments for automatically decomposing a mixture of normals [44]. His approach, however, required the solution of a ninth degree polynomial equation. Cohen in 1967 limited the problem to components with equal variances and presented a solution involving root finding and the first four moments [12]. Little attention was given to the likelihood maximization approach until 1972, when Tan and Chwang [57], and independently Fryer and Robertson [20] demonstrated that the method of moments is inferior to likelihood estimation for Cohen's limited problem and the more general cases.

In 1977, Dempster, Laird, and Rubin [16] presented an iterative scheme for handling missing data in maximization problems and established its theoretical convergence properties. This iterative scheme is called expectation maximization (EM). It has been and continues to be applied to likelihood maximization for the definition of finite mixture models.

A variety of alternate technologies have also received considerable attention for FGMM development: graphical methods, minimum distance techniques such as chi-squared and least squared minimization methods, Bayesian techniques, Newton-Raphson, and the method of scoring. Although no single method has been shown to be ideal for all situations, EM has several properties which make it an appealing technique [30, 62, 64]:

1. EM is simple to apply: no matrix inversion is required as with Newton-Raphson and the method of scoring.

2. EM is stable: Bayesian techniques develop numerical difficulties in high dimensional parameter spaces.

3. EM converges to singularities less often.

4. EM is monotonic: it is the only method for which likelihood is guaranteed to increase after each iteration.

One of the most common complaints in regard to EM is its slow convergence rate. Recently, however, Xu [62] has shown that although EM strictly has first order convergence properties, the matrix which characterizes the direction of its step with respect to the local gradient serves to reduce the condition number of the Hessian of the maximum likelihood surface so that nearly second order convergence rates are often achieved.

Maximum Likelihood Estimation

The standard likelihood equation for a collection of training instances, S, given a FGMM(Ψ) is

$$L(S|\Psi) = \prod_{j=1}^{|S|} p(\vec{x}_j|\Psi). \tag{3.17}$$

Equation (3.17) is maximized (or minimized) when

$$\frac{\partial L(S|\Psi)}{\partial \Psi} = 0. \tag{3.18}$$

Finding a maximum likelihood estimate of Ψ is equivalent to finding the maximum log likelihood estimate of Ψ:

$$LL(S|\Psi) = \sum_{j=1}^{|S|} log(p(\vec{x}_j|\Psi)). \tag{3.19}$$

By assuming that each sample, $\vec{x}_j$, arises from one of the component distributions, $G(\vec{x}_j|\mu_i, \Sigma_i)$, which exist in the mixture model in the portions ω_i, then a new variable $z_{j,i}$ can be introduced that captures the component membership for $\vec{x}_j$:

$$z_{j,i} = \begin{cases} 1, & \text{if } \vec{x}_j \in \text{ component } i; \\ 0, & \text{otherwise.} \end{cases} \tag{3.20}$$

As a result, the log likelihood equation can be rewritten as

$$LL(S|\Psi) = \sum_{j=1}^{|S|} \sum_{i=1}^{K} z_{j,i}(log(\omega_i) + log(G(\vec{x_j}, \phi_i))) \tag{3.21}$$

and this equation is maximal when

$$\frac{\partial LL(S|\Psi)}{\partial \Psi} = \sum_{j=1}^{|S|} \sum_{i=1}^{K} z_{j,i} \frac{\partial log(\omega_i)}{\partial \omega} + \sum_{j=1}^{|S|} \sum_{i=1}^{K} z_{j,i} \frac{\partial log(G(\vec{x_j}, \phi_i))}{\partial \phi} = 0. \tag{3.22}$$

This formulation of the likelihood equation requires the specification of the $z_{j,i}$, but during the development of the model, these terms are unknown. Thus, the use of this equation as the function to be solved converts the maximum like-lihood task to a "missing data" problem, which expectation maximization (EM) was developed to solve.

The EM algorithm was presented by Dempster, Laird, and Rubin [16] as an iterative maximization algorithm capable of handling missing data. The EM algo-rithm is applied to FGMMs by treating all $z_{j,i}$ as missing data. Each iteration of EM has two steps: an E-step and an M-step. The E-step, the expectation step, as-signs values to the missing data variables based on the current model parameters. The M-step, the maximization step, adjusts the parameters of the model using the current estimates of the missing data variables. The E-Step requires the calcula-tion of expected value of the log likelihood function conditional on the initially estimated parameters, $\Phi^{(0)}$, so

$$Q(\Phi|\Phi^{(0)}) = E[LL(S|\Phi)|\Phi^{(0)}]. \tag{3.23}$$

This is achieved by substituting the component conditional posterior probabilities of each sample for the component membership variables, i.e., the missing data. The component conditional posterior probabilities, $P_{j,i}$, of each instance, j, for each component, i, are

$$z_{j,i} = P_{j,i} = p(\overrightarrow{x_j}|\Phi;i) = \frac{\omega_i G(\overrightarrow{x_j};\phi_i)}{\sum_{k=1}^{K} \omega_k G(\overrightarrow{x_j};\phi_k)}. \tag{3.24}$$

The M-step adjusts the estimates of the parameters Φ based on this new log likelihood estimate. The weights are estimated by

$$\omega_i = \frac{1}{|S|} \sum_{j=1}^{|S|} P_{j,i} \forall i = 1...K, \tag{3.25}$$

and the remaining parameters of the model in the maximum log likelihood equa-tion are determined by substituting of $P_{j,i}$ for $z_{j,i}$ into the second term of Equation (3.22):

$$\sum_{j=1}^{|S|} \sum_{i=1}^{K} z_{j,i} \frac{\partial log(G(\overrightarrow{x_j};\phi_i))}{\partial \phi} = 0. \tag{3.26}$$

For a variety of component distribution shapes, solutions exist for this equation. Such is the case for FGMMs. Specifically, the mean and covariance of each com-ponent, $i = 1...K$, are

$$\overrightarrow{\mu_i} = \frac{1}{|S|} \sum_{j=1}^{|S|} \frac{P_{j,i}\overrightarrow{x_j}}{\omega_i}, \tag{3.27}$$

$$\Sigma_i = \frac{1}{|S|} \sum_{j=1}^{|S|} \frac{P_{j,i}(\overrightarrow{x_j} - \overrightarrow{\mu_i})^t(\overrightarrow{x_j} - \overrightarrow{\mu_i})}{\omega_i}. \tag{3.28}$$

Operation.   As with k-means, FGMMs can be used as classifiers or clusterers. In this chapter, as with k-means, FGMMs are used to model the populations independently and provide class conditional probabilities to a Bayesian classifier.

k-means was used to initialize the model's parameters. The EM algorithm was run to convergence. Convergence was indicated by less than 0.01% change in the likelihood over two iterations. Analysis of the training indicates that convergence was reached in most cases in less than 10 iterations, but occasionally as many as 250 were required.

Labeling Accuracy and Consistency.   It has been shown that FGMMs can be used to approximate arbitrary densities [19]. In practice, however, FGMMs via MLEM are reliant on the user specification of K and are subject to local maxima and non-optimal global maxima.

A variety of methods have been developed to automate the specification of K. None, however, have been generally accepted, and most make additional assumptions concerning the distributions being modeled [64] or introduce new parameters [38]. For generalized projective Gaussian distributions the automated specification of K can be a particularly difficult task. There is in actuality an infinite number of components, and the task becomes one of finding a finite number of components which well approximate that continuum.

Non-optimal global maxima generally occur on the fringes of a model's parameter space. They occur when a component becomes dedicated to a single sample. The variance of this component will tend toward zero while its maximum likelihood value will rapidly increase. Given a likelihood measure, the values at those maxima are actually unbounded. Bayesian methods have been applied in an effort to limit the effects of such maxima [61].

The existence of local maxima is revealed by starting from a different point in parameter space. Since k-means is being used to initialize the parameters, simply changing the order in which the data are supplied causes EM to begin with different initial values and a drastically different arrangement of components results. Figure 3.24 shows the results from FGMM via MLEM using $K = 7$ when the same data used to generate Figure 3.23 is presented in a different order. The effect of the change in the parameters of the components is obvious. A different local maximum has been reached.

The poor consistency resulting from local maxima is revealed by the graphs in Figure 3.25. They also illustrate the importance of the specification of the number of components in determining the accuracy as well as the consistency of the models.

Most researchers recommend developing several models of a set of data from different initial points in parameter space and using a range of K values and then choosing the "best" so as to avoid these problems. In practice such an approach can be time consuming, difficult to implement, and may require significant data. The approach to CGMM presented in this chapter addresses these issues.

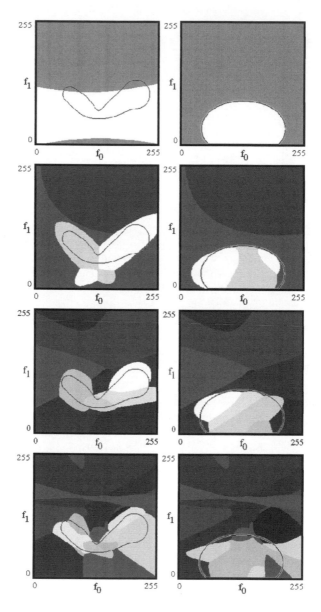

Figure 3.23. Class A/B (left column) and Class A$'$/B$'$ (right column) decision regions produced by finite Gaussian mixture modeling clustering per class using $K = 1$ (row 1), 2 (row 2), 4 (row 3), and 7 (row 4).

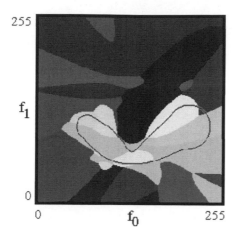

Figure 3.24. Different initialization produces different model; versus Figure 3.23, Problems 1, $K = 7$.

Method	K	TPR	FPR	oneSigma$_{TPR}$	oneSigma$_{FPR}$
FGMM	1	0.86124	0.29245	0.02382	0.03819
	2	0.92873	0.29276	0.02216	0.04280
	4	0.93117	0.28892	0.10046	0.07593
	7	0.91521	0.28005	0.16201	0.09829

Table 3.18. TPRs, FPRs, and consistency for Class B (Problem 1) from finite Gaussian mixture modeling classification.

Method	K	TPR	FPR	oneSigma$_{TPR}$	oneSigma$_{FPR}$
FGMM$'$	1	0.95209	0.11342	0.00972	0.01200
	2	0.95060	0.11444	0.01677	0.01742
	4	0.94488	0.11365	0.05717	0.02690
	7	0.94154	0.11764	0.09238	0.07252

Table 3.19. TPRs, FPRs, and consistency for Class B$'$ (Problem 2) from finite Gaussian mixture modeling classification.

Ease of Analysis. Despite the difficulties with FGMMs, they have received considerable attention and found their way into numerous practical applications [2, 3, 6, 22, 50, 59]. The most significant advance in terms of quantitative analysis has come from Louis [35]. His method for simultaneously defining Fisher's information matrix during the development of the model has enabled the application of a

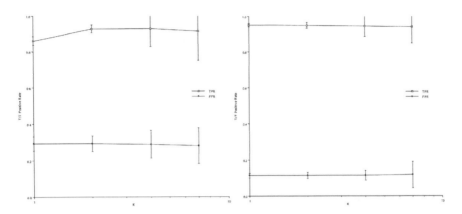

Figure 3.25. TPR/FPR/consistency versus k for Class B (left column) and for Class B$'$ (right column) for finite gaussian mixture modeling (FGMM) clustering for classification.

wide range of quantitative analytic techniques. Qualitative analysis is facilitated by the use of Gaussian components. Both the qualitative and quantitative analysis can be especially revealing if it is assumed that the population is actually a mixture of Gaussians. Such mixture models are referred to as direct mixture models. For extruded Gaussian distributions, since the number of components is actually infinite, only indirect FGMMs can be formed.

3.5   Summary of Classifiers' Performance

Figures 3.26 and 3.27 provide summaries of the performance of the classifiers discussed in this chapter for Problems 1 and 2, respectively. These figures are plots of TPR versus FPR. Each classifier's average TPR/FPR value is indicated by its abbreviated name. About each average is a circle whose area is proportional to the log of that classifier's combined oneSigma TPR/FPR variance. The larger the circle, the less consistent was that classification technique's accuracy throughout the Monte Carlo simulation.

This chapter has, in fact, presented and analyzed a range of approaches to and implementations of Gaussian mixture modeling. Linear classifiers imply GMMs having just one Gaussian component with a fixed covariance matrix. Gaussian classifiers use GMMs having just one component whose mean and covariance are fit to a class' training sample. k-means allows multiple Gaussians to represent a distribution, but the covariance matrices of those Gaussian components are fixed. Finite Gaussian mixture modeling uses multiple means to represent a distribution and allows their covariance matrices to adjust to the data. Parzen windowing goes

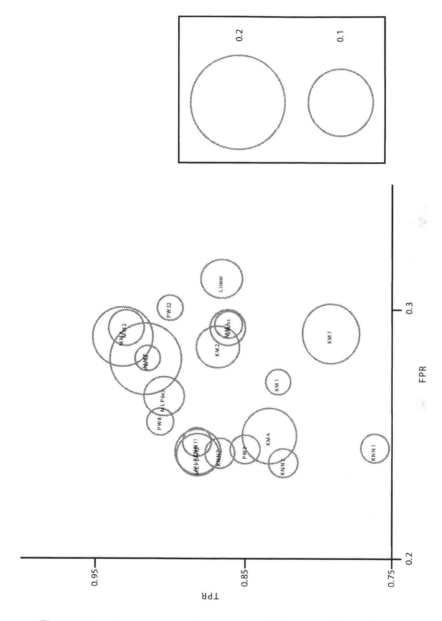

Figure 3.26. Average and oneSigma range of TPR versus FPR for Class B.

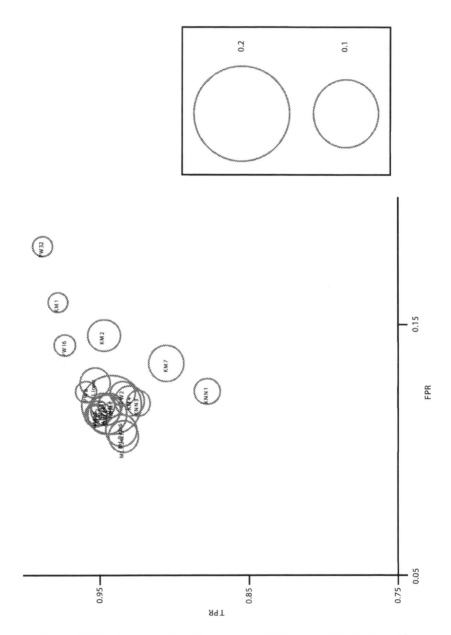

Figure 3.27. Average and oneSigma range of TPR versus FPR for Class B$'$.

Method	K	TPR	FPR	oneSigma$_{TPR}$	oneSigma$_{FPR}$
Gauss		0.86057	0.29396	0.01756	0.01737
Linear		0.86497	0.31235	0.06153	0.06942
KNN	1	0.76094	0.24433	0.02126	0.01861
	3	0.82346	0.23834	0.02152	0.01895
	7	0.86584	0.24230	0.02137	0.02049
	11	0.88107	0.24680	0.02085	0.02146
PW	2	0.84960	0.24390	0.02069	0.01960
	8	0.90534	0.25473	0.01700	0.01962
	16	0.91394	0.28018	0.01458	0.01801
	32	0.89889	0.30073	0.01420	0.01750
MLP	6x3	0.90327	0.26495	0.05007	0.05858
	12x6	0.88089	0.24267	0.06650	0.05939
	24x12	0.88044	0.24143	0.06509	0.06155
KM	1	0.82714	0.27073	0.01705	0.01739
	2	0.86754	0.28483	0.05093	0.03924
	4	0.83296	0.24890	0.06654	0.07657
	7	0.79113	0.29009	0.05728	0.12592
FGMM	1	0.86124	0.29245	0.02382	0.03819
	2	0.92873	0.29276	0.02216	0.04280
	4	0.93117	0.28892	0.10046	0.07593
	7	0.91521	0.28005	0.16201	0.09829

Table 3.20. Summary of the recorded Problem 1 TPR, FPR, and oneSigma ranges.

to the extreme of having one Gaussian per instance, but the covariance matricies of those Gaussians are fixed.

While these comparisons are very revealing, the ordering of the performance of the classifiers are specifically limited to the particular problems. It is often infeasible to run such extensive comparisons for every problem, but the question remains, which density estimation method is most correct for each population so that the most accurate and consistent classification results? Next, we present goodness-of-fit measures that help answer this question.

3.6 Goodness-of-Fit

As shown by the experiments in the previous section, the validity of the assumed distributions drives the accuracy and consistency of classifiers. The validity of a distribution model is assessed using goodness-of-fit measures to test the null hypothesis that there is no difference between two distributions except chance differences due to finite sampling. Methods for testing this hypothesis include:

Method	K	TPR	FPR	oneSigma$_{TPR}$	oneSigma$_{FPR}$
Gauss$'$		0.95211	0.11352	0.00966	0.01185
Linear$'$		0.95270	0.12630	0.02965	0.03227
KNN$'$	1	0.87729	0.12305	0.01832	0.01420
	3	0.92383	0.11823	0.01567	0.01395
	7	0.94232	0.11722	0.01426	0.01441
	11	0.94764	0.11751	0.01397	0.01481
PW$'$	2	0.93420	0.12209	0.01428	0.01476
	8	0.95858	0.12271	0.01032	0.01375
	16	0.97261	0.14125	0.00728	0.01416
	32	0.98743	0.18084	0.00448	0.01625
MLP$'$	6x3	0.94569	0.11549	0.03472	0.03211
	12x6	0.93519	0.10756	0.04273	0.03246
	24x12	0.93376	0.10504	0.03667	0.02764
KM$'$	1	0.97706	0.15863	0.00650	0.01508
	2	0.94669	0.14549	0.04098	0.01911
	4	0.92937	0.11924	0.02515	0.01614
	7	0.90416	0.13444	0.03270	0.02753
FGMM$'$	1	0.95209	0.11342	0.00972	0.01200
	2	0.95060	0.11444	0.01677	0.01742
	4	0.94488	0.11365	0.05717	0.02690
	7	0.94154	0.11764	0.09238	0.07252

Table 3.21. Summary of the recorded Problem 2 TPR, FPR, and oneSigma ranges.

1) omnibus procedures, 2) likelihood ratio tests involving specific alternatives, 3) measures of moments: skew, and/or kurtosis, and 4) graphical procedures. The most commonly applicable tests are omnibus tests geared to assess if a sample has a Gaussian distribution. We call these Gaussian goodness-of-fit (GGoF) functions. GGoF procedures can be grouped as those based on χ^2 measurements and those based on empirical distribution functions (EDF) [31].

χ^2 Measures. Pearson's idea when developing his (the original) χ^2 measure was to reduce the problem of GoF to the simpler problem of comparing observed bin frequencies, $O(j)$, with expected bin frequencies, $E(j)$. Three popular examples of χ^2 GoF functions are Pearson's χ^2 (Equation (3.29)), Read and Cressie's power divergent statistic (Equation (3.30)), and the log likelihood ratio (Equation (3.31)) [49]:

$$\chi_P^2 = \sum_{j=1}^{B} \frac{\left(O(j) - E(j)\right)^2}{E(j)}, \tag{3.29}$$

$$\chi^2_{R\&C} = \frac{9}{5} \sum_{j=1}^{B} O(j) \left(\left(\frac{O(j)}{E(j)} \right)^{2/3} - 1 \right),$$ (3.30)

$$\chi^2_{LLR} = 2 \sum_{j=1}^{B} O(j) \ln \left(\frac{O(j)}{E(j)} \right).$$ (3.31)

EDF Measures. EDF measures are based on the fact that "if one plots the ordered univariate sample versus the corresponding percentiles of the standard normal distribution, one should observe approximately a straight line if the sample indeed is normally distributed" [31]. Examples of EDF statistics include Cramer-von Mises based statistics such as the Cramer-von Mises statistic (Equation (3.32)) and the Anderson-Darling statistic (Equation (3.33)), the Shapiro-Wilk statistics, and the Kolmogorov-Smirnov statistics, e.g., Equations (3.34)-(3.36). These measures use the ordered $\overrightarrow{x_i}$ values z_i, the cumulative Gaussian distribution's value at the ordered sample values [56].

$$W^2 = \sum_{i=1}^{|S|} \left(z_i - \frac{2i-1}{2|S|} \right)^2 + \frac{1}{12|S|},$$ (3.32)

$$A^2 = \frac{1}{|S|} \sum_{i=1}^{|S|} (2i-1) \left(\ln(z_i) + \ln(1 - z_{(|S|+1-i)}) \right) - |S|,$$ (3.33)

$$D^+ = \max_{1 \le i \le |S|} \left(\frac{i}{|S|} - z_i \right),$$ (3.34)

$$D^- = \max_{1 \le i \le |S|} \left(z_i - \frac{i-1}{|S|} \right),$$ (3.35)

$$D = \max_{1 \le i \le |S|} (D^+, D^-).$$ (3.36)

Compared to χ^2 statistics, EDF statistics are generally more computationally expensive, require a problematic ordering of multivariate samples, and have not been as well-evaluated on discrete data. Since this chapter is concerned with large quantities of discrete data, further GGoF evaluation is limited to χ^2 methods.

χ^2 functions test the null hypothesis that the instances $\overrightarrow{x_i}$, $i = 1...|S|$, are well represented by the density function $F(\overrightarrow{x})$. Partitioning the random samples $\overrightarrow{x_i}$ into B cell ranges $R(j)$, $j = 1...B$, produces observed frequencies $O(j)$. The null hypothesis is true when the $O(j)$ have a binomial distribution with

$$E(j) = P(\overrightarrow{x_i} \text{ falls in } R(j)) = \int_{R(j)} dF(\overrightarrow{x}).$$ (3.37)

Thus, the GoF problem reduces to one of testing whether a multinomial distribution $O(j)$ has cell probabilities $E(j)$. Pearson showed that the quantities

$O(j) - E(j)$ have an approximately multivariate normal distribution and that the quadratic of Equation (3.29) is approximately χ^2 distributed with $B - 1$ degrees of freedom, χ^2_{B-1}. That is,

$$P(\chi^2_P \geq c) \to P(\chi^2_{B-1} \geq c) \text{ for any } c \geq 0, \tag{3.38}$$

and thus with confidence α

$$P(\chi^2_P \geq \chi^2_{B-1}(\alpha)) \to \alpha. \tag{3.39}$$

This fact is independent of whether F is univariate or multivariate, discrete or continuous. Fisher noted that the log likelihood ratio statistic, χ^2_{LLR}, is asymptotically equivalent to Pearson's χ^2_P. Read and Cressie have proven the same for their statistic, $\chi^2_{R\&C}$. Thus, rejection of model validity occurs when the observed value of a χ^2 function is greater than or equal to the χ^2 distribution value found in the $\chi^2_{B-1}(\alpha)$ tables for a pre-specified percentage point, i.e., $\alpha * 100\%$.

One important consideration is the "smoothness" of these functions. Smooth GoF functions are "constructed so as to have good power against alternatives whose probability density functions depart smoothly from the desired ..., Smooth changes include shifts in mean, variance, skew, and kurtosis" [48]. The smoothness criterion is important for the algorithms of this chapter so that the resulting GGoF space is differentiable and its extrema are well localized. Moment-based GoF functions utilizing Hermite polynomials directly address this smoothness criterion, but the use of such functions is computationally expensive [48]. The three tests being discussed have also been proven to be smooth functions [13]. Our experiments have shown that the binning technique also plays a significant role in the smoothness of the GGoF function.

3.6.1   Univariate Binning

The allocation of the samples to cells is "binning." A binning technique is defined by the number of bins and their feature space ranges.

Numerous researchers have devised formulas for suggesting the number of bins, B, for which the various χ^2 tests will provide optimal power. It is generally accepted that a test's power will increase as B is increased up to a point, and then for larger B values the power will begin to decrease. Dahiya and Gurland [14] suggest that frequently 4 or 5 bins are appropriate. Most researchers agree that the optimal number is usually quite small but may increase as the sample size increases [49]. We suggest $B = 6$ for most applications. However, this hyperparameter should be investigated for each application.

Bins can cover equal ranges in feature space, have equal probability, or deviate from equiprobable using a specified weighting [31]. This chapter also promotes

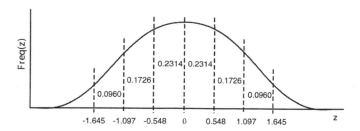

Figure 3.28. Allocation of the ±1.645σ extent of a Gaussian to 6 equirange bins. Expected bin probabilities are listed inside the bins.

an alternate binning strategy based on overlapping bins [25, 23] which has been shown to increase the accuracy of GoF estimates.

Bins sample within a fixed range of feature space; effectively, the sampled distributions are being compared to clipped Gaussians. It is important for this range to be sufficiently broad while limiting interference from neighboring clusters and speeding computations. Preliminary research indicates that a clipped Gaussian extent of 3σ to 4σ is sufficient for most work, but this hyperparameter should also be investigated for each application of GoF measures.

Equirange Bins.   Pearson's original work [44] used bins having equal feature space ranges. Such binning is independent of the distribution being considered. Consider the Gaussian distribution shown in Figure 3.28 which is partitioned into 6 equal spatial range bins. It has been suggested by numerous authors that for bins of unequal probabilities, enough instances should be considered so that each bin has an expected frequency greater than two. Given the expected probabilities shown, at least 21 instances are needed. This can be a limiting factor.

Equiprobable Bins.   It is generally accepted, following the work by Mann and Wald [37], that equiprobable bins provide the best power for most situations in which the parameters of the expected distribution are known, i.e., not estimated from the sample. The evaluation of a GGoF space is carried out using known parameters. That is, GGoF space defines the μ and σ, not the data. The allocation of a ±1.645σ extent of a Gaussian to 6 equiprobable bins is shown in Figure 3.29. This binning is dependent on the expected distribution. Because of the increase in power it is generally accepted that each bin should have an expected frequency of greater than one, requiring only 7 samples.

Overlapped Binning.   GoF measures generated using clipped Gaussians and hard limits between bins may be discontinuous due to sampling. These unwanted struc-

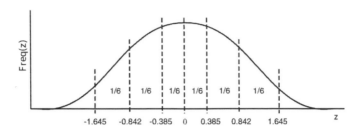

Figure 3.29. Allocation of the $\pm 1.645\sigma$ extent of a Gaussian to 6 equiprobable bins. Expected bin probabilities are listed inside the bins.

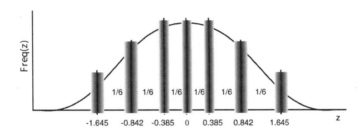

Figure 3.30. Bars indicate amount of overlap in overlapped binning with equiprobable bins.

tures are caused by the abrupt transition of instances from one bin to the next due to small changes in μ or σ being tested. As a result, a technique for smoothing the transition of the instances between the bins was considered. It operates by weighting each sample with respect to each bin [25, 23]. A variety of weighting techniques are possible, and overlapped binning can be applied in an equirange or an equiprobable binning manner. Overlapped-equiprobable binning is illustrated in Figure 3.30.

3.7 Conclusion

This chapter has introduced a range of classification systems, demonstrated their performance varies based on the validity of their underlying assumptions, and introduced goodness-of-fit measures to assess the validity of any assumed distribution shape, e.g., Gaussian. Scattergrams and other feature space visualizations are illustrated as tools for gaining an intuitive understanding of the data, distributions, and labelings of any multivariate classification problem. True positive and false positive rates are explained as quantifications of a classifier's accuracy given a set of training and testing data. Monte Carlo methods are introduced as a way of measuring classifier consistency given multiple sets of training and testing data. Most of the classifiers discussed are related to finite Gaussian mixture models with varying constraints and numbers of Gaussian components. χ^2 goodness-of-fit methods are introduced as one technique to assess if an assumed distribution shape matches the observed distribution shape; thereby aiding in the selection of a classification scheme for any given problem.

Again, we emphasis that few general truths are revealed in this chapter aside from the fact that a classifier's performance will vary based on the validity of its assumptions and the spread of the classes. We sought to present a range of popular classification schemes and provide the methods for qualitatively and quantitatively evaluating their utility for problems you may encounter and questions that may arise when conducting your future research in medical image analysis.

3.8 Appendix: Extruded Gaussian Distributions

Consider the proton density MR image shown again in Figure 3.31. It contains an inhomogeneity which is revealed by a dimming in the inferior cerebellum (lower portion of the brain). Locally within this image, the tissues have Gaussian distributions, but the means and covariances of those distributions change across the image because of the intensity inhomogeneity. Such spatially varying distributions produce *extruded Gaussian distributions*. A scatterplot of 984 hand-labeled white matter samples and 788 hand-labeled gray matter samples from this image are shown in Figure 3.32. The effect of the inhomogeneity is clear.

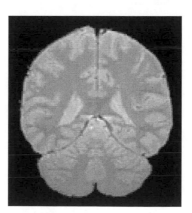

Figure 3.31. Proton density (PD) MR image. [The medical data used to generate this and all of the other images in this chapter were acquired in and provided by the Department of Radiology at The University of North Carolina at Chapel Hill]

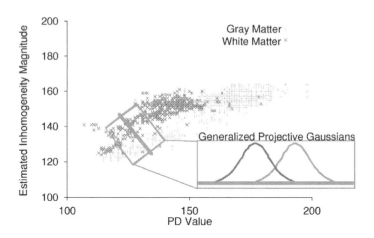

Figure 3.32. Scatterplot of hand-labeled instances from Figure 3.1. The distributions are not Gaussian as a whole, but have Gaussian cross-sections. They are referred to as extruded or generalized projective Gaussians.

For Problem 1, we define class B using splines to define the extrusion paths for the means and variances of a Gaussian. We use three cubic B-splines [47] and four isotropic control Gaussians, i.e., G_0, G_1, G_2, G_3. Each spline governs one of the three parameters of the continuum of Gaussians, i.e., one spline for each component of the mean vector and one for the variance. The parameters of the control Gaussians are given in Table 3.22. A visualization of the control Gaussians' density functions with the track of means $\vec{\mu}(t)$ for $t \in [0, 1]$ overlaid is shown in Figure 3.33.

The steps in generating an instance from Class B are as follows. A parametric value, t, is chosen from a uniform distribution, U[0,1]. The three splines are evaluated at that t value. The first two splines define the mean and the third defines the variance for an isotropic Gaussian distribution, i.e., $G(t)$ is defined by $\vec{\mu}(t)$ and $\sigma(t)$. The instance is then generated from that Gaussian distribution.

Class B

	μ_{f_0}	μ_{f_1}	σ^2
G_0	80	112	324
G_1	112	56	1
G_2	144	56	1
G_3	192	112	324

Table 3.22. The parameters of the control points of the splines which define the continuum of means and variances that are Class B.

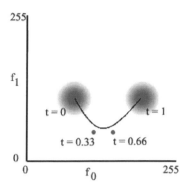

Figure 3.33. The four control Gaussians and the track of the continuum of means for Class B.

To illustrate the relationship between the Class A and Class B distributions in Problem 1 and thereby illustrate the unwanted relationships present in inhomogeneous medical images, the distributions can be used to generate a simulate medical image having two tissue types. Figure 3.34 is one possible pseudo-medical image based on these descriptions. Instances from Class A are uniformly spread across columns 0–64 and 196–256. Class B's instances are assigned to columns using a random pick from a Gaussian having a mean of 128 and a standard deviation of 32. The f_0 value generated for each sample specifies its row number, and the corresponding f_1 value specifies its intensity. One million instances were generated from each class to produce the image. Class B's intensity inhomogeneity is

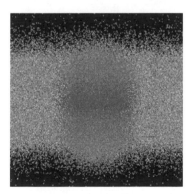

Figure 3.34. Pseudo-medical image depicting Problem 1: generated by interpreting f_0 as image row and f_1 as image intensity. Instances from Class A occupy the right and left portions of the image. Class B's instances occupy the central track.

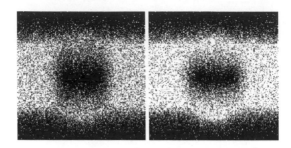

Figure 3.35. Two manual thresholdings of the pseudo-medical image in Figure 3.34. The effect of the intensity/row inhomogeneity of Class B is clearly visible. Ideally, the two outer bands would be colored white to indicate Class A, and no white would appear within the central band.

clearly visible as a non-linear correlation between its instances' intensity and row number. As a result of this inhomogeneity, near the top and bottom of this image, it is difficult to distinguish classes using only intensity. The effect of the inhomogeneity is accentuated when an intensity threshold is applied to distinguish the classes (Figure 3.35).

It can be imagined that such an image could result from SPECT or MR. The central region, i.e., Class B, might correspond with damaged tissue having reduced uptake of the isotope or contrast agent. The inhomogeneity may have resulted from attenuation from intervening structures or from MR field inhomogeneities. To accurately estimate the extent of the damaged tissue, the inhomogeneity/attenuation must be accurately modeled when the statistics of the classes are modeled, or the inhomogeneity can be removed via image processing to produce classes that have distributions that better resemble Gaussians. The processing of Gaussian data is the focus of Problem 2.

References

[1] Aylward, S. and J. Coggins. "Spatially invariant classification of tissues in MR images," in *Visualization in Biomedical Computing 1994* (R. A. Robb, ed.), (Mayo Clinic, Rochester, Minnesota) SPIE (1994).

[2] Aylward, S., and S. Pizer. "Continuous Gaussian mixture modeling," in *Proceedings of Information Processing in Medical Imaging*, Dordrecht, The Netherlands: Kluwer Academic Publishers, 1997. 176–189.

[3] Aylward, S. "Continuous Gaussian mixture modeling," *Pattern Recognition*, (September 2002): 1821–1833.

[4] Babich, G. and O. Camps. "Weighted Parzen windows for pattern classification," *IEEE Transactions on Pattern Analysis and Machine Intelligence* 18(5)(1996): 567-574.

[5] Bebis, G. and M. Georgiopoulos. "Feed-forward neural networks," *IEEE Potentials* (October/November 1994).

[6] Bellegarda, E., Bellegarda, J., *et al.* "A fast statistical mixture algorithm for on-line handwriting recognition," *IEEE Transactions on Pattern Analysis and Machine Intelligence* 16(12)(1994): 1227.

[7] Chen, T. and H. Chen. "Approximation capability to functions of several variables, nonlinear functionals, and operators by radial basis function neural networks," *IEEE Transactions on Neural Networks* 6(4)(1995): 904–910.

[8] Cheng, Y. "Mean shift, mode seeking, and clustering," *IEEE Transactions on Pattern Analysis and Machine Intelligence* 17(8)(1995): 790–799.

[9] Coggins, J. M. " A multiscale description of image structure for segmentation of biomedical images," in *Visualization in Biomedical Computing* (Atlanta, GA, IEEE, 1990).

[10] Coggins, J. M. "A statistical approach to multiscale medial vision," in *Mathematical Methods in Medical Imaging* (SPIE 1992).

[11] Coggins, J. M. and E. Graves. "Geometric image analysis using multiscale statistical features," in *AAAI Spring Symposium Series: Application of Computer Vision in Medical Image Processing* (Stanford University, AAAI Press, 1994).

[12] Cohen, A. C. "Estimation in mixtures of two normal distributions." *Technometrics* 9(1967): 15–28.

[13] Cressie, N. and T. Read. "Multinomial goodness-of-fit tests," *Journal of the Royal Statistical Society* 46(4)(1984): 440–464.

[14] Dahiya, R. and J. Gurland. "How many classes in the Pearson Chi-squared test?" *Journal of the American Statistical Association* 68(1973): 707–712.

[15] Dawant, B., Zijdenbos, A., *et al.* "Correction of intensity variations in MR images for computer-aided tissue classification." *IEEE Transactions on Medical Imaging* 12(4)(1993): 770–781.

[16] Dempster, A., Laird, N., *et al.* (1977). "Maximum likelihood for incomplete data via the EM algorithm," *Royal Statistical Society* 1(1)(1977).

[17] Duda, R. and P. Hart. *Pattern Classification and Scene Analysis*. New York: John Wiley and Sons, 1973.

[18] Duda, R., Hart, P., and D. Stork. *Pattern Classification*, second edition. New York: John Wiley and Sons, 2001.

[19] Ferguson, T. S. "Bayesian density estimation via mixtures of normal distributions," in *Recent Advances in Statistics*. New York: Academic Press, 1983. 287–302.

[20] Fryer, J. and C. Robertson. "A comparison of some methods for estimating mixed normal distributions," *Biometrika* 59(1972): 639–648.

[21] Geary, R. "Testing for normality," *Biometrika* 34(1947): 209–242.

[22] Gish, H. and M. Schmidt. "Text-independent speaker identification," *IEEE Signal Processing Magazine* 11(4)(1994): 18–32.

[23] Hall, P. "Tailor-made tests of goodness of fit," *Journal of the Royal Statistical Society* 47(1985): 125–131.

[24] Hornik, K., Stinchcombe, M., *et al.* "Multilayer feedforward networks are universal approximators," *Neural Networks* 2(1989): 359–366.

[25] Ivchenko, G. and S. Tsukanov. "On a new way of treating frequencies in the method of grouping observations, and the optimality of the X2 test," *Soviet Mathematics Doklady* 30(1984): 79–82.

[26] Jain, A. and R. Dudes. *Algorithms for Clustering Data*. Englewood Cliffs, NJ, Prentice Hall, 1988.

[27] Jain, A. K. *Fundamentals of Digital Image Processing*. Englewood Cliffs, NJ, Prentice Hall, 1989.

[28] Johnston, B., Atkins, M., *et al.* "Segmentation of multiple sclerosis lesions in intensity corrected multispectral MRI," *IEEE Transactions on Medical Imaging* 15(2)(1996): 154–169.

[29] Jones, M., Marron, J., *et al.* (1994). "A brief servey of bandwidth selection for density estimation," *Journal of the American Statistical Association* 91(1994): 401–407.

[30] Jordan, M. I. and L. Xu. *Convergence Results for the EM Approach to Mixtures of Experts Architectures*. (Massachusetts Institute of Technology, Artificial Intelligence Laboratory, 1993).

[31] Koziol, J. A. "Assessing multivariate normality: a compendium," *Communications in Statistics: Theories and Methods* 15(9)(1986): 2763–2783.

[32] Lecocq, C. and J.-G. Postaire. *Iterations of Morphological Transformations For Cluster Separation Pattern Classification*. (Symbolic-Numeric Data Analysis and Learning, Versailles, France, Nova Science Publishers, Inc, 1991).

[33] Lippmann, R. P. "An introduction to computing with neural nets." *IEEE ASSP* 3(4)(1987): 4–22.

[34] Loader, C. R. *Local Likelihood Density Estimation*. (ATT Bell Laboratories, 1995).

[35] Louis, T. "Finding the observed information matrix when using the EM algorithm," *Journal of the Royal Statistical Society* 44(2)(1982): 226–233.

[36] Mao, J. and A. K. Jain. "Artificial neural networks for feature extraction and multivariate data projection," *IEEE Transactions on Neural Networks* 6(2)(1995): 296–317.

[37] Mann, H. B. and A. Wald. "On the choice of the number of class intervals in the application of the chi-square test," *Annals of Mathematical Statistics* 13(1942): 306–317.

[38] McLachlan, G. and K. Basford. *Mixture Models*. New York, Marcel Dekker, Inc., 1988.

[39] Meyer, C., Bland, P., *et al.* "Retrospective correction of intensity inhomogeneities in MRI," *IEEE Transactions on Medical Imaging* 14(1)(1995): 36–41.

[40] Neter, J., Wasserman, W., *et al. Applied Statistics*. Boston, Allyn and Bacon, Inc., 1978.

[41] Newcomb, S. "A generalized theory of the combination of observations so as to obtain the best result," *American Journal of Mathematics* 8(1886): 343–366.

[42] Osman, H. and M. Fahmy. "On the discriminatory power of adaptive feedforward layered networks," *IEEE Transactions on Pattern Analysis and Machine Intelligence* 16(8)(1994): 837.

[43] Parzen, E. *Stochastic Processes*. San Francisco, Holden-Day, 1962.

[44] Pearson, K. "Contribution to the mathematical theory of evolution," *Phil. Transactions Roy. Soc.* A 185(1894): 71–110.

[45] Peterson, G., St. Clair, D., *et al.* "Using Taguchi's method of experimental design to control errors in layered perceptrons," *IEEE Transactions on Neural Networks* 6(4)(1995): 949–961.

[46] Poggio, T. and F. Girosi. "Networks for approximation and learning," *Proceedings of the IEEE* 78(9)(1990): 1481–1497.

[47] Press, W., Flannery, B., *et al. Numerical Recipes in C.* Cambridge, Cambridge University Press, 1990.

[48] Rayner, J. and D. Best. *Smooth Tests of Goodness of Fit.* Oxford, Oxford University Press, 1989.

[49] Read, T. and N. Cressie. *Goodness-of-Fit Statistics for Discrete Multivariate Data.* New York, Springer-Verlag, 1988.

[50] Samadani, R. "A finite mixtures algorithm for finding proportions in SAR images," *IEEE Transactions on Image Processing* 4(8)(1995): 1182–1186.

[51] Schalkoff, R. *Pattern Recognition: Statistical, Structural and Neural Approaches.* New York, John Wiley and Sons, Inc., 1992.

[52] Silverman, B. W. "Choosing the window width when estimating a density," *Biometrika* 65(1)(1978): 1–11.

[53] Silverman, B. W. *Density Estimation for Statistics and Data Analysis.* London, Chapman and Hall, 1986.

[54] Sobol', L. M. *A Primer for the Monte Carlo Method.* Boca Raton, CRC Press, 1994.

[55] Speckman, P. "Kernel smoothing in partial linear models," *Journal of the Royal Statistical Society* 50(3)(1988): 413–436.

[56] Stephens, M. "EDF statistics for goodness of fit and some comparisons," *Journal of the American Statistical Association* 69(347)(1974): 730–737.

[57] Tan, W. and W. Chang. "Some comparisons of the method of moments and the method of maximum likelihood in estimating parameters of a mixture of two normal densities," *Journal of the American Statistical Association* 67(1972): 702–708.

[58] Touzani, A. and J. Postaire. "Mode detection by relaxation," *IEEE Transactions on Pattern Analysis and Machine Intelligence* 10(6)(1988): 970–978.

[59] Waterhouse, S. and A. Robinson. "Non-linear prediction of acoustic vectors using hierarchical mixtures of experts. (*Neural Information Processing Systems, 7*, MIT Press Cambridge MA, 1995).

[60] Wells, W., Grimson, W., *et al.* "Adaptive segmentation of MRI data," *IEEE Transactions on Medical Imaging* 15(4)(1996): 429–442.

[61] West, M. "Approximating posterior distributions by mixtures," *Journal of the Royal Statistical Society* 55(2)(1993): 409–422.

[62] Xu, L. and M. Jordan. *On Convergence Properties of the EM Algorithm for Gaussian Mixtures.* (Massachusetts Institute of Technology, Artificial Intelligence Laboratory, 1995).

[63] Xu, L., Krzyzak, A., *et al.* "On radial basis function nets and kernel regression: statistical consistency, convergence rates, and receptive field size," *Neural Networks* 7(4)(1994): 609–628.

[64] Zhuang, X., Huang, Y., *et al.* "Gaussian mixture density modeling, decomposition, and applications," *IEEE Transactions on Image Processing* 5(9)(1996): 1293–1302.

Nonlinear Image Filtering with Partial Differential Equations

Ross Whitaker
University of Utah

4.1   Introduction

The idea of *filtering* is closely tied to traditional systems theory and its use of linear, stationary filters which can be represented, analyzed, and implemented in the frequency domain. To get beyond linear filtering, people have proposed various kinds of nonlinear filtering operations. Most nonlinear filtering methods are built upon one of three strategies: i) heuristics—set of rules on pixels that are designed specifically to achieve a particular result, ii) statistics—filters designed on robust statistics (e.g., median) or using properties from stochastic processes, iii) partial differential equations (PDEs)—the input is some initial value in a PDE and the output is from the solution of that PDE.

 This chapter introduces the idea of using PDEs to process images. The discussion begins with the mathematics of PDEs and filtering, introduces anisotropic diffusion for denoising images, describes some numerical schemes for solving the PDE, and then reviews several other families of PDEs for image processing.

4.2   Gaussian Blurring and the Heat Equation

The motivation for using partial PDEs for image processing begins with Gaussian blurring, which is a kind of low-pass transform. Low-pass transforms can be understood in the frequency (Fourier) domain as filters that reduce high frequencies more than low frequencies. Such transforms are useful for denoising, because often the noise in an image has more energy in high frequencies than does the *signal*, i.e., that part of the image which is useful for subsequent processing. Thus, most interesting images have (by far) most of their energy in the low frequencies.

For this discussion we will consider only 2D examples, because it keeps the notation simpler; the extension to higher-dimensional images falls directly out of the mathematical notation, which is quite general. We consider an image $f(x,y)$ to be a mapping from an image domain $\mathcal{D} \subset \mathbb{R}^2$, usually a small rectangle, to some real value, which is often called the *intensity*. That is, $f : \mathcal{D} \mapsto \mathbb{R}$.

The Gaussian is a kind of low-pass transform which has the form:

$$g(x,y) = \frac{1}{\sqrt{2\pi\sigma^2}} e^{\frac{x^2+y^2}{2\sigma^2}} \quad \text{and} \quad G(u,v) = \mathbb{F}\{g\} = e^{\frac{(x^2+y^2)\sigma^2}{2}}, \qquad (4.1)$$

where $\mathbb{F}$ denotes the Fourier transform, and σ is the standard deviation, which is the effective width of the kernel. Notice, this is the same as a normal distribution or *bell curve*, from statistics. As a low-pass transform the Gaussian kernel has several nice properties [6]. For instance it is smooth, does not ring, and is self-similar, which means that repeated applications of Gaussian filtering can be achieved with a single Gaussian filter. The relationship to PDEs is through the *diffusion equation*:

$$\frac{\partial f}{\partial t} = \frac{\partial^2 f}{\partial^2 x} + \frac{\partial^2 f}{\partial^2 y} = \nabla \cdot c \nabla f, \qquad (4.2)$$

where $f(x,y,t)$ can be viewed as a sequence of images that starts with $f(x,y,0)$ and *evolves* over time to become more smooth. The operator ∇ denotes the vector of partial derivatives, and we let $c = 1$ for the current discussion. Equation (4.2) is an *initial value problem* in that we treat it as a process where we put in an image I, i.e., $f(x,y,0) = I(x,y)$, and obtain a smoother image at some later time t by looking at the solution. The larger the value of t we choose the smoother the result.

Because Equation (4.2) is linear and stationary (shift invariant) the solution can be represented by convolution with a kernel, i.e., a linear filter. If we assume either an infinite or cyclical domain for the image and take the Fourier transform of f with respect to the spatial dimensions x and y, we can re-express Equation (4.2) in the frequency domain:

$$\frac{\partial F}{\partial t} = -(u^2 + v^2)F, \qquad (4.3)$$

where we have used the fact that $\mathbb{F}\{\partial f/\partial x\} = juF$, and $j^2 = -1$. Equation (4.3) is an ordinary differential equation, and its solution is an exponential of the form:

$$F(u,v,t) = e^{-(u^2+v^2)t} F(u,v,0). \qquad (4.4)$$

If we move the solution back to the spatial domain (it is valid in either domain), multiplication with G becomes a convolution with a Gaussian and we have the inverse Fourier transform of Equation (4.4) as

$$f(x,y,t) = \frac{1}{\sqrt{4\pi t}} e^{-(x^2+y^2)/4t} \otimes f(x,y,0). \qquad (4.5)$$

Thus, we see that the solution to the diffusion equation at a particular time is the same as convolving the initial conditions with a Gaussian of standard deviation $\sigma = \sqrt{2t}$. This analysis applies to continuous functions, but there are discrete versions of this formulation which give equivalent results.

Of course, convolving with a Gaussian can be quite fast, and is generally preferred over solving the diffusion equation in some more direct fashion. However, there are good motivations for solving the diffusion equation directly. One reason is boundary conditions. The relationship between diffusion and Gaussian blurring applies only if we have the proper boundary conditions. This is the familiar problem with the Fourier transform. In image processing, this means that we must either pad the image with some set of predetermined values or allow the image to simply *wrap around*, i.e., the left edge meets the right and the bottom meets the top to form a toroidal topology. Either of these choices can cause noticeable, sometimes undesirable artifacts at the edges. However, the PDE version of the problem allows one to define boundary conditions that are more appropriate for images.

Boundary conditions for PDEs are usually specified in terms of how the solution should behave at the boundaries of the domain, $\mathcal{D}$. The nature of the boundary conditions is related to the type of PDE; the diffusion equation is a second order PDE. The most common choice of boundary conditions for applications of diffusion equation to image processing is *Von Neumann conditions*, which is to specify the value of the derivatives of the solution across the boundary. If we denote the boundary of the $\mathcal{D}$ as Λ and the direction of the normal to the boundary as $\vec{n}$. We would typically choose $\partial f/\partial \vec{n} = 0$ for all points in Λ. This choice means that no image intensity (thinking of it as a physical quantity, e.g. density) flows across the boundaries of the image. Thus, the average grey scale value is preserved, and when applied to the problem of heat transfer, this preservation of energy is denoted as *adiabatic*. For images this means that the diffusion smoothes up to and along the boundary of the image but not across the boundary of the image. Figure 4.1 shows the effects of blurring with adiabatic diffusion as compared to simply padding the image with zeros.

4.2.1 Anisotropic Diffusion

One of the drawbacks of low-pass filtering to remove noise is that it also reduces high-frequency components of the image. Although images have most of their

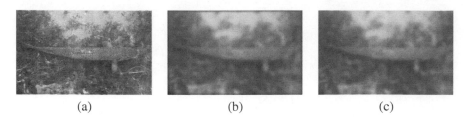

(a)   (b)   (c)

Figure 4.1. The use of PDEs for resolving boundary conditions: (a) original image; (b) Gaussian blurring with padding of 0 (black); (c) processing with diffusion equation and adiabatic boundary conditions.

energy in the low frequencies, the high-frequency components help create sharp boundaries in the image, or areas of locally high contrast, i.e., *edges*. Such image features are particularly important to human perception [7] and to a variety of higher image processing algorithms, such as edge detection. Unfortunately, smoothing an image with a low-pass filter typically has a disproportionate effect on such features.

To alleviate this problem, people have turned to nonlinear techniques, where the structure of the image drives the shape or influence of the filter. When filtering with the diffusion equation, information about edges can impact the smoothing through a variable conductance term:

$$f_t = \nabla \cdot c \nabla f, \text{ where } c : D \mapsto \Re^+, \tag{4.6}$$

where a subscript of a variable represents a partial derivative with respect to that variable. If we imagine diffusion as the flow of material or energy (e.g., heat) from places in the domain where there is a high concentration to places where there is a low concentration, then the term $c(x,y)$ controls the rate of flow and allows it to be defined differently in each part of the image.

If we control the conductance $c(x,y)$ and lower it near interesting image features, *e.g.*, edges, we can reduce the blurring in those areas. This is idea was proposed by Grossberg [5] in the context of modeling human vision and building artificial vision systems. In that work, he constructs a function $E(x,y)$ that quantifies edges in the initial image $I(x,y)$ and constructed $c(E)$ which is monotonically decreasing with E. That is, the conductance goes down at edges in the input. If we make c dependent only on the input data, I, and not the solution, f, we have a variable conductance system that is still linear.

Perona and Malik [10] proposed a nonlinear form of this scheme in which c depends on f. They proposed a PDE of the form

$$f_t = \nabla \cdot c(|\nabla f|)\nabla f, \tag{4.7}$$

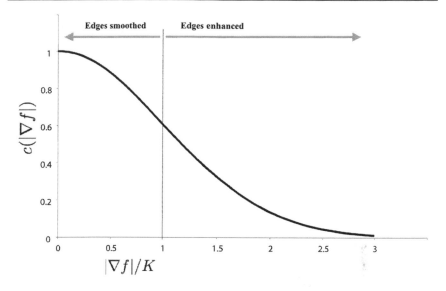

Figure 4.2. The conductance function is sigmoidal. A 1D analysis shows that edges with gradient values to the right of the inflection point get enhanced, while those to the left (lower values) get smoothed.

and called it *anisotropic diffusion*. The conductance function c is monotonically decreasing and should be sigmoidal in shape. A very popular form is

$$c(|\nabla f|) = e^{\frac{|\nabla f|^2}{2K^2}}, \qquad (4.8)$$

where K is a free parameter, with units of *intensity/space*, which controls the degree to which the gradient of f impacts the conductance. A graph of the conductance function is show in Figure 4.2.

The philosophy behind Equation (4.7) is that the edges in f, indicated by larger $|\nabla f|$, are not smoothed as the process progresses. Analysis shows that this equation can smooth away noise while preserving, or even enhancing, edges for which $|\nabla f| > K$. Figure 4.3 shows an example of a solution of this equation for a noisy input image.

When filtering an image by solving Equation (4.7) one must choose the amount of blurring t (related to σ in the linear case) and the contrast of the edges to be preserved K. Tuning both parameters correctly can be difficult, and depends on the contrast of the image, the level of noise, and the application. To make this easier many implementations express K in units of RMS gradient magnitude. Therefore

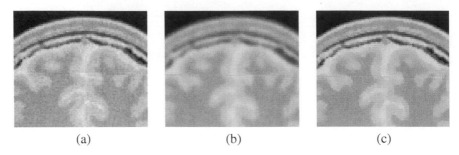

(a)   (b)   (c)

Figure 4.3. (a) MRI Image; (b) Gaussian blurring with $\sigma = 1.0$; (c) anisotropic diffusion with $t = 1.0$ and $k = 0.6$.

the conductance Equation (4.8) is typically computed as

$$c(|\nabla f|) = e^{\frac{|\nabla f|^2}{2k^2|\hat{\nabla}f|^2}}, \tag{4.9}$$

where $|\hat{\nabla}f|$ is the root mean squared of the gradient magnitude (across all of $\mathcal{D}$ or some neighborhood). In this case k is unitless, and depends less on the overall scaling or units of the input image. For most images good values of k range from 1.0 to 3.0.

The value of t to choose for the output depends on the application. If we use the relationship from the case where $c = 1$, a kernel width of $2\sqrt{2}$ pixels would require $t = 1$. For general purpose image denoising (e.g., medical images), $t = 1$ (4–5 iterations) often produces good results. For creating images that are very smooth with a few sharp boundaries (e.g., for segmentation) values of t up to 2.5 are sometimes appropriate.

This scheme and many variations of it have proven useful for denoising images. Several important improvements to the method are worth noting, but are otherwise beyond the scope if this paper. For instance, several authors have observed that the gradient used in the conductance term relies on derivatives of a noisy image [3, 25]. For this reason it is useful to compute the edge term in the conductance in conjunction with a smoothing kernel (e.g., Gaussian with standard deviation σ). In this way, Equation (4.7) becomes $f_t = \nabla \cdot c(|\nabla_\sigma f|)\nabla f$, where ∇_σ denotes a gradient obtained from derivative-of-Gaussian kernels with standard deviation σ. This smoothing term also provides well-posedness in the theoretical sense [3]. In practice few people smooth the gradient in the conductance term, and, because the image must be smoothed during every iteration, only small kernels (e.g., small σ) are actually practical.

An important generalization of anisotropic images is the extension to vector-valued data: that is, an image $\vec{f}: \mathcal{D} :\mapsto \mathbb{R}^m$, where m is the value of the output. For instance, $m = 3$ in the case of color images. Whitaker and Gerig [28] proposed treating vector-valued data through a system of coupled PDEs. If we think of $\vec{f}$ as a collection of scalar functions, i.e., $\vec{f} = f_1, f_2, \ldots f_m$, then the generalization of Equation (4.7) is

$$
\begin{aligned}
\frac{\partial f_1}{\partial t} &= \nabla \cdot c(D[\vec{f}]) \nabla f_1 \\
\frac{\partial f_2}{\partial t} &= \nabla \cdot c(D[\vec{f}]) \nabla \\
\vdots \quad &\quad \vdots \\
\frac{\partial f_m}{\partial t} &= \nabla \cdot c(D[\vec{f}]) \nabla f_m,
\end{aligned}
\tag{4.10}
$$

where $D[\vec{f}]$, the *dissimilarity*, is some generalization of the gradient magnitude to vector-valued functions. Equation (4.10) can be denoted $\vec{f}_t = \nabla \cdot c(D[\vec{f}]) \nabla \vec{f}$, where $\nabla \vec{f}$ is the matrix of partial derivatives, or the Jacobian. For dissimilarity, Whitaker [26] proposed the Frobenius norm (root sum of squares) of the Jacobian of $\vec{f}$, or some re-weighted Jacobian. More recently, Sapiro [12] proposed an approach, from color image processing, which uses the L_2 norm of $\nabla \vec{f}$, or combinations of the principle component values of $\nabla \vec{f}$. This norm has the advantage of favoring edges where the derivatives of the different components of $\vec{f}$ are aligned—which suggests it should be less sensitive to noise. Figure 4.4 shows the results of applying vector-valued anisotropic diffusion, using the L_2 norm of $\nabla \vec{f}$, to a color histology image.

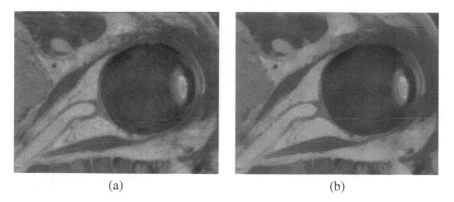

| (a) | (b) |

Figure 4.4. (See Plate VII) The result of vector valued anisotropic diffusion on a color image; (a) Original; (b) Filtering with dissimilarity defined as difference of the first two principal components of the Jacobian, with $t = 2.4$ and $K = 0.7$.

4.3   Numerical Implementations

In this section we describe some of the methods for solving these diffusion equations on a discrete grid. We assume, for each equation or mapping, that the image domain is a regular, isotropic grid with unit spacing. We denote the value of the function f at a grid point (i, j) and time step n as $f_{i,j,n}$, where the time step n is sometimes omitted when the context is clear. Equations (4.7) and (4.10) are usually solved with a finite difference scheme, using an Euler forward difference in time. The initial values are $f_{i,j,0} = I_{i,j}$ and the updates to the next discrete time step are given by

$$f_{i,j,n+1} = f_{i,j,n} + \Delta t \Delta f_{i,j,n}, \tag{4.11}$$

where $\Delta f_{i,j,n}$ is determined using finite differences on the spatial grid, (i, j).

To approximate the right side of Equation (4.3), we apply a sequence of finite difference operators, which means that each derivative is computed as a difference between a grid value and it neighbors. Different neighborhood configurations are possible and we denote the type of discrete difference using superscripts. For instance, a difference between a grid value and its forward neighbor (increasing grid index) is denoted $\delta^{(+)}$. We use $\delta^{(-)}$ to denote backward differences, and δ for central differences, i.e., the difference between forward and backward neighbors. For instance, differences in the x direction on a discrete grid are denoted

$$\delta_x^{(+)} f_{i,j} \equiv f_{i+1,j} - f_{i,j}, \tag{4.12}$$

$$\delta_x^{(-)} f_{i,j} \equiv f_{i,j} - f_{i-1,j}, \text{ and} \tag{4.13}$$

$$\delta_x f_{i,j} \equiv (f_{i+1,j} - f_{i-1,j})/2. \tag{4.14}$$

The result of each operator has a position on the grid, which is considered to be the average of the location of the grid points used to computed the difference. Thus the result of a $\delta_x^{(+)}$ operator is considered *staggered* because it is positioned at $(i + \frac{1}{2}, j)$, while the result of the central difference operator δ_x is situated at (i, j).

The anisotropic diffusion operator is approximated using a sequence of forward and backward differences (the result being centered):

$$\nabla \cdot c(|\nabla f|) \nabla f \approx \delta_x^- \left(c_{i+\frac{1}{2},j} \delta_x^+ f_{i,j} \right) + \delta_y^- \left(c_{i,j+\frac{1}{2}} \delta_y^+ f_{i,j} \right), \tag{4.15}$$

where $f_{i,j} = f_{i,j,n}$. Special care must be taken when computing the conductance at the *staggered* locations. The product of the conductance c and the inner-most derivative (in either x or y directions) is called the *flux*, and the update to a grid value is computed as a difference of fluxes. The fluxes must computed *between* grid points using an approximation to the $|\nabla f|$ that is centered over these stag-

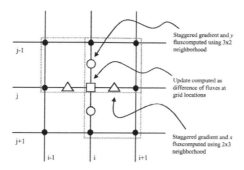

Figure 4.5. Fluxes, for both *x*- and *y*-directions, require gradients on a staggered grid. This is done using 6 neighboring values.

gered locations. We use the method proposed in [27]:

$$c_{i+\frac{1}{2},j} = c\left(\left[\delta_x^+ f_{i,j}\right]^2 + \frac{1}{4}\left[\delta_y f_{i,j} + \delta_y f_{i+1,j}\right]^2 \right), \qquad (4.16)$$

$$c_{i,j+\frac{1}{2}} = c\left(\left[\delta_y^+ f_{i,j}\right]^2 + \frac{1}{4}\left[\delta_x f_{i,j} + \delta_x f_{i,j+1}\right]^2 \right). \qquad (4.17)$$

This strategy is depicted in Figure 4.5.

When using the finite forward difference scheme given in Equation (4.11), Δt must be chosen to ensure stable solutions, using the diffusion number of the equation, which can be determined from a Von Neumann stability analysis [19]. For the finite difference scheme given in Equations (4.11)–(4.17), the stability condition gives $\Delta t \leq \frac{1}{4}$. These same conditions for M-dimensional domains require $\Delta t \leq \frac{1}{2M}$.

Several authors have proposed improved numerical schemes for anisotropic diffusion and related equations. For instance, Weickert [23] proposed a semi-implicit numerical scheme which allows significantly larger time steps. To achieve this, the conductance term c is fixed within each time step (i.e., a linear approximation to the equation). Therefore, Weickert acheives stability for very large time steps, but with reduced effectiveness for enhancement and significant directional artifacts. Acton [1] proposed a multi-grid scheme, which improves the computation time, especially when very long diffusion times are required, as in the case of reconstruction, described in Section 4.3.2.

4.3.1   Other PDE Methods

There are variety of other PDE-based methods for processing images. Here we give a brief description of a few of the more prevalent examples. For more comprehensive reviews see [21, 15, 14].

The Perona and Malik Equation (4.7) is based on the diffusion equation. Another strategy is to build PDEs that process the *level sets* or isophotes of an image. For 2D images these level sets are curves and for volumes they are surfaces (see Chapter 8) for more discussion of level sets and surface curvature). The strategy is to smooth an image by simultaneously shortening each level set (curve). The level-set curves are shortened (smoothed) by moving them in the direction and magnitude of their local curvature values. The curvatures of the level sets of an image f are given by

$$\kappa = \nabla \cdot \frac{\nabla f}{|\nabla f|}, \tag{4.18}$$

where the PDE for f that moves the level sets requires a multiplication with the gradient magnitude. That is,

$$\frac{\partial f}{\partial t} = |\nabla f| \nabla \cdot \frac{\nabla f}{|\nabla f|}. \tag{4.19}$$

This equation smoothes images in a way that tends to preserve straight edges, but it does not sharpen edges in the way that anisotropic diffusion does. The numerical implementation is often done using second derivatives with tightest-fitting kernels, which can be approximated by cascading forward and half derivatives. That is, $f_{xx} \approx \delta_{xx} f_{i,j} = \delta_x^{(-)} \delta_x^{(+)} f_{i,j}$. In 2D this gives

$$|\nabla f| \nabla \cdot \frac{\nabla f}{|\nabla f|} = \frac{f_{xx} f_y^2 + f_{yy} f_x^2 - 2 f_x f_y f_{xy}}{\left(f_x^2 + f_y^2\right)} \tag{4.20}$$

$$\approx \frac{d_{xx} f_{i,j} \left(d_y f_{i,j}\right)^2 + d_{yy} f_{i,j} \left(d_x f_{i,j}\right)^2 + d_y d_x f_{i,j} \left(d_x f_{i,j}\right) \left(d_y f_{i,j}\right)}{\left(\left(d_y f_{i,j}\right)^2 + \left(d_x f_{i,j}\right)^2 + \varepsilon\right)}, \tag{4.21}$$

where ε is chosen to avoid divide-by-zero problems near singularities. For stability $\Delta t < 1/4$ is required, but this scheme can still cause oscillations for any Δt. Some authors include a small amount of normal, homogeneous diffusion (Laplacian) in order alleviate this oscillation. An alternative implementation is a difference of normals, as described in [29]. In that scheme, the curvature is implemented by computing gradients of f on the staggered grid (in both x and y using a stencil like that in Equation (4.17)), normalizing those gradients, and computing the divergence of the resulting vectors. The gradient magnitude is applied afterward using the up-wind scheme, which is described in Chapter 8.

There are several interesting variations on this curvature-based smoothing approach. It can be applied to vector-valued images in the same way as anisotropic diffusion. Alvarez and Morel[2] include a variable weighting term in the front, which reduces the smoothing at edges:

$$\frac{\partial f}{\partial t} = c(|\nabla f|)|\nabla f|\nabla \cdot \frac{\nabla f}{|\nabla f|}, \tag{4.22}$$

where c is defined much like the conductance in Equation (4.7). This does improve the smoothing of edges, but still does not create the enhancement of edges associated with anisotropic diffusion.

A more general form of Equation (4.19) has been proposed [13, 16, 29] for image denoising:

$$\frac{\partial f}{\partial t} = |\nabla f|\nabla \cdot c(|\nabla f|)\frac{\nabla f}{|\nabla f|}, \tag{4.23}$$

where c is positive, bounded and sigmoidal, as in Equation (4.7). This *modified curvature flow* can enhance edges and has been shown to be less sensitive to local image contrast [29]. Therefore it is easier to choose an appropriate value of the parameter k. This equation must be solved using differences of normals. Examples of Equation (4.23) relative to anisotropic diffusion are shown in Figure 4.6.

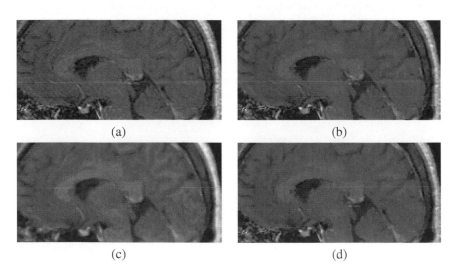

(a)　　　　　　　　　　(b)

(c)　　　　　　　　　　(d)

Figure 4.6. Processing MRI data with PDEs: (a) original data; (b) anisotropic diffusion ($t = 1.0$, $K = 0.6$); (c) curvature flow ($t = 1.0$); (d) modified curvature flow ($t = 6.0$, $K = 0.6$).

A variety of other possibilities have been explored, although must are related in some way to either anisotropic diffusion or curvature flow. For instance, Weickert [24] proposed a tensor for conductance rather than a scalar (in which case the conductance becomes anisotropic). This produces a richer set of possibilities, which can be used to enhance linear structures or texture. Several authors [18, 31] have proposed PDEs based on minimizing the surface area of manifolds that consist of graphics of functions f or $\vec{f}$. These PDEs generally have the form that resembles a curvature flow. Whitaker and Gerig [28] have proposed using a complex form of the gradient operator, which allows anisotropic diffusion to be tuned to specific frequencies. Several authors have proposed fourth-order PDEs for preserving piecewise smooth (rather than merely piecewise constant) functions [20, 22].

4.3.2   Variational Approaches

The previous discussion was motivated by Gaussian blurring, and it treated PDEs as filters. A PDE-based filter accepts an input image, uses it as initial conditions, and returns another image which has been smoothed, by an amount which is specified by the parameter t (how far we allow the evolution to progress). However, these PDEs are (mostly) minimizers of certain kinds of energy functionals, an interpretation which leads to other PDE-based processing strategies.

Anisotropic diffusion, curvature flow, and modified curvature flow are variational PDEs. This means that they represent the gradient descent, calculated from the first variation (in an appropriate metric space) of some penalty or energy functional. For instance, the Perona and Malik Equation (4.7), with the exponential conductance, has the corresponding energy functional:

$$E = \int_D e^{-|\nabla f|^2/2K^2} dx dy, \qquad (4.24)$$

where we omit constant factors for notational convenience. We will call such terms *smoothing penalties*. The level-set curvature flow has a corresponding energy functional

$$E = \int_D |\nabla f| dx dy, \qquad (4.25)$$

if one assumes an image metric based on level-set distance, which effectively introduces a factor of $|\nabla f|$ in the first variation [20].

Several authors [9, 17, 4] have noted that the variational form of anisotropic diffusion relates PDE-based filters to a large class of stochastic or Bayesian reconstruction algorithms. In this light, Nordstrom [9] proposed *biased anisotropic diffusion*,

$$\frac{\partial f}{\partial t} = (I - f) + \lambda \nabla \cdot c(|\nabla f|) \nabla f. \qquad (4.26)$$

This is the gradient descent on a smoothness penalty combined with a penalty for *image fidelity*, $\int (f - I)^2 dxdy$. The steady state of Equation (4.26) is, presumably, the desired solution. Thus, t is not a free parameter but should be as large as is necessary for this process to "settle down." The term λ controls the amount of smoothing. The initialization is an important issue, because the energy function might not be convex. Several authors have proposed methods, such as graduated nonconvexity [17, 8], for finding the global optimum. Most people simply use the input data I as the initial estimate, and satisfy themselves with local minima around that function.

Other related PDE reconstruction schemes have also been proposed. For instance, Rudin et al.[11] have proposed total variation reconstruction, which combines an image fidelity term with a curvature flow term (comes from the first variation of $\int |\nabla f| dxdy$). They also compute λ on automatically using the variance of the noise as a constraint. Another example is the reconstruction of height functions [30], which combines a projected form of surface smoothing with a fidelity to terrain or range data.

References

[1] Acton, S. "Multigrid anisotropic diffusion," *IEEE Transactions Image Processing* 7(3)(1998): 280-291.

[2] Alvarez, L. and J.-M. Morel. "A morphological approach to multiscale analysis: From principles to equations," in *Geometry-Driven Diffusion in Computer Vision* B. ter Haar Romeny, ed. Dordrecht, The Netherlands: Kluwer Academic Publishers, 1994. 229–254.

[3] Catte, F, Lions, P., Morel, J. and T. Coll. "Image selective smoothing and edge detection by nonlinear diffusion," *SIAM Numer. Anal.* 29(1)(1992): 182–193.

[4] Geiger, D. and A. Yuille, "A common framework for image segmentation," *International Journal of Computer Vision*, 6(3)(1991): 227–243.

[5] Grossberg, S. "Neural dynamics of brightness perception: Features, boundaries, diffusion, and resonance," *Perception and Psychophysics* 36(5),(1984) 428–456.

[6] Koenderink, J. "The structure of images," *Biol. Cybern.* 50, (1984): 363–370.

[7] Marr, D. *Vision*. San Francisco: Freeman, 1982.

[8] Nielsen, M. "Graduated non-convexity by functional focusing," *IEEE Transactions on Pattern Analysis and Machine Intelligence* 19(5)(1997): 521–525.

[9] Nordstrom, N. "Biased anisotropic diffusion—a unified regularization and diffusion approach to edge detection," *Image and Vision Comp.* 8(4)(1990): 318–327.

[10] Perona, P. and J. Malik, "Scale-space and edge detection using anisotropic diffusion," *IEEE Transactions on Pattern Analysis Machine Intelligence* 12 (1990): 629–639.

[11] Rudin, L., Osher, S., and C. Fatemi, "Nonlinear total variation based noise removal algorithms," *Physica D* 60 (1992): 259–268.

[12] Sapiro, G. and D. Ringach. "Anisotropic diffusion of multivalued images with applications to color filtering," *IEEE Transactions on Image Processing* 5 (1996): 1582–1586.

[13] Sapiro, G. "Color snakes," *CVIU*, 68(2)(1997): 247–253.

[14] Sapiro, G., Morel, J.-M., and A. Tannenbaum, "Special issue on differential equations in image processing," *IEEE Transactions Image Processing* 7(3)(1998).

[15] Sapiro, G. *Geometric Partial Differential Equations and Image Analysis.* Cambridge University Press, 2001.

[16] Shah, J., "Segmentation by nonlinear diffusion II," in *Proc. Conf. on Computer Vision and Pattern Recognition*(1992): 644–647.

[17] Snyder, W., Han, Y.-S., Bilbro, G., Whitaker, R. and S. Pizer, "Image relaxation: restoration and feature extraction," *IEEE Transactions on Pattern Analysis and Machine Intelligence* 17 (June 1995): 620–624.

[18] Sochen, N., Kimmel, R., and R. Malladi, "A general framework for low level vision," *IEEE Transactions on Image Processing* 7(3)(1998): 310–318.

[19] Strikwerda, J. *Finite Difference Schemes and Partial Differential Equations.* London, Chapman & Hall, 1989.

[20] Tasdizen, T., Whitaker, R., Burchard, P., and S. Osher. "Geometric surface processing via normal maps." *To appear: ACM Transactions on Graphics.*

[21] ter Haar Romeny, B. ed. *Geometry-Driven Diffusion in Computer Vision.* Dordrecht, The Netherlands: Kluwer Academic Publishers, 1994.

[22] Tumblin, J. and G. Turk. "Lcis: A boundary hierarchy for detail preserving contrast reduction," in *Proc. ACM SIGGRAPH* (1999): 83–90.

[23] Weickert, J. *Anisotropic Diffusion in Image Processing.* Teubner-Verlag, 1998.

[24] Weickert, J. "Coherence-enhancing diffusion of colour images," *Image and vision computing* 17 (1999): 201–212.

[25] Whitaker, R. and S. Pizer, "A multi-scale approach to nonuniform diffusion," *Computer Vision, Graphics, and Image Processing: Image Understanding* 57 (January 1993): 99–110.

[26] Whitaker, R. "Geometry-limited diffusion in the characterization of geometric patches in images," *Computer Vision, Graphics, and Image Processing: Image Understanding* 57 (January 1993): 111–120.

[27] Whitaker, R. "Volumetric deformable models: Active blobs," in *Visualization In Biomedical Computing 1994* (R. A. Robb, ed.), (Mayo Clinic, Rochester, Minnesota) SPIE (1994): 122–134.

[28] Whitaker, R. and G. Gerig. "Vector-valued diffusion," in *Geometry-Driven Diffusion in Computer Vision* (B. M. ter Haar Romeny, ed.), Dordrecht, The Netherlands: Kluwer Academic Publishers, 1994.

[29] Whitaker, R. and X. Xue. "Variable-conductance, level-set curvature for image denoising," in *Proceedings of IEEE International Conference on Image Processing*, (October 2001): 142–145.

[30] Whitaker, R. and E. Juarez-Valdes. "On the reconstruction of height functions and terrain maps from dense 3D data," *IEEE Transactions on Image Processing.* 11(7)(2002): 704–716.

[31] Yezzi, J. A. "Modified curvature motion for image smoothing and enhancement," *IEEE Transactions on Image Processing.* 7(3)(1998): 345–352.

Part Two

Segmentation

Segmentation Basics

Terry S. Yoo
National Library of Medicine, NIH

5.1 Introduction

Segmentation can be defined simply as the partitioning of a dataset into contiguous regions (or sub-volumes) whose member elements (e.g., pixels or voxels) have common, cohesive properties. This clustering of voxels or, conversely, the partitioning of volumes is a precursor to identifying these regions as objects and subsequently classifying or labeling them as anatomical structures with corresponding physiological properties. Segmentation is the means by which we impose structure on raw medical image data; we draw on this structure when we later visualize the anatomy or pathology in question as well as plan interventions and treatments to address the medical condition.

Just as there are many ways of interpreting an image (through statistics, differential geometry of intensity fields, or partial differential equations, etc.), there are many different approaches to the problem of segmentation derived from the many perspectives. This section of the book will briefly treat some of the concepts underlying these basic approaches and describe the foundations of the corresponding algorithms.

A common problem shared by many of the methods is that automated methods for partitioning structures within a dataset may not correspond to the object or structures that a human expert would identify. Often, the dataset is *oversegmented* or fractured into many small pieces that have to be aggregated into whole structures using expert, human intervention. Another common error is to *under-segment* a dataset, which requires later subdivision of mixed or connected objects into separate pieces. The result is that automated segmentation methods are often mixed with graphic user interfaces to allow clinicans and researchers to explore, manipulate, and edit the results created by computer segmentation algorithms.

5.1.1   A Basic Example

The example in Figure 5.1 (Courtesy of Ross Whitaker and Josh Cates, University of Utah) applies a watershed segmentation technique to the Visible Human Project Female dataset. This 24-bit color dataset often presents complex challenges in visualization arising from its density, size, and the multiple channels of the voxel elements. The watershed technique creates a series of "catchment basins," representing local minima of image features or metrics (an example is shown in Figure 5.1 (a)). The resulting watersheds can be hierarchically organized into succesively more global minima, and the resulting graph (a schematic of a simple graph is shown in Figure 5.1 (b)), linking all of the individual watershed regions, can be generated as a preprocessing step. Subsequently, the graph hierarchy can later be used to navigate the image, interactively exploring the dataset through dynamic segmentation of the volume by selectively building 3D watershed regions from the 4D height field. ITK can thus be linked to a graphical user interface as well as a visualization back-end in order to create a functioning interactive segmentation and visualization system for complex data (see Figure 5.2 (a)).

Figure 5.2 (b) shows the results of the semi-automated segmentation system. A user has defined the rectus muscle (highlighted on the left in grey or red), the left eye (highlighted in moderate shade), and the optic nerve (highlighted in a brighter shade). These tissue types can be compared with the correspond-

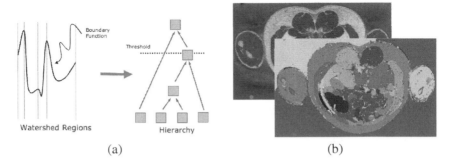

(a)   (b)

Figure 5.1. (See Plate I) Watershed segmentation: (a) a 1D height field (image) and its watersheds with the derived hierarchy; (b) a 2D color data slice from the NLM Visible Human Project and a derived watershed segmentation. [Example courtesy of Ross Whitaker and Josh Cates, Univ. of Utah]

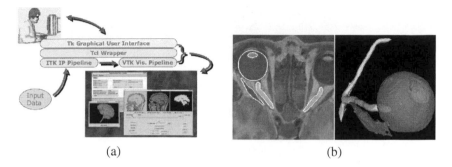

(a)   (b)

Figure 5.2. (See Plate II) An ITK watershed segmentation application with visualization provided by VTK and with the user interface provided by Tcl/TK. This example shows how ITK can be be integrated with multiple packages in a single application: (a) Data flow and system architecture for the segmentation user application; (b) anatomic structures (the rectus muscle, the optic nerve, and the eye) outlined in white (shown in 2D but extending in the third dimension) and the resulting visualization. [Example courtesy of Ross Whitaker and Josh Cates, Univ. of Utah]

ing anatomy reflected on the right side of the subject, revealing the difference in the highlighting color and the relative fidelity of the segmentation. It should also be noted that after preprocessing to find the primary watershed regions, navigating the hierarchy to build up regions is a natural and quick process with a graphical user interface. The watersheds and their related hierarchies extend in three dimensions making this segmentation example a native 3D application.

5.2   Statistical Pattern Recognition

Computer scientists often adopt related mathematics to solve complex problems. When dealing with medical image data, one powerful perspective is to treat the dataset, or a collection of datasets, as a statistical sampling of the human anatomy. Building upon statistics as a foundation for imaging research allows practioners to draw upon the powerful calculus associated with probabilites and distributions of pixel or voxel measurements of the patient's internal parts. Chapter 3 covered some of the basic concepts of statistical pattern recognition.

As a basic illustration of how images might be used to survey a problem of segmentation and visualization medical data, consider the example of rendering the bones of a human foot from CT data. If one considers the components of the dataset to be lumped into two or three types of tissue based on the x-ray attenuation of the material (i.e., skin, muscle, and bone), a simple threshold becomes a linear discriminant on the histogram, dividing the distribution of image intensity intensity values into two classes: "bone" and "not-bone."

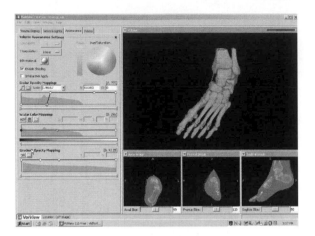

Figure 5.3. A simple threshold of intensity values can be used to create a binary discriminant, partitioning the distribution of voxel values into "bone" and "not-bone." The scale of this simple approach applies the labeling uniformly to the entire dataset. [Segmentation performed using the VolView software, courtesy of Kitware, Inc.]

Figure 5.3 shows the results of a simple division of intensity values applied to a portion of the Visible Human Project male dataset. This type of segmentation is particularly useful for x-ray CT data where intensity values have an intitutive mapping to physical density. More complex, multivalued data, such as MRI or color cryosections, require sophisticated statistical techniques, including Bayesian clustering and nonparametric methods such as Parzen windowing or k-nearest-neighbors. Chapter 3 covers these concepts in greater detail.

This segmentation effectively partitions the data into structures of interest to an orthopedist; however, the objects are undersegmented. That is, all the bones appear together as a single segment, and it is probably more useful to the clinician to be able to separate and articulate the bones individually. A single, simple statistical perspective treats the entire dataset simultaneously, affecting all parts of the image at once. No accomodation is made for the local connections among similar voxel values clustered in space. Separation of the many bones of the foot can be achieved using a user interface for expert editing of the segmented system in a tool similar to the one in the watershed example above.

5.3   Region Growing

A basic means of accounting for spatial proximity and connectedness as well as alikeness of intensity values is the use of a region-growing flood-fill technique. Figure 5.4 is a view of the same dataset in Figure 5.3, but with a 3D marker

Figure 5.4. Using a connected threshold approach helps define a contigious region of connected intensity values that maintain spatial as well as intensity coherence, defining specific structures or bones. This is an attempted segmentation of just the talus, or ankle bone, from the same dataset and view shown in Figure 5.3. The use of connected threshold segmentation meets with only mixed success. Flood-fill methods such as this one are prone to unwanted leaking across boundaries as well as noisy, rough surface descriptions. [Segmentation performed using the VolView software, courtesy of Kitware, Inc.]

placed in the talus (ankle bone) of the foot and a connected-threshold applied to recursively aggregate all voxels that are connected to the marked point by an unbroken lattice of voxels that appear within a user-defined interval of intensity values.

By applying statistical properties and including known or estimated values for intensity mean, variance, and scale, one can extend the concept of threshold-connected components to include confidence-connected components. These statistical methods of linking pixels or voxels in a dataset can be used to delineate regions or boundaries, depending on the application. This is a loose description of the idea of "fuzzy-connectedness"; for a complete and exhasutive treatment of this topic, see Chapter 6. More sophisticated approaches to region growing include a family of methods that use Markov random fields (see Chapter 7). Markov random fields models describe a class of region growing techniques built on statistical foundations that enable segmentations with similar properties to the methods of nonlinear diffusion built on partial differential equations described in Chapter 4.

5.4   Active Surfaces/Front Evolution

Alternate approaches to region growing include the family of adaptive surface definition methods which attempt to connect the margin voxels that circumscribe the region of interest. These methods can be built on statistical measures of alikeness or similarity or they can attempt to use energy-minimizing models that balance external boundary-following metrics with internal forces that attempt to promote smoothness. The fundamental mathematics of two such approaches are presented in this book: an implicit voxel-based method built on front evolution and level sets (see Chapter 8), and a parametric method built on finite element deformable meshes (see Chapter 9).

The principal advantages of using the adaptive surface definition techniques such as those described in this text are that the expected geometric properties of the segment boundaries can be enforced. In the example that we have been following, the segmentation of the ankle bone, we have seen that simple region growing, flood-fill approaches leave pits where the margins fall outside the expected interval, and they leak where nearby bones adjoin, connecting structures in unwanted ways. Bones are reasonably smooth, not pitted, and the ankle bone is a convex structure with few deep folds. By using an active surface or level set method, we can apply some differential geometry to the evolving front and limit the overall local curvature to recover a smooth object with no surface pits in the few deep folds, thus recreating the desired object while reducing the unwanted bridges between separate structures.

Figure 5.5 is a view of the same dataset in Figure 5.3, but with a 3D seed point marker placed in the ankle bone of the foot and a fast marching technique applied to grow a level set around the seed point, slowing the evolution of the implicit surface front where boundaries are detected while limiting the curvature or pits and folds in the surfaces. An isovalue of the front track can then be used to extract a surface that matches the desired structure.

The drawbacks of adaptive surface methods are that the very advantages of enforced smoothness limits them when defining small structures with high curvature such as blood vessels and nerve bundles. Initializing a level set close to the expected solution by placing multiple seeds or generating an approximate shape speeds the segmentation and enables the delineation of fine structures. Deformable finite element models also work best when they are initialized close to the target boundaries. These methods are essentially iterative, energy-minimizing, discrete solvers for partial differential equations whose parameters and coefficients govern the balance between smoothness and boundariness discovered in

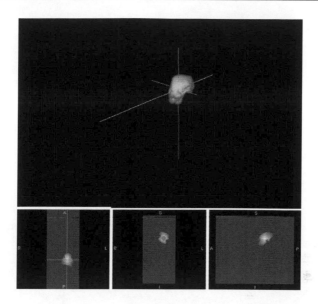

Figure 5.5. Applying a level set approach helps enable us to define a contigious region with expected geometric properties such as an enforced smoothness criteria. This is a better segmentation of the talus, or ankle bone, generating a smoother and better defined structure than that shown in Figure 5.4. Fast marching methods eliminate the pitted surfaces of flood fill techniques and reduce leaking between structures. [Fast marching methods available from ITK in VolView, courtesy of Kitware, Inc.]

the image. One pitfall of these techniques is that the curvature-limiting criteria not only misses some fine structures by rounding off corners, it has a tendency to create boundaries that are slightly smaller than the expected, perceived surface.

5.5   Combining Segmentation Techniques

Image statistics are powerful but often require the addition of surface geometry to generate effective results. How then can one initialize a deformable model or level set method? Figure 5.6 shows a pipeline that generates results from a fuzzy connectedness process to feed a Voronoi-based decomposition of the data, providing a useful structure for navigating the initial segmentation. The results of this pipeline applied to a portion of the Visible Human Project male dataset are shown in Figure 5.7. The entire process can be used to feed a deformable model or level set algorithm for further refining the model.

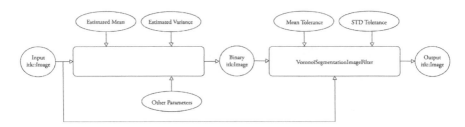

Figure 5.6. A pipeline showing the integration of fuzzy connectedness with Voronoi diagram classification. This process can be used as an input to a deformable model for additional processing and analysis. [Figure courtesy of Celina Imielinska, Columbia Univ., Jayaram Udupa, UPenn, and Dimitris Metaxas, Rutgers Univ.]

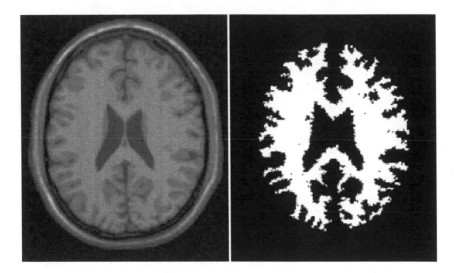

Figure 5.7. Segmentation results for the hybrid segmentation approach, showing the input image and the binary mask resulting from the segmentation pipeline. [Figure courtesy of Celina Imielinska, Columbia Univ., Jayaram Udupa, UPenn, and Dimitris Metaxas, Rutgers Univ.]

5.6   Looking Ahead

There is no perfect segmentation of any dataset; the correctness of the granularity of a partitioning of any picture or volume depends on the application and the insight desired. Just as there is no perfect segmentation, there is no preferred segmentation technique. Each type of dataset presents unique challenges that must be addressed, and the creative and resourceful programmer will carefully select from the many available tools, often combining them in a constructive process to achieve the desired goal.

This chapter is not intended as a complete overview of the research discipline of medical image segmentation, but rather as an introduction to the problem in general. We have covered very briefly some of the issues and approaches that are used in some common (and some uncommon) methods for medical data segmentation. The remainder of this section of the book covers the mathematics and fundamental algorithms of these approaches, providing a deeper and more comprehensive treatment of these tools.

Fuzzy Connectedness

Jayaram K. Udupa
University of Pennsylvania

Punam K. Saha
University of Pennsylvania

6.1 Background

6.1.1 Hard Connectedness

Two-dimensional (2D) digital picture processing activities started in the early 1960s [21, 53, 54]. Very early on, Rosenfeld [55] recognized the need and laid the foundation for digital counterparts of key topological and geometric concepts that have long been available for the continuous space (point sets). Working with a rectangular grid as the basis for defining a 2D digital picture, he defined almost all the fundamental concepts that are now ubiquitous, such as adjacency, neighborhood, path, connectedness, curves, boundaries, surroundedness, distance, perimeter, and area. These concepts were developed for hard (or crisp) sets of pixels which are usually treated as (hard) subsets of Z^2. For developing the theory underlying these concepts, it does not matter as to how these sets are generated. In the context of image processing and analysis, these hard sets result from image segmentation—the process of defining and delineating an object of interest in a given image. In particular, the concept of connectedness he developed found widespread use in image analysis such as in separating an already segmented object from other irrelevant structures, and in separating the component parts of a disconnected object that has also been segmented already. This concept was defined in a hard fashion in the sense that a set of pixels was considered to be either

131

connected or not connected. In the 1970s, three-dimensional (3D) digital images started becoming available with the advent of tomographic imaging, and attempts to develop the 3D (and the higher dimensional) counterparts of the above concepts started for hard subsets of Z^3 [5, 57, 69], motivated by the need to process, visualize, manipulate, and analyze the object information captured in these images. The generalization of the digital topological concepts, particularly of connectedness, which is the focus of this chapter, from hard sets to fuzzy subsets of Z^n (that is, Z^n together with a membership value in [0, 1] assigned to every element of Z^n) is not merely of academic interest, but has implications beyond just the generalization.

6.1.2   Fuzzy Connectedness

Why fuzzy? Images produced by any imaging device are inherently fuzzy. This fuzziness comes from several sources. Because of spatial and temporal resolution limitations, imaging devices introduce blur. They also introduce noise and other artifacts such as background intensity variation. Further, objects usually have material heterogeneity, and this introduces heterogeneity of intensities in the images. This is especially true in the biomedical context of the internal organs of living beings. The combined effect of all these factors is that object regions manifest themselves with a heterogeneity of intensity values in acquired images. The main rationale for taking fuzzy approaches in image analysis is to try to address and handle the uncertainties and intensity gradations that invariably exist in images as realistically as possible.

It appears to us that, unlike in the hard setting, the fuzzy treatment of the geometric and topological concepts can be made in two distinct manners in image analysis. The first approach is to carry out a fuzzy segmentation of the given image first so that we have a fuzzy subset of the image domain wherein every image element has a fuzzy object membership value assigned to it, and then to define the geometric and topological concepts on this fuzzy subset. The second approach is to develop these concepts directly on the given image, which implies that these concepts will have to be integrated somehow with the process of segmentation. Taking the first approach, Rosenfeld did some early work [56, 58] in developing fuzzy digital topological and geometric concepts. The second approach pursued by us and others has been mainly for developing the fuzzy connectedness concepts [75]. Rosenfeld [56] defined fuzzy connectedness by using a min-max construct on fuzzy subsets of 2D picture domains: the strength of connectedness of any path between any two pixels c and d is the smallest fuzzy pixel membership along the path and the strength of connectedness between c and d is the strength of the strongest of all paths between c and d. The process of obtaining the fuzzy subset in the first place is indeed the act of (fuzzy) segmentation. Through the introduction of a fundamental concept, a local fuzzy relation on image elements called *affinity* [75], it is possible to develop the concept of fuzzy connectedness directly on the given image and integrate it with, and actually utilize it for facilitating, image segmentation. As we already pointed out, image elements have a gradation of intensities within the same object region. In spite of this graded composition, they seem to hang together (form a *gestalt*) within the same object. We argue that this "hanging-togetherness" should be treated in a fuzzy manner, and an appropriate mathematical vehicle to realize this concept is fuzzy connectedness. If hard connectedness were to be used, then it would require a pre-segmentation. In other words, it cannot influence the result of segmentation itself, except in separating (hard) connected components. Fuzzy connectedness when utilized directly on the given image (via affinity), on the other hand, can influence the segmentation result by the spatio-topological consideration of how the image elements hang together, rather than by strategies of feature clustering that have used fuzzy clustering techniques [49] but have ignored the important topological property of hanging-togetherness. It is widely acknowledged that segmentation is one of the most crucial and difficult problems in image processing and analysis. Any fundamental advance that empowers the process of segmentation will take us closer to having a grip on this problem. We believe that fuzzy connectedness is one such recent development.

6.2   Outline of the Chapter

We have tried to make this chapter as self-contained as possible, leaving out a few details that are irrelevant for the main discussion. These details may be obtained from the cited references. We have attempted to describe the connectedness concepts in a unified way so that hard connectedness becomes a specific case of fuzzy connectedness. However, we shall give descriptions for the hard case also when it is appropriate and to emphasize the nuances of its relationship to the fuzzy case, or when more effective and efficient means are possible for the hard case. We shall see that this combined treatment opens interesting future avenues even for hard connectedness in the fuzzy setting. We encourage the reader to invoke mentally the process of particularization to the hard case while reading the general treatment.

By nature, this chapter is mathematical. The mathematical treatment may discourage some readers who are interested in only the main ideas embodied in fuzzy connectedness. Furthermore, in a casual reading, the mathematical details may shroud the central ideas that motivate the mathematics. For these reasons, in the rest of this section, we delineate the key ideas involved in fuzzy connectedness in an intuitive fashion utilizing Figures 6.1 and 6.2.

There are five key ideas underlying fuzzy connectedness: those related to (1) affinity, (2) scale, (3) generalized fuzzy connectedness, (4) relative fuzzy connectedness, and (5) iterative relative fuzzy connectedness. The latter three will

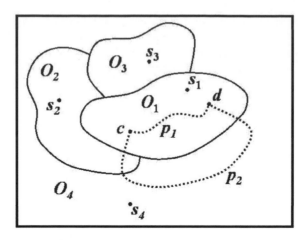

Figure 6.1. Illustration of the ideas underlying generalized fuzzy connectedness and relative fuzzy connectedness.

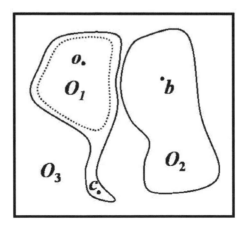

Figure 6.2. Illustration of the ideas underlying iterative relative fuzzy connectedness.

be abbreviated GFC, RFC, and IRFC, respectively. Fuzzy connectedness will be referred to as FC, and hard connectedness will be denoted by HC for ease of reference.

Affinity is intended to be a *local* relation between every two image elements u and v. That is, if u and v are far apart, the strength of this relationship is likely to be zero. If the elements are nearby, the strength of their affinity, lying between 0 and 1, depends on the distance between them, on the homogeneity of the intensities surrounding both, and on the closeness of their intensity-based features to the feature values expected for the object. In determining the homogeneity component and the features at u and v, the *scale* values at both u and v are utilized. The *scale* at any element is the radius of the largest hyperball centered at that element within which the intensities are homogeneous under some criterion of homogeneity. In determining the affinity between u and v, all elements in the scale regions of both u and v are consulted to make affinity (and subsequently, FC) less susceptible to noise and to intensity heterogeneity within the same object region.

Contrary to affinity, FC is intended to be a *global* fuzzy relation. FC strength between any two elements, such as c and d in Figure 6.1, is the strength of the strongest of all possible paths between c and d. A path such as p_1 is a sequence of nearby elements, and the strength of a path is the smallest affinity of pairwise elements along the path. Imagine Figure 6.1 to be a gray-level image with three object regions O_1, O_2, and O_3, and a background O_4. With respect to O_1, if an affinity relation is chosen appropriately, we expect path p_1 to have a higher strength than any path, such as p_2, that treads across multiple object regions.

In GFC, a fuzzy connected object of a certain strength θ and containing a given element is defined as the largest set of image elements containing that element such that the strength of connectedness between any two elements in that set is at least θ. The theoretical framework demonstrates that the problem of finding a fuzzy connected object of any strength can be solved via dynamic programming. In RFC, all co-objects of importance (in Figure 6.1, these are O_1, O_2, O_3, O_4) are let to compete for claiming membership of image elements via FC. Again, referring to Figure 6.1, let s_1, s_2, s_3, s_4 be seed elements in respective objects. An element such as d is considered to belong to object O_i if the strength of connectedness of d with respect to O_i's seed s_i is greater than the strength of connectedness of d with respect to each of the other seeds. The threshold θ required in GFC is thus eliminated in *RFC*. In IRFC, the idea is to refine iteratively the competition rules for RFC for different objects depending upon the results of the previous iteration. Referring to Figure 6.2, for an element such as c, its strength of FC from o and from b are likely to be similar due to the blurring that takes place where O_1 and O_2 come close together. In this case, the strongest path from b to c is therefore likely to pass through the "core" of O_1 which is indicated by the dotted curve in the figure. This core can be detected first and then excluded from consideration in a subsequent iteration for any path from b to c to pass through. Then, we can substantially weaken the strongest path from b to c compared to the strongest path from o to c which is still allowed to pass through the core. This leads us to an iterative strategy to separate O_1 and O_2 via RFC.

The organization of this chapter is as follows. In Section 6.3, we describe some basic terms and the notations used throughout the chapter. In Section 6.4, the theory of fuzzy connectedness is outlined quite independently of its connection to image segmentation. In Section 6.5, its implications in image segmenta-

tion are described. In Section 6.6, a variety of large applications that have utilized
fuzzy connectedness for image segmentation are described. In Section 6.7, our
conclusions are stated with some pointers for future work.

6.3   Basic Notations and Definitions

Let X be any reference set. A *fuzzy subset* [29] $\mathcal{A}$ of X is a set of ordered pairs
$\mathcal{A} = \{(x, \mu_{\mathcal{A}}(x)) \mid x \in X\}$ where $\mu_{\mathcal{A}} : X \to [0,1]$. $\mu_{\mathcal{A}}$ is called the *membership
function* of $\mathcal{A}$ in X. When $\mathcal{A}$ is a *hard subset* $\mu_{\mathcal{A}} : X \to \{0,1\}$ and is called
the *characteristic function* of $\mathcal{A}$ in X. The fuzzy subset $\mathcal{A}$ is called *nonempty* if
there exists an $x \in X$ such that $\mu_{\mathcal{A}}(x) \neq 0$; otherwise it is called the *empty fuzzy
subset* of X. We use Φ to denote an empty fuzzy or hard subset. The *fuzzy
intersection* and *fuzzy union* between two fuzzy subsets $\mathcal{A}$ and $\mathcal{B}$ of X are defined
as follows: $\mathcal{A} \cap \mathcal{B} = \{(x, \mu_{\mathcal{A} \cap \mathcal{B}}(x)) \mid x \in X\}$ where $\mu_{\mathcal{A} \cap \mathcal{B}}(x) = \min[\mu_{\mathcal{A}}(x), \mu_{\mathcal{B}}(x)]$,
$\mathcal{A} \cup \mathcal{B} = \{(x, \mu_{\mathcal{A} \cup \mathcal{B}}(x)) \mid x \in X\}$ where $\mu_{\mathcal{A} \cup \mathcal{B}}(x) = \max[\mu_{\mathcal{A}}(x), \mu_{\mathcal{B}}(x)]$.

A *2-ary fuzzy relation* ρ in X is a fuzzy subset of $X \times X$, $\rho = \{((x,y), \mu_{\rho}(x,y)) \mid$
$x, y \in X\}$ where $\mu_{\rho} : X \times X \to [0,1]$. Since we are not interested in fuzzy m-ary
relations for $m > 2$, we drop the qualifier "2-ary" for simplicity. We use μ sub-
scripted by the fuzzy (hard) subset under consideration to denote the membership
(characteristic) function of the fuzzy subset. In the hard case, ρ is called a *binary
relation* and $\mu_{\rho} : X \times X \to \{0,1\}$. Let ρ be any fuzzy relation in X. ρ is said to be
reflexive, if, $\forall x \in X$, $\mu_{\rho}(x,x) = 1$; *symmetric*, if, $\forall x,y \in X$, $\mu_{\rho}(x,y) = \mu_{\rho}(y,x)$; *transitive*, if, $\forall x, z \in X$, $\mu_{\rho}(x,z) = \max_{y \in X}[\min[\mu_{\rho}(x,y), \mu_{\rho}(y,z)]]$. A fuzzy relation ρ
is called a *similitude relation* in X if it is reflexive, symmetric and transitive. Note
that the above definitions of reflexivity, symmetry, and transitivity hold when ρ is
a hard binary relation. A binary (hard) relation in X that is reflexive, symmetric,
and transitive is called an *equivalence* relation in X. When ρ is an equivalence
relation in X, for any $o \in X$, the set $\{x \in X \mid \mu_{\rho}(x,o) = 1\}$ is called the *equivalence
class of o* in X. Similitudity and equivalence are analogous concepts in fuzzy and
hard subsets, respectively.

For any $n \geq 2$, let n-dimensional Euclidean space R^n be subdivided into hypercuboids by n mutually orthogonal families of parallel hyperplanes. Assume,
with no loss of generality, that the hyperplanes in each family have equal unit
spacing so that the hypercuboids are unit hypercubes, and we shall choose coordinates so that the center of each hypercube has integer coordinates. The hypercubes will be called *spels* (an abbreviation for "space elements"). When $n = 2$,
spels are called *pixels*, and when $n = 3$ they are called *voxels*. The coordinates
of the center of a spel are represented by an n-tuple of integers, defining a point
in Z^n. Z^n itself will be thought of as the set of all spels in R^n with the above
interpretation of spels, and the concepts of spels and points in Z^n will be used
interchangeably.

A fuzzy relation α in Z^n is said to be a *fuzzy adjacency* if it is both reflexive and symmetric. It is desirable that α be such that $\mu_\alpha(c,d)$ is a non increasing function of the distance $\|c-d\|$ between spels c and d. It is not difficult to see that the hard adjacency relations commonly used in digital topology [30] are special cases of fuzzy adjacencies. We shall denote these (hard adjacencies) collectively by α_h, for future reference. A generic definition of α_h is as follows. For any $1 \leq m \leq n$, the mth *order adjacency* of spels in Z^n is defined by

$$\mu_{\alpha_h}(c,d) = \begin{cases} 1, & \text{if } \|c-d\| \leq |\sqrt{m}|, \\ 0, & \text{otherwise.} \end{cases} \tag{6.1}$$

The 1st and 2nd order adjacencies correspond to the classical 4-way and 8-way adjacencies when $n = 2$, and when $n = 3$, they represent 6-way and 18-way adjacencies, and the 26-adjacency corresponds to the 3rd order adjacency for $n = 3$.

We call the pair (Z^n, α), where α is a fuzzy adjacency, a *fuzzy digital space*. Fuzzy digital space is a concept that characterizes the underlying digital grid system independent of any image-related concepts. We shall eventually tie this with image-related concepts to arrive at fuzzy object-related notions.

A *scene over* a fuzzy digital space (Z^n, α) is a pair $C = (C, f)$ where $C = \{c \mid -b_j \leq c_j \leq b_j \text{ for some } b \in Z_+^n\}$; Z_+^n is the set of n-tuples of positive integers; f, called *scene intensity*, is a function whose domain is C, called the *scene domain*, and whose range $[L,H]$ is a set of numbers (usually integers). C is a *binary scene over* (Z^n, α) if the range of f is $\{0,1\}$. In this case, we refer to a spel c in C such that $f(c) = 0$ as a *0-spel* of C, and if $f(c) = 1$, we refer to c as a *1-spel* of C. Binary scenes usually result from a hard segmentation of a given scene. When the result of segmentation is a fuzzy subset of the scene domain, this fuzzy subset can be equivalently represented by a scene wherein the scene intensity represents the fuzzy membership value in [0,1]. We call such scenes *membership scenes* over (Z^n, α). Unless specified otherwise, whenever we refer to a scene in this chapter, it encompasses all three types: gray-level scene, membership scene, and binary scene. We assume in this chapter that $n \geq 2$. We also note that f may be vector-valued (when C is not a binary scene); for example, C may represent a color scene in which case f has three (R,G,B) components. All concepts described in this chapter can be readily generalized from scalar scenes to vectorial scenes. For simplicity and ease of understanding of the concepts, we shall assume from now on that f is scalar-valued. We shall remark on any derivations needed to handle the vectorial case at appropriate points in the chapter.

A nonempty *path* p_{cd} in a scene C over (Z^n, α) from c to d is a sequence $\langle c = c_1, c_2, \cdots, c_l = d \rangle$ of $l \geq 1$ spels in C; l is called the *length* of the path. Note that the successive spels in a path need not be "adjacent" in the sense hard adjacency is defined in digital topology as in Equation (6.1). When they are indeed α_h-

adjacent, we shall refer to this as a *contiguous* (or, α_h-) *path*. The set of all spels in the path p_{cd} is denoted by $\mathcal{E}(p_{cd})$. An empty path in C, denoted $\langle \rangle$, is a sequence of no spels. Paths of length 2 will be referred to as *links*. The set of all paths in C from c to d is denoted by P_{cd} (Note that c and d are not necessarily distinct.). The set of all paths in C, defined as $\bigcup_{c,d \in C} P_{cd}$, is denoted by P_C. We define a binary *join to* operation on P_C, denoted "+" as follows: for every two nonempty paths $p_{cd} = \langle c_1, c_2, \cdots, c_{l_1} \rangle \in P_C$ and $p_{de} = \langle d_1, d_2, \cdots, d_{l_2} \rangle \in P_C$ [note that $c_{l_1} = d_1 = d$],

$$p_{cd} + p_{de} = \langle c_1, c_2, \cdots, c_{l_1}, d_2, \cdots, d_{l_2} \rangle, \tag{6.2}$$
$$p_{cd} + \langle \rangle = p_{cd}, \tag{6.3}$$
$$\langle \rangle + p_{de} = p_{de}, \tag{6.4}$$
$$\langle \rangle + \langle \rangle = \langle \rangle. \tag{6.5}$$

Note that the join to operation between $p_{c_1 c_2}$ and $p_{d_1 d_2}$ is undefined if $c_2 \neq d_1$. It is shown in [75] (Proposition 2.1) that for any scene $C = (C, f)$ over any fuzzy digital space (Z^n, α), the following relation holds for any two spels $c, e \in C$:

$$P_{ce} = \{ p_{cd} + p_{de} \mid d \in C \text{ and } p_{cd} \in P_{cd} \text{ and } p_{de} \in P_{de} \}. \tag{6.6}$$

We define a binary relation *greater than* on P_C, denoted ">", as follows. Let $p = \langle c_1, c_2, \cdots c_{l_p} \rangle$ and $q = \langle d_1, d_2, \cdots d_{l_q} \rangle$ be any paths in C. We say that $p > q$ if and only if we can find a mapping g from the set of spels in q to the set of spels in p that satisfies all of the following conditions:

1. $g(d_i) = c_j$ only if $d_i = c_j$.

2. There exists some $1 \leq m \leq l_p$, for which $g(d_1) = c_m$.

3. For all $1 \leq j < l_q$, whenever $g(d_j) = c_i$, $g(d_{j+1}) = c_k$ for some $k \geq i$ and $c_i = c_{i+1} = \cdots = c_{k-1}$.

Some examples follow: $\langle c_1, c_2, c_3, c_3, c_4, c_5 \rangle > \langle c_3, c_4, c_4, c_5 \rangle$; $\langle c_1, c_2, c_3, c_4 \rangle > \langle c_3, c_3, c_3, c_3, c_3, c_3 \rangle$. Trivially, every non-empty path in C is greater than the empty path $\langle \rangle$ in C.

6.4 Theory

6.4.1 Early Works and History

Hard Connectedness

In many applications (e.g., lesion detection [76] in MR images of Multiple Sclerosis patients' heads, character recognition [11]), labeling in terms of connected sets after hard segmentation of a scene into a binary scene becomes useful. There are image processing operations wherein it is useful to label both the set of 0-spels and 1-spels in this manner into connected components. Rosenfeld [55] was the first to formulate systematically HC and to describe for labeling connected components. He also observed that if 8-way adjacency is used for 1-spels, 4-way adjacency must be used for 0-spels (and vice versa) to satisfy certain properties such as the digital Jordon curve theorem that a simple closed curve is necessary and sufficient to split a plane into an exterior and an interior. In [57], he formulated different topological concepts such as "surroundedness," "background," and "cavities." Kong and Rosenfeld [30] studied valid adjacency pairs for 3D spaces and Herman and Udupa [23, 73] studied these issues for n-dimensional spaces.

A classical two-scan algorithm for computing binary connected components was presented by Rosenfeld and Pfaltz [53] in the 1960s. The method required a large global table to record equivalence of initial component labels. A space-efficient two-scan algorithm for component labeling was presented in [40]. The method used a small local equivalence table that stored only the equivalence between the current and the preceding rows. Using local equivalence tables it partially performed the job of equivalencing in the first top-down scan and completed it in a subsequent bottom-up scan. This approach was generalized to 3D spaces by Lumia [41]. Ronse and Devijver [52] suggested the use of special hardware to compute a run-length coding (list of contiguous, typically, horizontal runs of 1-spels) of a binary scene and then apply component labeling on these compact data. Several parallel algorithms for binary component labeling have been reported in the literature [16, 42]. A recursive algorithm to label simultaneously connected components of 0-spels, 1-spels, and boundaries separating them is described in [73]. A detailed survey of connected component labeling algorithms for binary scenes has been presented in [67]. A classical algorithm [53] of hard component labeling in binary images is presented in Section 6.5.2.

Fuzzy Connectedness

The notion of the "degree of connectedness" of two spels was first introduced by Rosenfeld [56, 58] in the context of studying the topology of fuzzy subsets of Z^2. He considered contiguous paths in Z^2 and defined the *strength of connectedness* of a contiguous path p_{cd} from c to d in a fuzzy subset $\mathcal{A}$ of Z^2 as the smallest membership value along the path, and the *degree of connectedness*, denoted by $\mu_R(c,d)$ ("R" to denote Rosenfeld's approach and to distinguish it from more general approaches described later), as the strength of the strongest path between the two spels. Therefore,

$$\mu_R(c,d) = \max_{p \in P_{cd}} \left[\min_{e \in \mathcal{E}(p)} \mu_{\mathcal{A}}(e) \right]. \tag{6.7}$$

He used the notion of the degree of connectedness to define certain topological and geometrical entities which had been previously defined for hard sets of spels in Z^2. Two spels $p, q \in Z^2$ are said to be *connected* if $\mu_R(c,d) = \min(\mu_{\mathcal{A}}(c), \mu_{\mathcal{A}}(d))$. He showed that this binary (hard) relation of connectedness in a fuzzy subset $\mathcal{A}$, denoted by $\rho_{\mathcal{A}}$, is reflexive and symmetric, but not transitive, and consequently, $\rho_{\mathcal{A}}$ is not an equivalence relation. Therefore, the components defined by $\rho_{\mathcal{A}}$ may not partition Z^2, *i.e.*, they may have nonempty overlap. In the same paper, he defined the number of components and genus in a fuzzy subset and the membership value of a spel in a component using the notions of plateaus, tops, and bottoms and associating three different connected sets with each top. Also, he introduced the notions of separation and surroundedness in a fuzzy subset and showed that surroundedness describes a weak partial ordering, i.e., the hard binary relation defined by surroundedness is reflexive, antisymmetric, and transitive.

Rosenfeld's "degree of connectedness" [56] was further studied to understand the topological, geometrical, and morphological properties of fuzzy subsets [58]. However, there was no consideration in these works as to how this concept could be exploited to facilitate image segmentation. Dellepiane and Fontana [17, 18] and Udupa and Samarasekera [75, 74] were the first to suggest this use. Dellepiane *et al.* utilized Rosenfeld's "degree of connectedness' to arrive at a segmentation method (to be discussed in Section 6.5.2). Udupa *et al.* [75] simultaneously introduced a different framework, bringing in a key concept of a local fuzzy relation called *affinity* on spels to capture local hanging-togetherness of spels. They showed how affinity can incorporate various image features in defining fuzzy connectedness, presented a general framework for the theory of fuzzy connectedness, and demonstrated how dynamic programming can be utilized to bring the otherwise seemingly intractable notion of fuzzy connectedness into segmentation. They also advanced the theory of fuzzy connectedness considerably [62, 63, 80, 64, 77], bringing in notions of relative fuzzy connectedness [80, 77] which were further extended by Herman and De Carvalho [24], iteratively defined relative fuzzy connectedness [64, 80], and addressed the virtues of the basic min-max construct used in fuzzy connectedness [63]. Affinity forms the link between fuzzy connectedness and segmentation. Saha *et al.* [62] studied the issue of how to construct effective affinities and the use of local scale for this purpose. Aspects related to the computational efficiency of fuzzy connectedness algorithms have also been studied [9, 46].

The general framework of fuzzy connectedness based on affinity has been utilized in conjunction with other methods of segmentation [27, 28, 3, 26, 8, 15], particularly with deformable models [27, 28], a Voronoi diagram-based method [3], and level set methods [26]. The fuzzy connectedness methods have been utilized for image segmentation extensively in several applications, including Multiple Sclerosis lesion quantification [76, 65, 43, 44, 20, 78, 2, 22], late life depression [31, 32], MR angiography [51, 36], CT angiography [59, 1], brain tumor assessment [38, 45], breast density quantification via mammograms [60], understanding upper airway disorders in children [39], and CT colonography [79].

In the rest of Section 6.4, we shall first describe the theory of HC and then delineate the general FC theory based on affinity and its extensions to RFC and IRFC. We shall indicate, when appropriate, the particularization to the hard case and its implications. We shall present all pertinent details to understand the structure of the theory but only state the main results as theorems. For details including proofs, please refer to [75, 62, 63, 80, 64, 77].

6.4.2 Hard Connectedness

Let $C = (C, f)$ be a binary scene over (Z^n, α_h) where α_h is a hard adjacency relation defined as in Equation (6.32). For $k \in \{0, 1\}$, let $U_k = \{c \in C \,||\, f(c) = k\}$. Two spels $c, d \in C$ are said to be α_h-*connected in* U_k if and only if there exists an α_h-path p_{cd} in C such that $\mathcal{E}(p_{cd}) \subset U_k$. Any subset Q_k of U_k is said to be α_k-*connected* if every two spels of Q_k are α_k-connected in Q_k. Q_k is said to be an α-*component of k-spels in* C if Q_k is α_h-connected and for any $c \in U_k - Q_k$, $Q_k \cup \{c\}$ is not α-connected. Clearly "α_h-connected in," denoted by "A_h", is a binary (hard) relation in C.

The following theorem, which is straightforward to prove, is the basis for all algorithms that track or label hard components in binary scenes.

Theorem 1 *For any binary scene* $C = (C, f)$ *over* (Z^n, α_h) *and for any* $k \in \{0, 1\}$, A_h *is an equivalence relation in* U_k, *and every* α_h-*component of k-spels in* C *is an equivalence class of o in* U_k *for some* $o \in C$.

The proof of this theorem follows as a particular case of the more general theorems (Theorems 2 and 3) formulated for FC in the next section. We note that the above theorem is valid even when α_h (as defined in Equation 6.32)) is replaced by any hard binary relation that is reflexive and symmetric (such as 6-adjacency in 2D spaces achieved by taking all 4-adjacencies and any two diagonal adjacencies). An algorithm for labeling α_h-components of 1-spels in C is presented in Section 6.5.

6.4.3 Generalized Fuzzy Connectedness

Fuzzy Affinity

Let $C = (C, f)$ be any scene over (Z^n, α). Any fuzzy relation κ in C is said to be a *fuzzy spel affinity* (or, *affinity* for short) *in* C if it is reflexive and symmetric. In practice, for κ to lead to meaningful segmentation, it should be such that, for any $c, d \in C$, $\mu_\kappa(c, d)$ is a function of (i) the fuzzy adjacency between c and d; (ii) the

homogeneity of the spel intensities at c and d; (iii) the closeness of the spel intensities and of the intensity-based features of c and d to some expected intensity and
feature values for the object. Further, $\mu_\kappa(c,d)$ may depend on the actual location
of c and d (i.e., μ_κ is shift variant). A detailed description of a generic functional
form for μ_κ and how to select the right affinity for a specific application are presented in Section 6.5. Throughout this chapter, κ with appropriate subscripts and
superscripts will be used to denote fuzzy spel affinities.

Note that affinity is intended to be a "local" relation. This is reflected in the
dependence of $\mu_\kappa(c,d)$ on the degree of adjacency of c and d. If c and d are far
apart, $\mu_\kappa(c,d)$ will be zero. How the extent of the "local" nature of κ is to be
determined will be addressed in Section 6.5. We note that the concept of affinity
is applicable to membership and binary scenes also. If the scene is a membership
scene, then affinity should depend perhaps only on adjacency and homogeneity of spel intensities. And then, FC analysis becomes a fuzzy analog of hard
connected component analysis. In the case of binary scenes, to realize classical
connected component analysis [67], affinity should be chosen to be simply one
of the hard adjacency relations α_h of Equation (6.1) (or, something similar) for
like-valued spels.

Path Strength and Fuzzy Connectedness

Our intent is for fuzzy connectedness to be a global fuzzy relation which assigns a
strength to every pair (c,d) of spels in C. It makes sense to consider this strength
of connectedness to be the largest of the strengths assigned to all paths between
c and d as originally suggested by Rosenfeld. (The physical analogy one may
consider is to think of c and d as being connected by many strings, each with its
own strength. When c and d are pulled apart, the strongest string will break at
the end, which should be the determining factor for the strength of connectedness
between c and d). However, it is not so obvious as to how the strength of each path
should be defined. Several measures based on the affinities of spel pairs along the
path, including their sum, product, and minimum, all seem plausible. (For binary
scenes, minimum and product are plausible and have the same meaning if κ_0 is
used as affinity. But we note that other forms of affinity that assign values in $[0,1]$
instead of from $\{0,1\}$ are also possible for binary scenes.) Saha and Udupa [63]
have shown that the minimum of affinities is the only valid choice for path strength
(under a set of reasonable assumptions stated in Axioms 1–4 below) so as to arrive
at important results that facilitate computing fuzzy connectedness via dynamic
programming. We shall examine some details pertaining to this tenet now.

Let $C = (C,f)$ be a scene over a fuzzy digital space (Z^n,α) and let κ be
a fuzzy affinity in C. A *fuzzy κ-net* $\mathcal{N}$ in C is a fuzzy subset of P_C with its
membership function $\mu_\mathcal{N} : P_C \rightarrow [0,1]$. $\mu_\mathcal{N}$ assigns a strength to every path of
P_C. For any spels $c,d \in C$, $p_{cd} \in P_{cd}$ is called a *strongest path* from c to d if

$\mu_{\mathcal{N}}(p_{cd}) = \max_{p \in P_{cd}}[\mu_{\mathcal{N}}(p)]$. The idea of κ-net is that it represents a network of all possible paths between all possible pairs of spels in C with a strength assigned to each path.

Axiom 1 *For any scene C over (Z^n, α), for any affinity κ and κ-net $\mathcal{N}$ in C, for any two spels $c, d \in C$, the strength of the link from c to d is the affinity between them; i.e., $\mu_{\mathcal{N}}(\langle c, d \rangle) = \mu_{\kappa}(c, d)$.*

Axiom 2 *For any scene C over (Z^n, α), for any affinity κ and κ-net $\mathcal{N}$ in C, for any two paths $p_1, p_2 \in P_C$, $p_1 > p_2$ implies that $\mu_{\mathcal{N}}(p_1) \leq \mu_{\mathcal{N}}(p_2)$.*

Axiom 3 *For any scene C over (Z^n, α), for any affinity κ and κ-net $\mathcal{N}$ in C, fuzzy κ-connectedness K in C is a fuzzy relation in C defined by the following membership function. For any $c, d \in C$,*

$$\mu_K(c, d) = \max_{p \in P_{cd}}[\mu_{\mathcal{N}}(p)]. \tag{6.8}$$

For fuzzy (and hard) connectedness, we shall always use the upper case form of the symbol used to represent the corresponding fuzzy affinity.

Axiom 4 *For any scene C over (Z^n, α), for any affinity κ and κ-net $\mathcal{N}$ in C, fuzzy κ-connectedness in C is a symmetric and transitive relation.*

Axiom 1 says that a link is a basic unit in any path, and that the strength of a link (which will be utilized in defining path strength) should be simply the affinity between the two component spels of the link. This is the fundamental way in which affinity is brought into the definition of path strength. Note that in a link $\langle c_i, c_{i+1} \rangle$ in a path (unless it is a contiguous path), c_i and c_{i+1} may not always be adjacent (in the sense of the α_h relation). That is, c_i and c_{i+1} may be far apart differing in some of their coordinates by more than 1. In such cases, Axiom 1 guarantees that the strength of $\langle c_i, c_{i+1} \rangle$ is determined by $\mu_{\kappa}(c, d)$ and not by "tighter" paths of the form $\langle c_i = c_{i,1}, c_{i,2}, \cdots, c_{i,m} = c_{i+1} \rangle$ wherein the successive spels are indeed adjacent as per α_h. Since κ is by definition reflexive and symmetric, this axiom guarantees that link strength is also a reflexive and symmetric relation in C. Axiom 2 guarantees that the strength of any path changes in a non-increasing manner along the path. This property is sensible and becomes essential in casting fuzzy connected-object tracking (whether in a given scene, membership scene, or a binary scene) as a dynamic programming problem. Axiom 3 says essentially that the strength of connectedness between c and d should be the strength of the strongest path between them. Its reasonableness has already been discussed. Finally, Axiom 4 guarantees that fuzzy connectedness is a similitude relation (equivalence relation, in the hard case) in C. This property is essential in devising a dynamic programming solution to the otherwise seemingly prohibitive combinatorial optimization problem of determining a fuzzy connected object. (The commonly-used region growing approaches to find a connected component in a binary scene over (Z^n, α_h) constitute a particular case of the dynamic programming strategy purported here for FC. However, as we shall see later, by using fuzzy affinity relations (instead of just α_h), interesting operations other than hard connected-component labeling can be performed even on binary scenes).

Using the above axioms, the following main theorem was proved in [63]:

Theorem 2 *For any scene* $C = (C, f)$ *over* (Z^n, α), *and for any affinity* κ *and* κ-*net* $\mathcal{N}$ *in* C, *fuzzy* κ-*connectedness* K *in* C *is a similitude relation in* C *if and only if*

$$\mu_K(c,d) = \max_{p \in P_{cd}} \left[\min_{1 < i \leq l_p} [\mu_\kappa(c_{i-1}, c_i)] \right],$$

where p *is the path* $\langle c_1, c_2, \cdots, c_{l_p} \rangle$.

Following the spirit of the above theorem, path strength from now on will be defined as follows:

$$\mu_{\mathcal{N}}(p) = \min_{1 < i \leq l_p} [\mu_\kappa(c_{i-1}, c_i)], \tag{6.9}$$

where p is the path $\langle c_1, c_2, \cdots, c_{l_p} \rangle$. For the remainder of this chapter, we shall assume the above definition of path strength. If we interpret equivalence to be the hard-case analog of similitudity for the fuzzy case, the above theorem and its proof follow for the hard case also.

Fuzzy Connected Objects

Let $C = (C, f)$ be any scene over (Z^n, α), let κ be any affinity in C, and let θ be a fixed number in $[0, 1]$. Let S be any subset of C. We shall refer to S as the set of *reference spels* or *seed spels* and assume throughout that $S \neq \phi$. A *fuzzy* $\kappa\theta$-*object* $O_{K\theta}(s)$ of C containing a seed spel $s \in S$ of C is a fuzzy subset of C whose membership function is

$$\mu_{O_{K\theta}(s)}(c) = \begin{cases} \eta(f(c)), & \text{if } c \in O_{K\theta}(s) \subset C, \\ 0, & \text{otherwise.} \end{cases} \tag{6.10}$$

In this expression, η is an *objectness function* whose domain is the range of f and whose range is $[0,1]$. It maps imaged scene intensity values into objectness values. For most segmentation purposes, η may be chosen to be a Gaussian whose mean and standard deviation correspond to the intensity value expected for the object region and its standard deviation (or some multiple thereof). The choice of η should depend on the particular imaging modality that generated C and the actual physical object under consideration. (When a hard segmentation is desired,

$O_{K\theta}(s)$ (defined below) will constitute the (hard) set of spels that represents the extent of the physical object and η will simply be the characteristic function of $O_{K\theta}(s)$.) $O_{K\theta}(s)$ is a subset of C satisfying all of the following conditions:

(i) $$s \in O_{K\theta}(s); \tag{6.11}$$

(ii) $$\text{for any spels } c,d \in O_{K\theta}(s), \mu_K(c,d) \geq \theta; \tag{6.12}$$

(iii) $$\text{for any spel } c \in O_{K\theta}(s) \text{ and any spel } d \notin O_{K\theta}(s), \mu_K(c,d) < \theta. \tag{6.13}$$

$O_{K\theta}(s)$ is called the *support of* $O_{K\theta}(s)$, i.e., a maximal subset of C such that it includes s and the strength of connectedness between any two of its spels is at least θ. Note that in the hard case when C is a binary scene over (Z^n, α_h), for any $0 < \theta \leq 1$, $O_{K\theta}(s)$ constitutes an α_h-component of k-spels in C containing s for some $k \in \{0, 1\}$.

A *fuzzy $\kappa\theta$-object* $O_{K\theta}(S)$ *of C containing a set S of seed spels of C* is a fuzzy subset of C whose membership function is

$$\mu_{O_{K\theta}(S)}(c) = \begin{cases} \eta(f(c)), & \text{if } c \in O_{K\theta}(S) = \bigcup_{s \in S} O_{K\theta}(s), \\ 0, & \text{otherwise.} \end{cases} \tag{6.14}$$

$O_{K\theta}(S)$ is the *support of* $O_{K\theta}(S)$, i.e., the union of the supports of the fuzzy connected objects containing the individual seed spels of S. The motivation for defining $O_{K\theta}(S)$ is the following. Often it is difficult to make sure through a single seed spel that all spels that loosely hang together and that form the support of the fuzzy object of interest can be captured through the specification of a single spel. Further, when we use methods to generate seed spels automatically, it is often easier to generate a set of spels sprinkled over the object region than to generate exactly one seed spel placed strategically in the object region.

Several important properties of fuzzy connected objects have been established [75, 63]. The following theorem gives a guidance as to how a fuzzy $\kappa\theta$-object $O_{K\theta}(S)$ of C should be computed. It is not practical to use the definition directly for this purpose because of the combinatorial complexity. The following theorem provides a practical way of computing $O_{K\theta}(S)$.

Theorem 3 *For any scene $C = (C, f)$ over (Z^n, α), for any affinity κ in C, for any $\theta \in [0, 1]$, for any objectness function η, and for any non-empty set $S \subset C$, the support $O_{K\theta}(S)$ of the fuzzy $\kappa\theta$-object of C containing S equals*

$$O_{K\theta}(S) = \{c \mid c \in C \text{ and } \max_{s \in S}[\mu_K(s,c)] \geq \theta\}. \tag{6.15}$$

The above theorem says that, instead of utilizing Equations (6.11)–(6.14) that define the support $O_{K\theta}(S)$ of the fuzzy $\kappa\theta$-object containing S and having to check the strength of connectedness of every pair (c,d) of spels for each seed $s \in S$, we need to check the maximum strength of connectedness of every spel c with respect to the seed spels. This is a vast simplification in the combinatorial complexity. $O_{K\theta}(S)$ as shown in Equation (6.15) can be computed via dynamic programming, given C, S, κ, and θ, as described in Section 6.5. In the hard case and when S is a singleton set, this theorem formally asserts the possibility of finding an α_h-component via region growing starting from a seed contained in the component, a result which is intuitively fairly obvious and that has been used extensively in image processing. What we have done here is to cast this result in the context of a vastly more general scenario of FC.

The following theorem characterizes the robustness of specifying fuzzy $\kappa\theta$-objects through sets of seed spels.

Theorem 4 *For any scene $C = (C, f)$ over (Z^n, α), for any affinity κ in C, for any $\theta \in [0, 1]$, for any objectness function η, and for any non empty sets $S_1, S_2 \subset C$, $O_{K\theta}(S_1) = O_{K\theta}(S_2)$ if and only if $S_1 \subset O_{K\theta}(S_2)$ and $S_2 \subset O_{K\theta}(S_1)$.*

The above theorem has important consequences in the practical utilization of the fuzzy connectedness algorithms for image segmentation. It states that the seeds must be selected from the same physical object and at least one seed must be selected from each physically connected region. High precision (reproducibility) of any segmentation algorithm with regard to subjective operator actions (and with regard to automatic operations minimizing these actions), such as specification of seeds, is essential for their practical utility. Generally, it is easy for human operators to specify spels within a region in the scene corresponding to the same physical object in a repeatable fashion. Theorem 4 guarantees that, even though the sets of spels specified in repeated trials (by the same or other human operators) may not be the same, as long as these sets are within the region of the same physical object in the scene, the resulting segmentations will be identical. It is important to point out that the intensity-based connectedness method as proposed by Dellepiane et al. [17, 18] (see Section 6.5) fails to satisfy this robustness property. Many other region growing algorithms [48] which adaptively change the spel inclusion criteria during the growth process cannot guarantee this robustness property. The validity of this property for the hard case is obvious. However, to examine this property as a particularization of a more general scenario of FC is instructive.

Other interesting properties of fuzzy connectedness and of fuzzy connected objects have been studied in [75, 63]. It has been shown that the support $O_{K\theta}(S)$ of the fuzzy connected object $O_{K\theta}(S)$ monotonically decreases in size with the strength of connectedness threshold θ while it monotonically increases with the set S of seed spels. It has also been shown that $O_{K\theta}(S)$ is connected (in the hard sense) if and only if the strength of connectedness between every two seed spels is at least θ.

6.4.4   Relative Fuzzy Connectedness

In the original fuzzy connectedness theory [75] as described in Section 6.4.3, an object is defined on its own based on the strength of connectedness utilizing a threshold θ. The key idea of relative fuzzy connectedness is to consider all co-objects of importance that are present in the scene and to let them compete among themselves in having spels as their members. Consider a 2D scene composed of multiple regions corresponding to multiple objects as illustrated in Figure 6.1. If O_1 is the object of interest, then the rest of the objects O_2, O_3, and O_4 may be thought of as constituting the background as far as O_1 is concerned. With such a thinking, Figure 6.3 shows just two objects O, the object of interest, and B, the background, equivalent to the scene in Figure 6.1. Although the theory for RFC can be developed for simultaneously considering multiple objects [61], for simplicity, we shall describe here the two-object case, but keep in mind the treatment of object grouping mentioned above.

Suppose that the path p_{co} shown in Figure 6.3 represents the strongest path between o and c. The basic idea in relative fuzzy connectedness is to first select reference spels o and b, one in each object, and then to determine to which object any given spel belongs based on its relative strength of connectedness with the reference spels. A spel c, for example, would belong to O since its strength of connectedness with o is likely to be greater than that with b. This relative strength of connectedness offers a natural mechanism for partitioning spels into regions based on how the spels hang together among themselves relative to others. A spel such as a in the boundary between O and B will be grabbed by that object with whom a hangs together most strongly. This mechanism not only eliminates

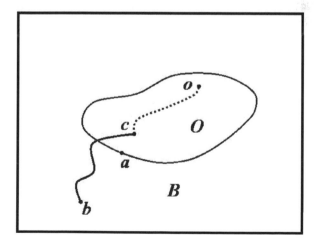

Figure 6.3.   Illustration used for defining relative fuzzy connectedness.

the need for a threshold required in the GFC method but also offers potentially more powerful segmentation strategies for two reasons. First, it allows more direct utilization of the information about all objects in the scene in determining the segmentation of a given object. Second, considering from a point of view of thresholding the strength of connectedness, it allows adaptively changing the threshold depending on the strength of connectedness of objects that surround the object of interest. The formal definition of relative fuzzy connected objects along with their important properties are presented in this section.

For any spels o, b in C, define

$$P_{ob_\kappa} = \{c \mid c \in C \text{ and } \mu_K(o,c) > \mu_K(b,c)\}. \tag{6.16}$$

The idea here is that o and b are typical spels specified in "object" and "background", respectively. Note that $P_{ob_\kappa} = \phi$, if $b = o$.

A *fuzzy κ-object O of a scene $C = (C, f)$ containing a spel o relative to a background containing a spel b* is the fuzzy subset of C defined by the following membership function. For any $c \in C$,

$$\mu_O(c) = \begin{cases} \eta(f(c)), & \text{if } c \in P_{ob_\kappa}, \\ 0, & \text{otherwise.} \end{cases} \tag{6.17}$$

We will refer to O simply as a *relative fuzzy κ-object* of C. In the hard case (when hard affinity relations are used on hard scenes), it is easy to see that GFC and RFC merge and result in the same object definition.

The following proposition states the cohesiveness of spels within the same relative fuzzy connected object.

Proposition 5 *For any scene $C = (C, f)$ over (Z^n, α), for any affinity κ in C, and for any spels o, b, q and c in C such that $q \in P_{ob_\kappa}$,*

$$\mu_K(q,c) > \mu_K(b,c) \tag{6.18}$$

if, and only if, $c \in P_{ob_\kappa}$.

Note that the above result is not valid if "$\geq$" is used in Equation (6.16) instead of "$>$".

This proposition asserts the robustness of the relative fuzzy κ-object to the reference spel selected in the object region. However, constancy of the κ-object with respect to reference spels specified in the background requires more constraints, as indicated by the following theorem.

Theorem 6 *For any scene $C = (C, f)$ over (Z^n, α), for any affinity κ in C, and for any spels o, b, q and r in C such that $q \in P_{ob_\kappa}$,*

$$P_{ob_\kappa} = P_{qr_\kappa}, \text{ if } r \in P_{bo_\kappa}. \tag{6.19}$$

Note that the condition in Equation (6.19) is sufficient (for $P_{ob_\kappa} = P_{qr_\kappa}$) but not necessary. The necessary and sufficient condition is expressed in the following theorem.

Theorem 7 *For any scene $C = (C, f)$ over (Z^n, α), for any affinity κ in C, and for any spels o, b, q and r in C such that $q \in P_{ob_\kappa}$,*

$$P_{ob_\kappa} = P_{qr_\kappa} \text{ if, and only if, } \mu_K(b, o) = \mu_K(r, o). \tag{6.20}$$

The above two theorems have different implications in the practical computation of relative fuzzy κ-objects in a given scene in a repeatable, consistent manner. Although less specific, and therefore more restrictive, Theorem 6 offers practically a more relevant guidance than Theorem 7 for selecting spels in the object and background so that the relative fuzzy κ-object defined is independent of the reference spels. Note how the above theorems are valid even for the hard case although their implications are rather straightforward.

An algorithm for determining relative fuzzy connected objects follows directly from algorithms for determining objects based on generalized fuzzy connectedness as described in Section 6.5.

6.4.5   Iterative Relative Fuzzy Connectedness

The principle behind this strategy is iteratively to refine the competition rules for different objects depending upon the results of the previous iteration. Consider the situation illustrated in Figure 6.2 which demonstrates three objects O_1, O_2 and O_3. It is very likely that, for a spel such as c, $\mu_K(o, c) \approx \mu_K(b, c)$ because of the blurring that takes place in those parts where O_1 and O_2 come close together. In this case, the strongest path from b to c is therefore likely to pass through the "core" of O_1 which is indicated by the dotted curve in the figure. This core (which is roughly P_{ob_κ}, defined in Section 6.4.4) can be detected first and then excluded from consideration in a subsequent iteration for any path from b to c to pass through. Then, we can substantially weaken the strongest path from b to c compared to the strongest path from o to c which is still allowed to pass through the core. This leads us to an iterative strategy to grow from o (and so complementarily from b) to more accurately capture O_1 (and O_2) than if a single-shot relative connectedness strategy is used. The phenomenon illustrated in Figure 6.2 is general and may be characterized in the following way. Most objects have a core part, which is relatively easy to segment after specifying the seed spel, and other diffused, subtle and fine parts that spread off from the core, which pose segmentation challenges. Although the latter seem to hang together fairly strongly with the core from a visual perceptual point of view, because of the ubiquitous noise and blurring, it is difficult to devise computational means to capture them as part of the same object by using a single-shot strategy. The iterative strategy captures these loose parts in an incremental and reinforcing manner. This formulation is described in this section.

For any fuzzy affinity κ and any two spels $c, d \in C$, define

$$\mu_{\kappa^0_{ob}}(c,d) = \mu_{\kappa}(c,d), \qquad (6.21)$$

$$P^0_{ob_{\kappa}} = \{c \,|\, c \in C \text{ and } \mu_K(o,c) > \mu_{K^0_{ob}}(b,c)\}. \qquad (6.22)$$

Note that $P^0_{ob_{\kappa}}$ is exactly the same as $P_{ob_{\kappa}}$, defined in Equation (6.16). Assuming that $P^{i-1}_{ob_{\kappa}}$ and κ^{i-1}_{ob} are defined for any positive integer i, $P^i_{ob_{\kappa}}$ and κ^i_{ob} are defined as follows. For any $c, d \in C$,

$$\mu_{\kappa^i_{ob}}(c,d) = \begin{cases} 1, & \text{if } c = d, \\ 0, & \text{if } c \text{ or } d \in P^{i-1}_{ob_{\kappa}}, \\ \mu_{\kappa}(c,d), & \text{otherwise,} \end{cases} \qquad (6.23)$$

$$P^i_{ob_{\kappa}} = \{c \,|\, c \in C \text{ and } \mu_K(o,c) > \mu_{K^i_{ob}}(b,c)\}. \qquad (6.24)$$

An *iteratively defined fuzzy κ^i-object O^i of a scene $C = (C, f)$ containing a spel o relative to a background containing a spel b* is a fuzzy subset of C defined by the membership function

$$\mu_{O^i}(c) = \begin{cases} \eta(f(c)), & \text{if } c \in P^i_{ob_{\kappa}}, \\ 0, & \text{otherwise.} \end{cases} \qquad (6.25)$$

We will refer to O^i simply as an *iterative relative fuzzy κ^i-object of C*. Below, we state a few important properties of iterative relative fuzzy connectedness. Note that, in the hard case, IRFC and RFC are equivalent which in turn merge with GFC.

Proposition 8 *For any scene $C = (C, f)$ over (Z^n, α), for any affinity κ in C, for any spels o and b in C, and for any non-negative integers $i < j$, $P^i_{ob_{\kappa}} \subseteq P^j_{ob_{\kappa}}$.*

The above proposition states essentially that $P^i_{ob_{\kappa}}$ is non-contracting as iterations continue. The following proposition states that $P^i_{ob_{\kappa}}$ and $P^j_{bo_{\kappa}}$ maintain their disjoincy at every iteration for any i and j.

Proposition 9 *For any scene $C = (C, f)$ over (Z^n, α), for any affinity κ in C, for any spels o, b in C, and for any non-negative integers i, j, $P^i_{ob_{\kappa}} \cap P^j_{bo_{\kappa}} = \phi$.*

The following theorem states an important property of iterative relative fuzzy κ^i-objects and shows their robustness with respect to reference seed spels. It is analogous to Theorem 6 on non-iterative relative fuzzy κ-objects.

Theorem 10 *For any scene* $C = (C, f)$ *over* (Z^n, α), *for any affinity* κ *in* C, *for any spels* $o, b, q,$ *and* r *in* C *such that* $q \in P_{ob_\kappa}$, *and for any non-negative integer* i, $P^i_{ob_\kappa} = P^i_{qr_\kappa}$ *if* $r \in P_{bo_\kappa}$.

Algorithms for determining iterative relative fuzzy objects are more complicated. See Section 6.5 for a further discussion on this topic.

6.5　Methods and Algorithms

6.5.1　Affinity Definition

As pointed out earlier, affinity is the link between the topological concept of fuzzy connectedness and its utility in image segmentation. As such, its definition determines the effectiveness of segmentation.

A detailed formulation of affinity definition was presented in [62]. In the same paper, it was first explicitly pointed out that, besides adjacency (a purely spatial concept), affinity should consist of two additional and distinct components: homogeneity-based and object feature-based components. This formulation can also be identified in the original affinity-based fuzzy connectedness work [75] although this was not explicitly stated. These two components are quite independent and there exist certain dichotomies between them. The object feature-based component does not capture the notion of path homogeneity. To clarify this, consider two spels c and d that are in the same object region but that are far apart. Assume that there is a slow varying background intensity component as found in MR images due to magnetic field inhomogeneities. Spels c and d are likely to have very different scene intensities although they belong to the same object. Nevertheless, one can find a contiguous path from c to d in the scene domain such that the intensities of each pair (c', d') of successive spels in the path are very similar. That is, c' and d' have a strong homogeneity-based affinity. Conversely, the homogeneity-based component alone cannot adequately capture the notion of a global agreement to known object intensity characteristics. Again consider a path from c to d along which spel intensity changes very slowly. The path may pass through different object regions. Without an object feature-based component, this path will indicate a strong connectedness between c and d and therefore may merge the different objects it passes through into one object. The following is a general functional form for μ_κ as proposed in [75, 62]:

$$\mu_\kappa(c, d) = \mu_\alpha(c, d) \, g(\mu_\psi(c, d), \mu_\phi(c, d)), \tag{6.26}$$

where μ_ψ and μ_ϕ represent the homogeneity-based and object-feature-based components of affinity, respectively. ψ and ϕ may themselves be considered as fuzzy relations in C. The strength of relation ψ indicates the degree of local hanging-togetherness of spels because of their similarities of intensities. The strength of relation ϕ indicates the degree of local hanging-togetherness of spels because of the similarity of their feature value to some (specified) object feature. General constraints on the functional form of g were described in [62] and the following examples were presented:

$$\mu_\kappa = \mu_\alpha \mu_\psi, \tag{6.27}$$

$$\mu_\kappa = \mu_\alpha \mu_\phi, \tag{6.28}$$

$$\mu_\kappa = \mu_\alpha \sqrt{\mu_\psi \mu_\phi}, \tag{6.29}$$

$$\mu_\kappa = \frac{1}{2} \mu_\alpha (\mu_\psi + \mu_\phi). \tag{6.30}$$

In the hard case,

$$\psi(c,d) = \phi(c,d) = \begin{cases} 1, & \text{if } f(c) = f(d), \\ 0, & \text{otherwise,} \end{cases} \tag{6.31}$$

Thus, the above equations are identical.

A fundamental question that arises in defining affinity (both ψ and ϕ) is how to determine the extent of the neighborhood of spels. This question also arises in all image processing operations that entail a local process. If the intensities, or features derived from them, at the level of the individual spels are considered for defining ψ and ϕ, this definition will be sensitive to noise. The idea then is to consult a neighborhood around each of c and d in defining $\mu_\psi(c,d)$ and $\mu_\phi(c,d)$ such that the neighborhood represents a region within the same object as in which c or d is situated.

"Scale" is a fundamental, well-established concept in image processing [37, 50, 68]. The premise behind this concept is to consider the local size of the object as the neighborhood size in carrying out whatever local operations that are done on the image. Saha *et al.* [62] introduced a new notion of location-specific object scale. Roughly, *object scale in* C at any spel $c \in C$ is the radius $r(c)$ of the largest hyperball centered at c which lies entirely in the same object region. Ironically, all this is done exactly for the purpose of defining the object in the first place and it appears that object definition is needed first to define scale. A simple and effective algorithm is presented in [62] to estimate object scale at each spel without explicit object definition but based on the continuity of intensity homogeneity alone. To estimate scale at a spel c, a *fraction of object* $FO_k(c)$ on the periphery of a digital ball $B_k(c)$ with center at c and radius k is computed as follows:

$$FO_k(c) = \frac{\sum_{d \in B_k(c) - B_{k-1}(c)} W_\psi(|f(c) - f(d)|)}{|B_k(c) - B_{k-1}(c)|}, \tag{6.32}$$

where W_ψ is a homogeneity function. As demonstrated in [62], a zero-mean un-normalized Gaussian is generally a good choice for W_ψ. The standard deviation parameter σ_ψ depends on the overall noise level in the image. An automatic method to determine this parameter for a specific image is described in [62]. The scale computation algorithm iteratively increases the ball radius k by 1, starting from 1, and checks for the fraction of the object $FO_k(c)$ containing c that is contained in the ball. The first time this fraction falls below a threshold level, the ball is considered to have entered into a significantly different region. See [62] for details.

In the rest of this section, we describe how scale is utilized in defining ψ and ϕ. Let $B_{xy}(z)$ denote the digital ball defined by:

$$B_{xy}(z) = \{e \in C \mid \|z - e\| \leq \min[r(x), r(y)]\}. \tag{6.33}$$

First, to define ψ, consider any two spels $c, d \in C$ such that $\mu_\alpha(c, d) > 0$. Consider any spels $e \in B_{cd}(c)$ and $e' \in B_{cd}(d)$ such that they represent the corresponding spels within $B_{cd}(c)$ and $B_{cd}(d)$, that is, $c - e = d - e'$. We will define two weighted sums $D^+(c, d)$ and $D^-(c, d)$ of differences of intensities between the two balls as follows.

$$\delta_{cd}^+(e, e') = \begin{cases} f(e) - f(e'), & \text{if } f(e) - f(e') > 0, \\ 0, & \text{otherwise,} \end{cases} \tag{6.34}$$

$$\delta_{cd}^-(e, e') = \begin{cases} f(e') - f(e), & \text{if } f(e) - f(e') < 0, \\ 0, & \text{otherwise,} \end{cases} \tag{6.35}$$

and

$$D^+(c, d) = \sum_{\substack{e \in B_{cd}(c); e' \in B_{cd}(d) \\ \text{s.t. } c - e = d - e'}} [1 - W_\psi(\delta_{cd}^+(e, e')) \, \omega_{cd}(\|c - e\|)], \tag{6.36}$$

$$D^-(c, d) = \sum_{\substack{e \in B_{cd}(c); e' \in B_{cd}(d) \\ \text{s.t. } c - e = d - e'}} [1 - W_\psi(\delta_{cd}^-(e, e')) \, \omega_{cd}(\|c - e\|)]. \tag{6.37}$$

W_ψ and ω_{cd} are window functions (such as a Gaussian). We note that the parameters of ω_{cd} depend on $r(c)$ and $r(d)$.

The connection of the above equations to the homogeneity-based affinity ψ is as follows. There are two types of intensity variations surrounding c and d: intra-object and inter-object variations. The intra-object component is generally random, and therefore, is likely to be near zero overall. The inter-object component, however, has a direction. It either increases or decreases along the direction given by $c - d$, and is likely to be larger than the intra-object variation. It is reasonable, therefore, to assume that the smaller of $D^+(c, d)$ and $D^-(c, d)$ represents the intra-object component, and the other represents the combined effect of the two components. (Note that when the values of δ_{cd}^+ (respectively, δ_{cd}^-) are small, $D^+(c, d)$ (respectively, $D^-(c, d)$) also becomes small). Note that, if there is a slow background component of variation, within the small neighborhood considered, this component is unlikely to cause a variation comparable to the inter-object component. This strategy leads us to the following functional form for μ_ψ:

$$\mu_\psi(c,d) = 1 - \frac{|D^+(c,d) - D^-(c,d)|}{\sum_{e \in B_{cd}(c)} \omega_{cd}(||c - e||)}. \tag{6.38}$$

Note that $|D^+(c,d) - D^-(c,d)|$ represents the degree of local inhomogeneity of the regions containing c and d. Its value is low when both c and d are inside an (homogeneous) object region. Its value is high when c and d are in the vicinity of (or across) a boundary. The denominator in Equation (6.38) is a normalization factor.

In the formulation of ϕ, instead of considering directly the spel intensities $f(c)$, we will consider a filtered version of it that takes into account the ball $B_r(c)$ at c defined by

$$B_r(c) = \{e \in C \mid ||c - e|| \leq r(c)\}. \tag{6.39}$$

The filtered value at any $c \in C$ is given by

$$f_a(c) = \frac{\sum_{e \in B_r(c)} f(e) \omega_{cd}(||c - e||)}{\sum_{e \in B_r(c)} \omega_{cd}(||c - e||)}, \tag{6.40}$$

where ω_{cd} (another window function) and its parameters depend on $r(c)$. The functional form of μ_ϕ is given by

$$\mu_\phi(c,d) = \begin{cases} \frac{\min[W_o(c), W_o(d)]}{\max[W_b(c), W_b(d)] + \min[W_o(c), W_o(d)]}, & \text{if } \min[W_o(c), W_o(d)] \neq 0, \\ & \text{and } c \neq d, \\ 1, & \text{if } c = d, \\ 0, & \text{otherwise}, \end{cases} \tag{6.41}$$

where W_o and W_b are intensity distribution functions for the object and background, respectively. See [62] for details on how the parameters of these and other functions can be estimated for a given application. We note that, when C is a binary scene, it is possible to define affinity relations κ in C in a fuzzy manner so that meaningful fuzzy connectedness analysis can be carried out by the appropriate choice of κ. The scale scene of C becomes a distance transform of C, and various (hard) morphological operations followed by (hard) connectivity analysis can be realized by a single FC analysis carried out on the scale scene.

6.5.2   Algorithms

Dellepiane *et al.* [17, 18] used the formulation of fuzzy connectedness proposed by Rosenfeld [56, 58] for image segmentation. Briefly, their method works as follows. Let $C = (C, f)$ be a given scene and let H be the maximum intensity over the scene. A membership scene $C_M = (C, f_M)$ is first created, where, for all $c \in C$, $f_M(c) = f(c)/H$. Let s denote a seed spel; another membership scene $C_M^s = (C, f_M^s)$ is computed from C_M and s, where, for all $c \in C$,

$$f_M^s(c) = 1 - |f_M(c) - f_M(s)|.$$

Finally, the intensity-based connectedness of C with respect to the seed s is expressed as another membership scene $C_{MK}^s = (C, f_{MK}^s)$, where, for all $c \in C$:

$$
\begin{aligned}
f_{MK}^s(c) &= \max_{p \in P_{cs}} \left[\min_{d \in \mathcal{E}(p)} f_M^s(d) \right] \\
&= 1 - \min_{p \in P_{cs}} \left[\max_{d \in \mathcal{E}(p)} |f_M(d) - f_M(s)| \right].
\end{aligned}
$$

A hard segmentation is obtained by thresholding C_{MK}^s. The authors presented a non-iterative algorithm using a tree expansion mechanism for computing C_{MK}^s given C. Note that the above intensity-based connectedness is sensitive to seed selection as the seed intensity is used for computing C_M^s, and subsequently for the connectedness C_{MK}^s.

The affinity-based approach of Udupa and Samarasekera [75, 74] not only overcomes this difficulty of dependence on seed points (Theorems 3,5,6,9) but also makes it possible to extend the fuzzy connectedness concept to relative and iterative relative fuzzy connectedness (with much relevance in practical image segmentation) and to demonstrate (Theorem 2) that the well-developed computational framework of dynamic programming [14] can be utilized to carry out fuzzy connectedness computations efficiently.

In this section we describe two algorithms [75] for extracting a fuzzy $\kappa\theta$-object containing a set S of spels in a given scene C for a given affinity κ in C, both based on dynamic programming, and another for computing iterative relative fuzzy connected objects. A fourth algorithm for labeling hard components is also presented. The first algorithm, named $\kappa\theta FOEMS$ ($\kappa\theta$-*fuzzy object extraction for multiple seeds*), extracts a fuzzy $\kappa\theta$-object of C of strength θ generated by the set S of reference spels. In this algorithm, the value of θ is assumed to be given as input and the algorithm uses this knowledge to achieve an efficient determination of the $\kappa\theta$-object. The second algorithm, named $\kappa FOEMS$, creates as output what we call a κ-*connectivity scene* $C_{KS} = (C, f_{KS})$ of C *generated by* the set S of reference spels defined by $f_{KS}(c) = \max_{s \in S}[\mu_K(s, c)]$.

Algorithm $\kappa\theta FOEMS$ terminates faster than $\kappa FOEMS$ for two reasons. First, $\kappa\theta FOEMS$ produces the hard set based on Equation (6.15). Therefore, for any spel $c \in C$, once we find a path of strength θ or greater from any of the reference spels to c, we do not need to search for a better path up to c, and hence, can avoid further processing for c. This allows us to reduce computation. Second, certain computations are avoided for those spels $d \in C$ for which $\max_{s \in S}[\mu_K(s, d)] < \theta$.

Unlike $\kappa\theta FOEMS$, $\kappa FOEMS$ computes the best path from the reference spels of S to every spel c in C. Therefore, every time the algorithm finds a better path up to c, it modifies the connectivity value at c and subsequently processes other spels which are affected by this modification. The algorithm generates a connectivity scene $C_{KS} = (C, f_{KS})$ of C. Although, $\kappa FOEMS$ terminates slower, it has a practical advantage. After the algorithm terminates, one can interactively specify θ and thereby examine various $\kappa\theta$-objects and interactively select the best θ. The connectivity scene has interesting properties relevant to classification and in shell rendering and manipulation [71] of $\kappa\theta$-objects.

Algorithm $\kappa\theta FOEMS$
Input: C, S, κ, η, and θ as defined in Sections 6.3 and 6.4.
Output: $O_{K\theta}(S)$ as defined in Section 6.4.
Auxiliary Data Structures: An n-D array representing a temporary scene $C' = (C, f')$ such that f' corresponds to the characteristic function of $O_{K\theta}(S)$, and a queue, Q, of spels. We refer to the array itself by C' for the purpose of the algorithm.

begin
    1. set all elements of C' to 0 except those spels $s \in S$ which are set to 1;
    2. push all spels $c \in C$ such that for some $s \in S$ $\mu_K(s, c) \geq \theta$ to Q;
    3. *while* Q is not empty *do*
    4.   remove a spel c from Q;
    5.   *if* $f'(c) \neq 1$ *then*
    6.   set $f'(c) = 1$;
    7.   push all spels d such that $\mu_K(c, d) \geq \theta$ to Q;
        endif;
    endwhile;
    8. create and output $O_{K\theta}(S)$ by assigning the value $\eta(f(c))$ to all $c \in C$ for which $f'(c) > 0$ and 0 to the rest of the spels;
end

Algorithm $\kappa FOEMS$
Input: C, S, and κ as defined Sections 6.3 and 6.4.
Output: A scene $C' = (C, f')$ representing the κ-connectivity scene C_{KS} of C generated by S.
Auxiliary Data Structures: An n-D array representing the connectivity scene $C' = (C, f')$ and a queue, Q, of spels. We refer to the array itself by C' for the purpose of the algorithm.

begin
    1. set all elements of C' to 0 except those spels $s \in S$ which are set to 1;
    2. push all spels $c \in C$ such that, for some $s \in S$, $\mu_\kappa(s,c) > 0$ to Q;
    3. *while* Q is not empty *do*
    4.   remove a spel c from Q;
    5.   find $f_{max} = \max_{d \in C}[\min(f'(d), \mu_\kappa(c,d))]$;
    6.   *if* $f_{max} > f'(c)$ *then*
    7.   set $f'(c) = f_{max}$;
    8.   push all spels e such that $\min[f_{max}, \mu_\kappa(c,e)] > f'(e)$ to Q;
        endif;
    endwhile;
    9. output the connectivity scene C';
end

Both algorithms work in an iterative fashion, and in each iteration in the *while-do* loop, they remove exactly one spel from the queue Q and check if the strength of connectedness at that spel needs to be changed. In case a change is needed, the change is passed on to the rest of the scene through the neighbors by entering them into Q. It may be noted that both algorithms start after initialization of strengths of connectedness at the seed spels. Also, note that the strength of connectedness at any spel never reduces during runtime of these algorithms. It was shown [75, 63] that both algorithms terminate in a finite number of steps, and when they do so, they produce the expected results. Recently, studies [9, 46] on the speed-up of these algorithms have been reported. A preliminary effort was made by Carvalho *et al.* [9] by using Dijkstra's algorithm [19] in place of the dynamic programming algorithm suggested in [75]. Based on 2D phantoms and two 2D MR slice scenes, they suggested that a 7-fold to 8-fold speed-up in fuzzy connectedness object computation can be achieved. Nyúl and Udupa [46] made an extensive study on the speed of a set of 18 algorithms utilizing a variety of 3D scenes. They presented two groups of algorithms: label-correcting and label-setting, and a variety of strategies of implementation under each group. They tested these algorithms in several real medical applications and concluded that there is no "always optimal" algorithm. The optimality of an algorithm depends on the input data as well as on the type of affinity relations used. In general, it was found that, for blob-like objects with a spread-out distribution of their intensities or other properties (such as brain tissues in MRI), label-setting algorithms with affinity-based hashing outperform others, while, for more sparse objects (vessels, bone) with high contrast (MRA, CTA), label setting algorithms with geometric hashing are more speed-efficient. Also, they demonstrated that using the right algorithms on fast hardware (1.5 GHz Pentium PC), interactive speed (less than one second per 3D scene) of fuzzy object segmentation is achievable.

We note that, in the hard case, when S contains a single seed spel, for any $0 < \theta \leq 1$, both algorithms behave in a similar manner. There is considerable redundancy in these algorithms when used for finding hard connected components. More efficient algorithms are available in the literature for tracking and labeling hard connected components [67]. The beauty of the above algorithms is that they can track fuzzy or hard connected objects in a gray or a binary scene. An efficient algorithm, termed $\alpha_h CE$ (α_h-*component extraction*), expressly for tracking and labeling hard connected objects in a binary scene is given below [53]. The method labels all connected components of 1-spels using only two scans. During its first scan, it assigns a new label when a (locally) new region of 1-spels is encountered and propagates all labeled regions along the scan direction. During this process when it finds two or more regions with different labels to merge, it saves the information in an equivalence table which keeps track of equivalent labels in the form of a graph. After completing the full first scan of the entire scene in this fashion, it parses the equivalence table to determine the maximal sets of globally connected labels and assigns a new label to each such maximal set. Finally, it revisits the scene and assigns to each 1-spel the new label which corresponds to the maximal set to which its old label belongs. Although, improvements of this classical method have been made subsequently, we present the old method because of its simplicity. More advanced methods may be found in [67].

Algorithm $\alpha_h CE$

Input: C (binary) and α_h as defined in Sections 6.3 and 6.5.

Output: A gray scene $O_{\alpha_h} = (C, f_{O_{\alpha_h}})$ where $f_{O_{\alpha_h}} = 0$ for all 0-spels and its value for a 1-spel indicates the label of the α_h-component to which the spel belongs; thus the set of 1-spels with the same $f_{O_{\alpha_h}}$ value constitutes one α_h-component.

Auxiliary Data Structures: $EQTABLE$ is a table with each row containing a pair of equivalent initial component labels; $CUR\_COMPONENT$ is a list of initial labels constituting the same α_h-component of C; $PARSED\_TABLE$ is a function mapping an initial label to a final component label.

begin

1.   initialize $f_{O_{\alpha_h}}(c) = 0$ for all $c \in C$;

2.   initialize $new\_label = 1$ and $EQTABLE = PARSED\_TABLE = empty$;

3.   *for* all $c \in C$ visit in a raster scan fashion *do*

4.   *if* $f(c) = 1$ *then*

5.   set $N_{\alpha_h}^-(c)$ = set of pre-visited 1-spels that are α_h-adjacent to c;

6.   *if* $N_{\alpha_h}^-(c) = $ empty *then*

7.   set $f_{O_{\alpha_h}}(c) = new\_label$;

8.   increase *new_label* by '1';

9.   *else*

10.   set $M = \min[f_{O_{\alpha_h}}(d) \mid d \in N_{\alpha_h}^-(c)]$;

11.   set $f_{O_{\alpha_h}}(c) = M$;

12.   set $S' = $ set of distinct $f_{O_{\alpha_h}}$ values for the spels in $N_{\alpha_h}^-(c)$;

13.   *for* $x \in S'$ and $x \neq M$ *do*

14.   Add_row(*x*,*M*,*EQTABLE*);

                             /* Append a row in *EQTABLE* containing */

                             /* the two entries *x* and *M*. */

                 endfor;

             endif;

         endif;

     endfor;

16.   initialize all rows in *EQTABLE* as "unmarked";

17.   set *new_label* = 1;

18.   *while EQTABLE* has an "unmarked" row *do*

19.   set *CUR_COMPONENT* = empty;

20.   set *N* = first "unmarked" row in *EQTABLE*;

21.   set *CUR_COMPONENT* = DFS(*N*, *EQTABLE*);

                     /* DFS performs a depth-first search [25] of the graph induced by *EQTABLE* starting from the node representing the second entry of the row *N* and marks all rows visited during the search process. The method returns a list of all labels belonging to one connected group. */

22.   *for* $x \in CUR\_COMPONENT$ *do*

23.   set $PARSED\_TABLE(x) = new\_label$;

         endfor;

24.   increase *new_label* by '1';

     endwhile;

25.   *for* all $c \in C$ visit in a raster scan fashion *do*

26.   *if* $f(c) = 1$ *then*

27.   set $f_{O_{\alpha_h}}(c) = PARSED\_TABLE(f_{O_{\alpha_h}}(c))$;

         endif;

     endfor;

end

Given $\kappa FOEMS$, an algorithm for relative fuzzy connectedness is easy to realize. First, the connectivity scenes $C_{K_o} = (C, f_{K_o})$ and $C_{K_b} = (C, f_{K_b})$ are computed using $\kappa FOEMS$ for seeds o and b, respectively. Subsequently, a spel $c \in C$ is included in the support P_{ob_κ} of the fuzzy connected object of o relative to the background containing b if $f_{K_o}(c) > f_{K_b}(c)$. Iterative relative fuzzy connectedness is an iterative method; the connectivity scene $C_{K_o} = (C, f_{K_o})$ is computed once, and in each iteration, the connectivity scene $C_{K_b^i} = (C, f_{K_b^i})$ is computed for seed spel b and affinity κ_{ob}^i as defined in Equation (6.23) in each iteration i. The algorithm for computing iterative relative fuzzy connectedness [64, 80] is presented below.

Algorithm $\kappa IRFOE$
Input: $C = (C, f)$, κ, spels o and b, and η as defined in Section 6.3.
Output: Iteratively defined fuzzy κ-object O^i containing o relative to a background containing b.
Auxiliary Data Structures: The κ-connectivity scene $C_{K_o} = (C, f_{K_o})$ of o in C; the κ_{ob}^i-connectivity scene $C_{K_{ob}^i} = (C, f_{K_{ob}^i})$ of b in C.

begin
   0.   compute C_{K_o} using $\kappa FOEM$ with the input as C, o, and κ;
   1.   set $i = 0$ and $\kappa_{ob}^i = \kappa$;
   2.   set $flag$ = true, and $count$ = 0;
   3.   *while* $flag$ is true *do*
   4.   set $flag$ = false, $old\_count = count$, and $count$ = 0;
   5.   compute $C_{K_{ob}^i}$ using $\kappa FOEM$ with the input as C, b, and κ_{ob}^i;
   6.   set $\kappa_{ob}^i = \kappa$;
   7.   *for* all $c \in C$ *do*
   8.   *if* $f_{K_o}(c) > f_{K_{ob}^i}(c)$ *then*
   9.   set $\mu_{O^i}(c) = \eta(f(c))$;
  10.   increment $count$ by 1;
  11.   *for* all $d \in C$ *do*
  12.   set $\mu_{\kappa_{ob}^i}(c, d) = 0$;
            endfor;
  13.   *else* set $\mu_{O^i}(c) = 0$;
         endif;
       endfor;
  14   *if* $count \neq old\_count$ *then* set $flag$ = true;
  15.   increment i by 1;
     endwhile;
  15. output O^i;
end

6.6   Applications

In this section, we shall describe seven applications in which the fuzzy connectedness algorithms have been utilized and tested extensively. It is one thing to have a powerful segmentation framework and quite another matter to put it to practical use, such as in clinical radiology, for routinely processing hundreds and thousands of scenes. The latter usually involves considerable work in (1) engineering the proper utilization of the framework, and (2) evaluating the efficacy of the method in each application. For the fuzzy connectedness framework, the first task entails the determination of the affinity relation appropriate for the application, choosing one among generalized, relative, and iterative relative fuzzy connectedness strategies, and devising other auxiliary processing and visualization methods needed under the application. The second task consists of assessing the precision, accuracy, and efficiency of the complete method in the application [81]. Precision here refers to the reliability (reproducibility) of the complete method taking into consideration all subjective actions that enter into the entire process including possible variations due to repeat acquisition of the scenes for the same subject on the same and different scanners, and any operator help and input required in segmentation. Accuracy relates to the agreement of the segmented result to truth. Efficiency refers to the practical viability of the segmentation method. This has to consider pure computational time and the time required in producing any operator help needed in segmentation. The above two tasks have been accomplished under each application (except evaluative studies for the last application which are very different in their requirement from other applications). For reasons of space, we shall not present here details of either the engineering or the evaluation studies, but refer the reader to references given under each application. In the rest of this section, we give a cursory description of each of the seven application areas outlining the segmentation need and methodology.

6.6.1   Multiple Sclerosis

Multiple sclerosis (MS) is an acquired disease of the central nervous system characterized primarily by multi-focal inflammation and destruction of myelin [10]. Inflammation and edema are accompanied by different degrees of demyelination and destruction of oligodendrocytes, and may be followed by remyelination, axonal loss and/or gliosis. The highest frequency of MS occurs in northern and central Europe, Canada, the USA, and New Zealand and South of Australia [33]. In the US, it affects approximately 350,000 adults and stands out as the most frequent cause of non-traumatic neurologic disability in young and middle-aged adults. In its advanced state, the disease may severely impair the ambulatory ability and may even cause death. The most commonly used scale to clinically measure disease progression in MS is the expanded disability status scale (EDSS)

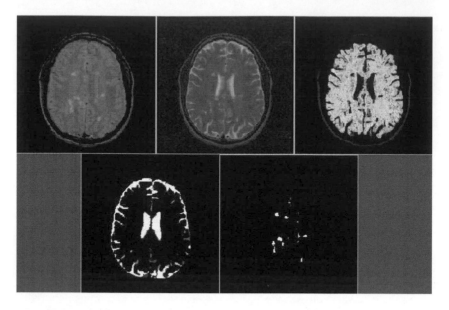

Figure 6.4. Top row: A slice of a T2 and proton weighted scene of an MS patient's head and the segmented brain parenchymal region. Bottom row: the segmented CSF region and the brain lesions (which appear hyper intense in both T2 and PD scenes).

introduced by Kurtzke [34]. The clinical quantification of disease severity is subjective and sometimes equivocal. The development of new treatments demands objective outcome measures for relatively short trials. Therefore, MR imaging has become one of the most important paraclinical tests for diagnosing MS and in monitoring disease progression in MS.

Various MRI protocols have been used in studying MS [78]. At present they include dual-echo T2-weighted imaging, T1-weighted imaging with and without Gadolinium contrast, magnetization transfer imaging, diffusion-weighted imaging, and MR spectroscopy. The different protocols convey information about different aspects of the disease. Current MRI-based research on MS is focused on (1) characterizing the natural course of the disease and to distinguish among disease subgroups through images, (2) characterizing treatment effects to assess different treatment strategies. In achieving these goals, several image processing operations are utilized, most crucial among them being segmentation of MR brain images in the different protocols into different tissue regions and lesions. Our efforts in this direction utilizing fuzzy connectedness and the ensuing clinical results are summarized in [78]. Figure 6.4 shows an example of a T2-weighted and proton-density-weighted scene of an MS patient's head and the segmented tissue regions.

A somewhat similar problem from the image processing perspective arises in the study of late-life depression [31, 32]. The aim here is to understand the disease and its treatment effects in the case of white matter lesions occurring in elderly subjects due to age and the associated effects on the function of their brain. In MS (and other neurological applications), a variety of MRI protocols are utilized. The actual imaging parameters used in these protocols vary among institutions. In spite of numerous brain MR image segmentation methods developed during the past 15 years, none of them is capable of handling variations within the same protocol, and much less, the variation among protocols. What we need is a segmentation "workshop" wherein a protocol-specific segmentation method can be fabricated quickly. In [47], one such workshop is described utilizing the fuzzy connectedness framework.

6.6.2   MR Angiography

MR angiography (MRA) plays an increasingly important role in the diagnosis and treatment of vascular diseases. Many methods are available to acquire MRA scenes; the most common is perhaps acquisition with an intravascular contrast agent such as Gadolinium. An image processing challenge in this application is how to visualize optimally the imaged vascular tree from the acquired scene. In the Maximum Intensity Projection (MIP) method commonly used for this purpose, the maximum voxel intensity encountered along each projection ray is assigned to a pixel in the viewing plane to create a rendition as illustrated in Figure 6.5. Although, the MIP method does not require prior segmentation of the scene, it obscures the display with clutter due to high intensities coming from various artifacts as shown in Figure 6.5. By segmenting those aspects of the vessels containing bright intensities via fuzzy connectedness, the clutter can be mostly eliminated as illustrated in Figure 6.5 [51]. Since the intensity of the image varies over the scene (due to various reasons including magnetic field inhomogeneities and different strengths of the contrast agent in different parts of the vessels), simple thresholding and connectivity analysis fail to achieve a proper segmentation of the vessels.

Recently, new imaging techniques with contrast agents that stay in the blood much longer and that cause far less extravessation than Gadolinium have been devised [35]. Although, these agents stay in the blood longer and hence provide a more uniform enhancement of even significantly thinner vessels, their side effect is that they enter into venous circulation also because of the long decay time. Consequently, both the arteries and veins enhance in the acquired scenes and it becomes very difficult to separate mentally the arterial tree from the venous tree in a display such as the one created by a MIP method (see Figure 6.6) [36]. Iterative relative fuzzy connectedness can be effectively utilized in this situation [36] to separate the arterial tree from the venous from a scene that is created by

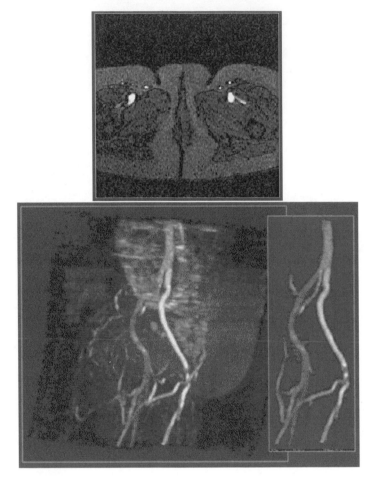

Figure 6.5. A slice of an MRA scene acquired with Gadolinium administration (top); a MIP rendition of this scene which shows much clutter, and a MIP rendition of the region segmented by using fuzzy connectedness (bottom).

segmenting the entire vascular tree first by using generalized fuzzy connectedness. This is illustrated in Figure 6.6.

6.6.3   CT Angiography

Currently, in the study of vascular diseases, CT angiography (CTA) [1] is a serious contender with MRA to eventually replace conventional x-ray angiography. A major drawback of CTA, however, has been the obscuration of vessels by bone

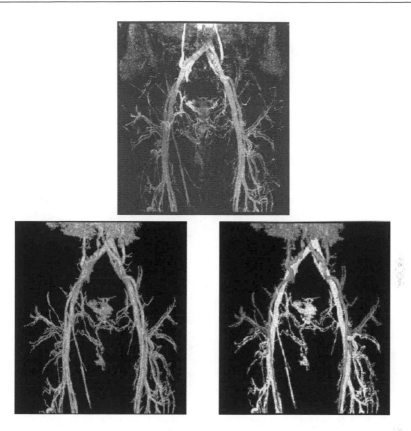

Figure 6.6. (See Plate VI) A MIP rendition of an MRA scene acquired with the blood-pool contrast agent MS325 (top), the entire vascular tree segmented via generalized fuzzy connectedness and volume rendered (bottom-left), and the arteries and veins separated by using iterative relative fuzzy connectedness (bottom-right).

in the 3D display of CTA images. The problem is caused by blood as well as bone assuming high and comparable intensity in the CTA scene and often by vessels running close to osseous structures. Partial volume and other artifacts make it really difficult to suppress bone in the latter situation. This scenario is particularly true for cerebral vessels since bone surrounds the brain and vessels come close to the inner surface of the calvarium at many sites. Interactive clipping of bone in three dimensions is not always feasible or would produce unsatisfactory results and slice-by-slice outlining is simply not practical since most studies involve hundreds of slices comprising a CTA scene.

A method employing fuzzy connectedness has been proposed in [59] for the suppression of osseous structures, and its application in neuro-surgery, particu-

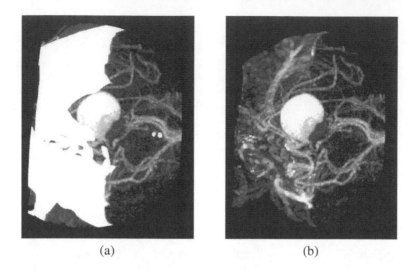

(a)   (b)

Figure 6.7. A MIP rendition of CTA of the head of a subject, and a MIP rendition after bone has been suppressed via relative fuzzy connectedness.

larly for managing cerebral aneurysms, is described in [1]. The method uses both generalized and iterative relative fuzzy connectedness along with some morphological operations to separate the vessels from osseous structures. Particularly challenging in this context are the suppression of thin bones and of bones juxtaposed close to vessels. Figure 6.7 shows an example of a MIP rendition of a cerebral CTA scene before and after suppression of bones by using the method.

6.6.4   Brain Tumor

The management of patients with brain tumors (particularly gliomas) is at present difficult in the clinic since no objective information is available as to the morphometric measures of the tumor. Multiprotocol MR images are utilized in the radiologic clinic to determine the presence of tumor and to discern subjectively the extent of the tumor [38]. Both subjective and objective quantification of brain tumor are difficult due to several factors. First, most tumors appear highly diffused and fuzzy with visually indeterminable borders. Second, the tumor pathology has different meaning in images obtained by using different MRI protocols. Finally, a proper definition of what is meant by "tumor" in terms of a specific pathology and a determination of what definition is appropriate and useful for the task on hand are yet to be determined. In the presence of these difficulties, we arrived at the following operational definition for the purpose of quantification. We seek two measures to characterize the size of the tumor, the first being the

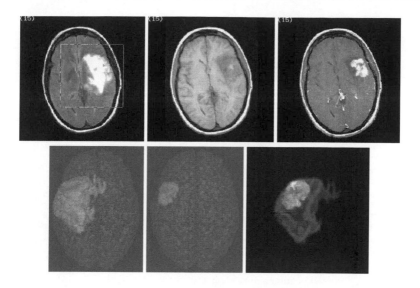

Figure 6.8. (See Plate V) Top row: A slice each of a FLAIR scene; a T1-weighted scene; a T1 with Gadolinium scene. Bottom row: A surface rendition of brain plus edema segmented from the FLAIR scene; brain plus edema and enhancing aspect of tumor segmented from the two T1 scenes; edema plus enhancing tumor volume rendered.

volume of the edema and the second being the active part of the tumor. Edema is defined to be constituted by the regions of high signal intensity in scenes acquired via the FLAIR sequence. Active aspects of the tumor are considered to be formed of regions that have high signal intensity in T1-weighted scenes acquired with Gadolinium administration but not in T1-weighted scenes acquired without Gadolinium.

Our approach to derive these measures from the MR scenes is described in [45]. Briefly, it consists of applying the generalized fuzzy connectedness method to the FLAIR scene to segment the hyper-intense regions and then estimating the volume. Further, the two T1 scenes are registered, then one scene is subtracted from the other, and the hyper-intense region in the subtracted scene is segmented and its volume is computed. The reason for subtracting the scenes is to remove any post-surgical scar/blood appearing as hyper-intense regions in the T1 scenes (without Gadolinium). Figure 6.8 shows an example of this application.

6.6.5 Breast Density

Breast density as measured from the volume of dense tissue in the breast is considered to indicate a risk factor for breast cancer [83]. X-ray mammography is

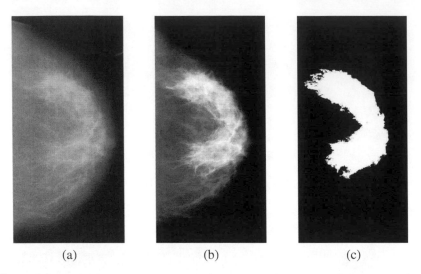

(a)   (b)   (c)

Figure 6.9. A digitized mammogram; the fuzzy connectivity scene of the dense region; the segmented binary region.

one of the most commonly used imaging modalities for breast cancer screening. Estimation of breast density from mammograms is therefore a useful exercise in managing patients at high risk for breast cancer. It may also be useful in assessing how this risk factor is affected by various treatment procedures such as hormone therapy.

In the Wolfe classification scheme [84, 85], a mammographer assigns a grade of density based on a visual inspection of the mammogram. This method, besides being very rough and subjective, may have considerable inter-reader and intra-reader variations and variation from one projective image to another of the same breast. We have developed a method utilizing generalized fuzzy connectedness to segment a digitized mammogram into dense fibroglandular and fatty regions and to provide several measures of density [60]. The fuzzy connectedness framework seems to be especially suitable in this application since the dense regions appear fuzzy with a wide range of scene intensities. Figure 6.9 shows an example.

6.6.6 Upper Airway in Children

MRI has been used to visualize the upper airway of adult humans. This enables imaging of the air column as well as of the surrounding soft tissue structures including lateral pharyngeal wall musculature, fat pads, tongue and soft palate [66]. With detailed images of the upper airway and its nearby structures, researchers can now use image processing methods to segment efficiently, visual-

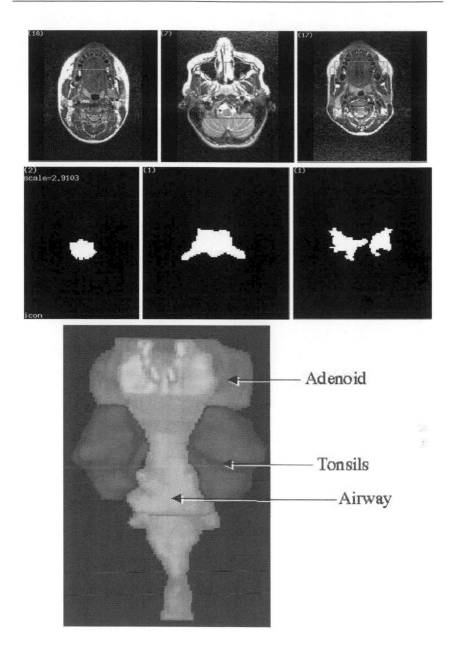

Figure 6.10. A slice of a T2 scene of the upper airway region of a child; the corresponding segmented airway and adenoid and palatine tonsils; a 3D rendition of the three structures.

ize, and measure these structures. Measurements of upper airway geometry and the architecture of the surrounding structures can be used to investigate the reasons for obstructive sleep apnea (OSA), and to assess the efficacy of medical treatment [66].

Similarly, MRI has recently been introduced to study the upper airway structure of children during growth and development and to assess the anatomic role in pathological conditions such as OSA [4]. MRI is particularly useful in children since it can delineate the lymphoid tissues surrounding the upper airway very accurately. These tissues, the tonsils and adenoids, figure predominately in childhood OSA and are a common cause for pediatric medical visits and surgery.

Our methodology [39] uses axial T2 images. Generalized fuzzy connectedness is used for segmenting the upper airway and adenoid and palatine tonsils. The system performs all tasks from transforming data to giving a final result report. To achieve accurate quantification and assure quality, the system allows clinicians to interact with critical data processing steps as minimally as possible. Figure 6.10 shows an example.

6.6.7   CT Colonoscopy

An established imaging method for detecting colon cancer has been x-ray projection imaging. A more fool-proof method has been the endoscopic examination through a colonoscopy that is commonly employed in the clinic as a screening process. Recently, in view of the invasiveness and attendant patient discomfort of this latter procedure, methods utilizing tomographic images have been investigated [82]. Dubbed virtual colonoscopy or CT colonoscopy, the basic premise of these methods is to utilize the 3D information available in a CT scene of the colon, acquired upon careful cleaning and preparation of the colon, to produce 3D displays depicting views that would be obtained by an endoscope. Since full 3D information is available, displays depicting colons that are cut or split open or unraveled in other forms can also be created.

In the system we have developed [79], generalized fuzzy connectedness is used for segmenting the colon. A 3D display of the entire colon is then created and the user selects several points along the colon on the outer surface so displayed to help the system create a central line running through the colon. Subsequently, along this central line view points are selected at uniform intervals, and at each such point, three renditions are created of the inner wall. These renditions represent the views of the inner wall of the colon that would be obtained by viewing the wall surface enface in three directions that are 120° apart. The speed of our surface rendering method is high enough to produce in real-time the set of three views that will be obtained as the view point moves along the central line automatically. Figure 6.11 shows an example.

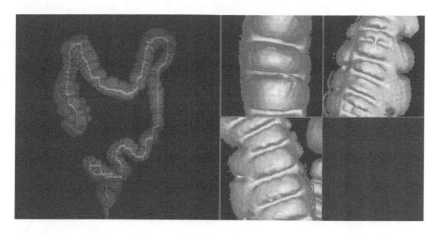

Figure 6.11. A surface rendition of the whole colon with the central line shown. Three enface views of the colon wall obtained from a view point situated on the central line with the viewing directions that are 120° apart.

6.7 Concluding Remarks

Although the idea of fuzzy connectedness existed in the literature since the late 1970s [56], its full-fledged development and application to image segmentation started in the 1990s [74]. Apart from generalizing the well known concept of hard connectedness to fuzzy subsets, fuzzy connectedness allows capturing the spatio-topological concept of hanging-togetherness of image elements even in the presence of a gradation of their intensities stemming from natural material heterogeneities, blurring and other imaging-phenomenon-related artifacts. In this chapter, we have examined in a tutorial fashion the different approaches that exist for defining fuzzy connectedness in fuzzy subsets of the discrete space as well as directly in given scenes. The latter, in particular, allow us to tie the notion of fuzzy connectedness with scene segmentation, leading to an immediate practical utilization of fuzzy connectedness in scene analysis. Starting from the fundamental notion of fuzzy connectedness, we have demonstrated how more advanced concepts such as relative and iterative relative fuzzy connectedness can be developed which have practical consequences in scene segmentation. We have also indicated how hard connectedness becomes a particularization of fuzzy connectedness in the case of binary scenes. We have also examined seven major medical application areas to which the fuzzy connectedness principles have been applied for delineating objects.

Fuzzy connectedness has several interesting characteristics that enhance its effectiveness in scene segmentation. The scale-based versions of fuzzy connectedness make it less sensitive to noise. It is also possible to consider texture in the definition of affinity to arrive at connectedness and segmentation of regions in terms of texture. We note that fuzzy connectedness is to some extent immune to a slow varying component of background scene intensity as demonstrated in [74]. This is not difficult to explain. Such slow varying components do not affect affinities much, as such the strengths of connectedness of paths are not significantly affected, and therefore, so also the overall fuzzy connectedness.

Many methods exist for fuzzy clustering of feature vectors in a feature space [7, 6]. It would be worth investigating if the added hanging-togetherness character of fuzzy connectedness can lead to improved clustering strategies in this area. Another direction for future research is to use fuzzy connectedness, relative and iterative relative fuzzy connectedness on binary scenes and on membership scenes for separating components and for doing morphological operations on them. The aspect of the size of the object at various spels can be captured via scale values and separation or connectedness of binary objects can be analyzed via fuzzy connectedness of scale scenes. It should be feasible to obtain directly medial line and skeletal representations of objects from binary scenes by using fuzzy connectedness on scale scenes.

The iterative relative fuzzy connectedness theory and algorithms currently available [80, 64] are for only two objects. Considering the fact that most scenes contain more than two objects, for these algorithms to be useful, they and the associated theory should be extended to multiple objects.

One of the current limitations (stemming from a theoretical requirement) in the design of affinities [61] is that their functional form should be shift-invariant over the whole scene domain, although there are ways of tailoring affinities to variations in the characteristics of different object regions (see [61]). It will be practically useful to investigate ways to allow different functional forms for different object regions and their theoretical and practical consequences.

A fundamental question arises in relative fuzzy connectedness as to how to do object groupings. Even when multiple objects are considered as in Figure 6.1, each object region may consist of a group of objects. From the consideration of the effectiveness of segmentation, certain groupings may be better than others or even better than the consideration of *all* object regions that may exist in the scene. Should the object groupings be considered based on separability? That is, the scene domain is first conceptually divided into a set of most separable groups of objects. These groups are then segmented via RFC (or IRFC). Next, within each group, further conceptual groupings are made based on separability, and each such subgroup is segmented via RFC. This process is continued until the desired objects are reached and segmented. Such a hierarchical RFC (or IRFC) has interesting theoretical, computational, and application-specific questions to be studied.

Fuzzy connectedness in its present form attempts to extract as much information as possible entirely from the given scene. It, however, does not attempt to use in its framework prior shape and appearance knowledge about object boundaries to be segmented. These entities have been shown to be useful in scene segmentation [12, 13]. When information is missing in certain segments of the boundary, fuzzy connectedness often goes astray establishing connection with other objects surrounding the object of interest. Ways to incorporate prior boundary shape appearance knowledge into the fuzzy connectedness framework may mitigate this problem. This is a challenging and worthwhile direction to pursue. The consideration of multiple objects in the model and utilization of such models in conjunction with multiple object relative fuzzy connectedness opens many new avenues for segmentation research.

Acknowledgements

The authors are grateful to Dr. Tianhu Lei and Jianguo Liu for Figures 6.6, 6.8 and 6.10 and to Dr. Harvey Eisenberg for data pertaining to Figure 6.11. The research reported here is funded by a DHHS grant NS37172 and contract LM-0-3502.

References

[1] Abrahams, J., Saha, P., Hurst, R., LeRoux, P., and J. Udupa. "Three-dimensional bone-free rendering of the cerebral circulation using computed tomographic angiography and fuzzy connectedness," *Neurosurgery* 51(2002): 264–269.

[2] Alderliesten, T., *et al.* "Objective and reproducible segmentation and quantification of tuberous sclerosis lesions in FLAIR brain MR images," in *Proceedings of SPIE: Medical Imaging* 4322(2001): 1509–1518.

[3] Angelini, E., Imielinska, C., Jin, Y., and A. Laine. "Improving statistics for hybrid segmentation of high-resolution multichannel images," in Proceedings of SPIE: Medical Imaging 4684(2002): 401–411.

[4] Arens R., *et al.* "Linear dimensions of the upper airway structure during development: assessment by magnetic resonance imaging", *American Journal of Respiratory Critical Care in Medicine*, 165(2002): 117–122.

[5] Artzy, E., Frieder, G., and G. Herman. "The theory, design, implementation, and evaluation of a three-dimensional surface detection algorithm," *Computer Graphics and Image Processing* 15(1981): 1–24.

[6] Backer, E. and A. Jain. "A clustering performance measure based on fuzzy set decomposition," *IEEE Transactions on Pattern Analysis and Machine Intelligence* 3(1981): 66–74.

[7] Bezdek, J. *Fuzzy mathematics in pattern classification*, Ph.D. dissertation, (Cornell University, Ithaca, NY, 1973).

[8] Bonnet, N. and S. Maegherman. "Algorithme de segmentation semi-automatique par la ligne de partage des eaux," *Internal Report* (In French, LERI, University of Reims, 1998).

[9] Carvalho, B., Gau, C., Herman, G., and T. Kong. "Algorithms for fuzzy segmentation," *Pattern Analysis and Applications* 2(1999): 73–81.

[10] Charcot, J. *Lectures on the Diseases of the Nervous System*, (The New Sydenham Society, London, 1877).

[11] Connell, S. and A. K. Jain. "Template-based online character recognition," *Pattern Recognition* 34(2001): 1–14.

[12] Cootes, T., Taylor, C., Cooper, D., and J. Graham. "Active shape models–their training and application," *Computer Vision and Image Understanding* 61(1995): 38–59.

[13] Cootes, T., Edwards, G., and C. Taylor. "Active appearance models," *IEEE Transactions on Pattern Analysis and Machine Intelligence* 23(2001): 681–685.

[14] Cormen, T., Leiserson, C., and R. Rivest. *Introduction to Algorithms*. McGraw-Hill, New York, 1991.

[15] Cutrona, J. and N. Bonnet. "Two methods for semi-automatic image segmentation based on fuzzy connectedness and watershed," *France-Iberic Mecroscopy Congress*, (Barcelona 2001): 23–24.

[16] Danielsson, P.-E. and S. Tanimoto. "Time complexity for serial and parallel propagation in images," in *Advanced and Algorithms for Digital Image Processing*, A. Oosterlinck and P.-E. Danielsson, (eds.), *Proceedings of the SPIE* 435(1983): 60–67.

[17] Dellepiane, S., and F. Fontana. "Extraction of intensity connectedness for image processing," *Pattern Recognition Letters*, 16(1995): 313–324.

[18] Dellepiane, S., Fontana, F., and G. Vernazza. "Nonlinear image labelling for multivalued segmentation," *IEEE Transactions on Image Processing* 5(1996): 429–446.

[19] Dijkstra, E. "A note on two problems in connection with graphs," *Numeric Mathematics* 1(1959): 269–271.

[20] Ge, Y., *et al.* "Brain atropy in relapsing-remitting multiple sclerosis and secondary progressive multiple sclerosis: longitudinal quantitative analysis," *Radiology* 213(2000): 665–670.

[21] Grimsdale, R., Summer, F., Tunis, C. and T. Kilburn. "A system for automatic recognition of patterns," *Proceedings of the IEEE* 106B(1959): 210–221.

[22] He, R. and P. A. Narayana. "Automatic delineation of Gd enhancements on magnetic resonance images in multiple sclerosis," *Medical Physics* 29(2002): 1536–1546.

[23] Herman, G. "Discrete multidimensional Jordan surfaces," *Graphical Models and Image Processing* 54(1992): 507–515.

[24] Herman, G. and B. De Carvalho. "Multiseeded segmentation using fuzzy connectedness," *IEEE Transactions on Pattern Analysis and Machine Intelligence* 23(2001): 460–474.

[25] Horowitz, E. and S. Sahani. *Fundamentals of Data Structures.* Rockville, MD: Computer Science Press, 1982

[26] Jin, Y., Laine, A., and C. Imielinska. "An adaptive speed term based on homogeneity for level-set segmentation," in Proceedings of SPIE: Medical Imaging 4684(2002): 383–400.

[27] Jones, T. and D. Metaxas. "Automated 3D segmentation using deformable models and fuzzy affinity," in *Proceedings of Information Processing in Medical Imaging 1997* Berlin: Springer (1997): 113–126.

[28] Jones, T. *Image-Based Ventricular Blood Flow Analysis*, Doctoral Dissertation (University of Pennsylvania, 1998).

[29] Kaufmann, A. *Introduction to the Theory of Fuzzy Subsets*, Vol. 1, (Academic Press, New York, 1975).

[30] Kong, T. and A. Rosenfeld, "Digital topology: introduction and survey," *Computer Vision, Graphics, and Image Processing* 48(1989): 357–393.

[31] Kumar, A., Bilker, W., Jin, Z., Udupa, J.,, and G. Gottlieb. "Age of onset of depression and quantitative neuroanatomic measures: absence of specific correlations," *Psychiatry Research Neuroimaging* 91(1999): 101–110.

[32] Kumar, A., Bilker, W., Jin, Z., and J. Udupa. "Atropy and high intensity lesions: complementary neurobiological mechanisms in late-life major depression," *Neuropsychopharmacology* 22(2000): 264–274.

[33] Kurtzke, J. "Geographic distribution of multiple sclerosis: an update with special reference to Europe and the Mediterranean region," *Acta Neurologica Scandinavica* 62(1980): 65–80.

[34] Kurtzke, J. "Rating neurologic imapairment in multiple sclerosis: an expanded disability status scale (EDSS)," *Neurology* 33(1983): 1444–1452.

[35] Lauffer, R., *et al.* "MS-325: Albumin-targeted contrast agent for MR angiography," *Radiology* 207(1998): 529–538.

[36] Lei, T., Udupa, J., Saha, P., and D. Odhner. "Artery-vein separation via MRA—an image processing approach," *IEEE Transactions on Medical Imaging*, 20(2001): 689–703.

[37] Lindeberg, T. *Scale-Space Theory in Computer Vision*. Boston, MA: Kluwer, 1994.

[38] Liu, J., Udupa, J., Hackney, D., and G. Moonis. "Estimation of tumor volume with fuzzy connectedness segmentation of MRI," in *Proceedings of SPIE: Medical Imaging* 4322(2001): 1455–1465.

[39] Liu, J., Udupa, J., Dewey, D., McDonough, J., and R. Arens. "A system for upper airway segmentation and measurement via MR imaging and fuzzy connectedness," *Academic Radiology* 10(2003): 13–24.

[40] Lumia, R., Shapiro, L., and O. Zuniga. "A new connected component algorithm for virtual memory computers," *Computer, Vision, Graphics, and Image Processing*, 22(1983): 287–300.

[41] Lumia, R. "A new three-dimensional connected components algorithm," *Computer, Vision, Graphics, and Image Processing*, 23(1983): 207–217.

[42] Manohar, M. and H. K. Ramaprian. "Connected component labelling of binary images on a mesh connected massively parallel processor," *Computer, Vision, Graphics, and Image Processing* 45(1989): 143–149.

[43] Miki, Y., *et al.* "Differences between relapsing remitting and chronic progressive multiple sclerosis as determined with quantitative MR imaging," *Radiology* 210(1999): 769–774.

[44] Miki, Y., *et al.* "Relapsing-remitting multiple sclerosis: longitudinal analysis of MR images–lack of correlation between changes in T2 lesion volume and clinical findings," *Radiology*, 213(1999): 395–399.

[45] Moonis, G., Liu, J., Udupa, J., and D. Hackney. "Estimation of tumor volume using fuzzy connectedness segmentation of MR images," *American Journal of Neuroradiology* 23(2002): 356–363.

[46] Nyúl, L. and J. Udupa. "Fuzzy-connected 3D image segmentation at interactive speeds," *Graphical Models*, in press.

[47] Nyúl, L. and J. Udupa. "Protocol-independent brain MRI segmentation method," in *Proceedings of SPIE: Medical Imaging* 4684(2002): 1588–1599.

[48] Pal, N. and S. Pal. "A review of image segmentation techniques," *Pattern Recognition* 26(1993): 1277–1294.

[49] Pham, D. "Spatial models for fuzzy clustering," *Computer Vision and Image Understanding* 84(2001): 285–297.

[50] Pizer, S., Eberly, D., Fritsch, D., and B. Morse. "Zoom-Invariant vision of figural shape: the mathematics of cores," *Computer Vision and Image Understanding* 69(1)(1998): 55–71.

[51] Rice, B. and J. Udupa. "Clutter-free volume rendering for magnetic resonance angiography using fuzzy connectedness," *International Journal of Imaging Systems and Technology* 11(2000): 62–70.

[52] Ronse, C. and P. A. Devijver. *Connected Components in Binary Images: The Detection Problem*. Research Studies Press, Ltchworth, England, 1984.

[53] Rosenfeld, A. and J. Pfatz. "Sequential operations in digital picture processing," *Journal of the Association of Computing Machinery* 14(1966): 471–494.

[54] Rosenfeld, A. and J. Pfatz. "Distance functions in digital pictures," *Pattern Recognition* 1(1968): 33–61.

[55] Rosenfeld, A. "Connectivity in digital pictures", *Journal of Association for Computing Machinery* 17(1970): 146–160.

[56] Rosenfeld, A. "Fuzzy digital topology," *Information and Control* 40(1)(1979): 76–87.

[57] Rosenfeld, A. "Three dimensional digital topology," *Information and Control* 50(1981): 119–127.

[58] Rosenfeld, A. "The fuzzy geometry of image subsets," *Pattern Recognition Letters* 2(1991): 311–317.

[59] Saha, P., Udupa, J., and J. Abrahams. "Automatic bone-free rendering of cerebral aneurysms via 3D-CTA," in *Proceedings of SPIE: Medical Imaging* 4322(2001): 1264–1272.

[60] Saha, P., Udupa, J., Conant, E., Chakraborty, D., and D. Sullivan. "Breast tissue density quantification via digitized mammograms," *IEEE Transactions on Medical Imaging* 20(2001): 792–803.

[61] Saha, P. and J. Udupa. "Relative fuzzy connectedness among multiple objects: theory, algorithms, and applications in image segmentation," *Computer Vision and Image Understanding* 82(2001): 42–56.

[62] Saha, P., Udupa, J., and D. Odhner. "Scale-based fuzzy connected image segmentation: theory, algorithms, and validation," *Computer Vision and Image Understanding* 77(2000): 145–174.

[63] Saha, P. and J. Udupa. "Fuzzy connected object delineation: axiomatic path strength definition and the case of multiple seeds," *Computer Vision and Image Understanding* 83(2001): 275–295.

[64] Saha, P. and J. Udupa. "Iterative relative fuzzy connectedness and object definition: theory, algorithms, and applications in image segmentation," in *Proceedings of IEEE Workshop on Mathematical Methods in Biomedical Image Analysis*, (Hilton Head, South Carolina, 2000): 28–35.

[65] Samarasekera, S., Udupa, J., Miki, Y., and R. Grossman. "A new computer-assisted method for enhancing lesion quantification in multiple sclerosis," *Journal of Computer Assisted Tomography* 21(1997): 145–151.

[66] Schwab, R., Gupta, K., Gefter, W., Hoffman, E., and A. Pack. "Upper airway soft tissue anatomy in normals and patients with sleep disordered breathing: significance of the lateral pharyngeal walls," *American Journal of Respiratory Critical Care in Medicine* 152(1995): 1673–1689.

[67] Shapiro, L. "Connected component labelling and adjacency graph construction," in *Topological Algorithms for Digital Image Processing*, T. Y. Kong and A. Rosenfeld, (eds.), (Elsevier Science B. V., Amsterdam, The Netherlands, 1996).

[68] Tabb, M. and N. Ahuja. "Multiscale image segmentation by integrated edge and region detection," *IEEE Trans. Image Processing* 6(5)(1997): 642–655.

[69] Udupa, J., Srihari, S. and G. Herman. "Boundary detection in multidimensions," *IEEE Transactions on Pattern Analysis and Machine Intelligence* 4(1982): 41–50.

[70] Udupa, J. and V. Ajjanagadde. "Boundary and object labelling in three-dimensional images," *Computer Vision, Graphics and Image Processing* 51(1994): 355–369.

[71] Udupa, J. and D. Odhner. "Shell Rendering," *IEEE Computer Graphics and Applications* 13(6)(1993): 58–67.

[72] Udupa, J., Odhner, D., Samarasekera, S., Goncalves, R., Iyer, K., Venu-gopal, K., and S. Furuie. "3DVIEWNIX: a open, transportable, multidimen-sional, multimodality, multiparametric imaging system," in *Proceedings of SPIE: Medical Imaging* 2164(1994): 58–73.

[73] Udupa, J. "Multidimensional digital surfaces," *Graphical Models and Image Processing*, 56(1994): 311–323.

[74] Udupa, J. and S. Samarasekera. "Fuzzy connectedness and object defini-tion," in *Proceedings of SPIE: Medical Imaging* 2431(1995): 2–11.

[75] Udupa, J. and S. Samarasekera. "Fuzzy connectedness and object defini-tion: Theory, algorithms, and applications in image segmentation," *Graphi-cal Models and Image Processing* 58(1996): 246–261.

[76] Udupa, J., *et al.*. "Multiple sclerosis lesion quantification using fuzzy con-nectedness principles," *IEEE Transactions on Medical Imaging*, 16(1997): 598–609.

[77] Udupa, J., Saha, P., and R. A. Lotufo. "Fuzzy connected object definition in images with respect to co-objects", in *Proceedings of SPIE: Medical Imag-ing* 3661(1999): 236–245.

[78] Udupa, J., Nyúl, L., Ge, Y., and R. Grossman. "Multiprotocol MR image segmentation in multiple sclerosis: experience with over 1000 studies," *Aca-demic Radiology* 8(2001): 1116–1126.

[79] Udupa, J., Odhner, D., and H. Eisenberg. "New automatic mode of vi-sualizing the colon via CT," in *Proceedings of SPIE: Medical Imaging* 4319(2001): 237–243.

[80] Udupa, J., Saha, P., and R. Lotufo. "Relative fuzzy connectedness and ob-ject definition: theory, algorithms, and applications in image segmentation," *IEEE Transactions on Pattern Analysis and Machine Analysis* 24(2002): 1485–1500.

[81] Udupa, J., *et al.* "Methodology for evaluating image segmentation algo-rithms," in *Proceedings of SPIE: Medical Imaging* 4684(2002): 266–277.

[82] Vining, D. "Virtual colonoscopy," *Gastrointest Endosc Clinics of North America* 7(1997): 285–291.

[83] Warner, E. *et al.* "The risk of breast-cancer associated with mammographic parenchymal patterns: a meta-analysis of the published literature to examine the effect of method of classification," *Cancer Detection and Prevention* 16(1992): 67–72.

[84] Wolfe, J. "Breast pattern as an index of risk for developing breast cancer," *AJR* 126(1976): 1130–1139.

[85] Wolfe, J. "Breast parenchymal patterns and their changes with age," *Radiology* 121(1976): 545–552.

Markov Random Field Models

Ting Chen
Rutgers University

Dimitris N. Metaxas
Rutgers University

7.1 Markov Random Field Models: Introduction and Previous Work

The region-based segmentation methods such as those in [1, 10], make use of the homogeneity of inner regions in images. The advantage of these methods is that the image information inside the object is considered. However, they often fail to capture the global boundary and shape properties of the object because they do not consider the boundary information explicitly. This may lead to noisy boundaries and holes inside the object.

Markov random field (MRF) models have been used in segmentation since the early 80s ([13, 5, 18, 9, 12]). MRF models are probabilistic models that use the correlation between pixels in a neighborhood to decide the object region. Compared to typical region growing methods in [1], MRF models include more information in the pixel's neighborhood. Pixels in the image are no longer simply assigned to objects or the background only according to their own values. Instead, we calculate the probability of a pixel belonging to the object using its pixel value and values of its neighboring pixels.

One big problem for those region-based methods using MRF models is that the process of achieving the *maximum a posteriori* (MAP) estimator of the object region in the image volume can be computationally expensive. One solution to the problem is to use the Gibbs distribution. Geman and Geman proposed an

181

image restoration and segmentation application in the Bayesian framework using the Gibbs distribution [10]. Some other groups later converted their work into segmentation applications ([10, 16, 8]). The main framework of Geman and Geman's approach is: For an image $X = (F,L)$, where F is the matrix of the observable pixel intensities and L denotes the (dual) matrix of the unobservable edge elements, a relaxation algorithm is developed to maximize the conditional probability $P(L|G)$ given the observation data $G = g$. This form of Bayesian estimation is known as the MAP estimation. Geman and Geman used the MRF property of the image and a Gibbs sampler to avoid the formidable computational burden of MAP.

The main purpose of using Gibbs sampler here is to create an energy function that can be minimized. By using the Gibbs prior model, any image with an MRF property can have its conditional probability simplified to be computationally efficient. This makes it possible for us to get a MAP estimation of the object region. However, in [10] and other later segmentation applications using the Gibbs prior model, no boundary information has been used to improve the segmentation quality.

In [2, 3], Chan used a Gibbs prior model which integrates the object boundary information to reconstruct medical images from 2D projections. Chan considered the image reconstruction from projection as estimating the original image from a distorted observation. In [3], a Gibbs distribution served as the prior distribution. A Bayesian framework was then used in the MAP estimation procedure. The Gibbs distribution contains prior region and boundary information so that the reconstruction of the object based on this distribution is of high quality. The advantage of this approach is to use a Gibbs distribution that integrates the boundary information instead of the Gibbs prior model that only considers pairwise pixel similarities. By using the new Gibbs distribution, it is possible to use more information from the image. We are not only capable of modelling the homogeneity of neighboring pixels, but we are also able to capture the object edge features in the region of interest.

In our approach in [4], we construct a new energy function for the Gibbs prior model. The new Gibbs prior model still integrates the boundary information and it is more suitable for segmentation.

7.2   Gibbs Prior Model Theories

Most medical images are Markov random field images, that is, the statistics of a pixel in a medical image are related to the statistics of pixels in its neighborhood. According to the equivalence theorem proved by Hammersley and Clifford [11], a Markov random field is equivalent to a Gibbs field under certain restrictions. Therefore the joint distribution of a medical image with MRF property can be written in the Gibbsian form as follows:

$$\Pi(X) = Z^{-1} \exp(-H(X)), \tag{7.1}$$

where $Z = \sum_{z \in \mathbf{X}} \exp(-H(z))$, $H(X)$ models the energy function of image X, $\mathbf{X}$ is the set of all possible configurations of the image X, Z is a normalizing factor, and z is an image in the set of $\mathbf{X}$. The local and global properties of MRF images will be incorporated into the model by designing an appropriate energy function $H(X)$, and minimize it. The lower the value of the energy function, the higher the value of the Gibbs distribution and the image fits better to the prior distribution. Therefore the segmentation procedure corresponds to the minimization of the energy function. Our Gibbs prior model has the following energy function:

$$H_{prior}(X) = H_1(X) + H_2(X), \tag{7.2}$$

where $H_1(X)$ models the piecewise pixel homogeneity statistics and $H_2(X)$ models the continuity of the boundary. $H_1(X)$ takes the form of

$$H_1(X) = \vartheta_1 \sum_{s \in X} \sum_{t \in \partial s} (\Phi(\Delta_{s,t})(1 - \Psi_{st}) + \alpha^2 \Psi_{st}), \tag{7.3}$$

where

$$\Psi_{st} = \{ \begin{matrix} 1 & \Delta_{s,t} > \alpha^2, \\ 0 & \Delta_{s,t} < \alpha^2, \end{matrix} \tag{7.4}$$

where s and t are neighboring pixels, ϑ_1 is the weight for the homogeneity term $H_1(X)$, and α is the threshold for the object boundary. α is very important in the energy function definition since it explicitly defines the edge in the image. $\Delta_{s,t}$ is the variance between pixels s and t. Φ is a function based on the variance. For simplicity, in our Gibbs model we define $\Phi(\Delta_{s,t}) = \Delta_{s,t}^2$.

In general, the homogeneity term $H_1(X)$ has a smoothing effect on pixels inside the object and will leave boundary features beyond the threshold unchanged. In image regions without boundary features, Ψ_{st} will be assigned to 0. We can rewrite Equation (7.3) as

$$H_1(X) = \vartheta_1 \sum_{s \in X} \sum_{t \in \partial s} \Delta^2. \tag{7.5}$$

When we minimize $H_1(X)$, the image region will be further smoothed. However, for pixels that have different gray level intensities (the variance between two pixels beyond the threshold α) from their neighbors, Ψ_{st} will be assigned to 1. We can rewrite Equation (7.3) as

$$H_1(X) = \vartheta_1 \sum_{s \in X} \sum_{t \in \partial s} \alpha^2. \tag{7.6}$$

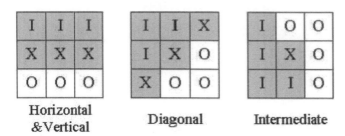

Horizontal
&Vertical   Diagonal   Intermediate

Figure 7.1. Pixels labelled "I" are inside the object; pixels labelled "O" are outside the object; pixels labelled "X" all have gray values similar to either "I" or "O". The shading area is the object region when assuming "X" is similar to "I".

$H_1(X)$ has a constant value in the neighborhood of s so the minimization of $H_1(X)$ does not have any smoothing effect in this region of the image. The variance between these two pixels will remain during the minimization procedure.

The continuity term of the energy function $H_2(X)$ models the continuity of the object boundaries. $H_2(X)$ has the following form:

$$H_2(X) = \vartheta_2 \sum_{x \in X} \sum_{i=1}^{N} W_i(x), \qquad (7.7)$$

where x is a pixel, ϑ_2 is the weight term for the boundary continuity, N is the number of local configurations, and $W_i(x)$ are weight functions (also called the potential functions) of local configurations. The weights of local characteristics are defined based on high-order cliques. In our model, the potential functions are defined on a neighborhood system based on 3×3 cliques. These high-order cliques are capable of modelling the boundary continuity. We depict clique definitions in Figure 7.1.

When we assign "1" to pixels belonging to the object and "0" to pixels in the background, for a 3×3 clique, there are altogether 2^9 possible "1,0" configurations. We group these configurations into six types of cliques. All these local configurations are symmetries with respect to the x and y axes, so vertical and horizontal configurations are assumed to be the same. The other three types of cliques not shown in Figure 7.1 are: cliques consisting only of pixels inside the object (type IV), cliques consisting only of pixels in the background (type V), and cliques that do not belong to any of those five types (type VI).

We assign different values to potentials of different clique configurations. Cliques that belong to the same clique type have similar properties and share the same potential value. We assign lower values to clique configurations that locate at smooth and continuous object boundaries than those assigned to other configurations. Therefore, when we minimize $H_2(X)$, the pixels in the image (especially those near the boundary) will alter their gray values to form clique configurations that are assigned a low value. The results of these alternations are: 1) the boundary will become smoother, and 2) the boundary will extend into image regions without obvious gradient steps, which will lead to a continuous object boundary.

7.3   Bayesian Framework and Posterior Energy Function

We use the Bayesian framework to get a MAP estimation of the object region. According to Bayes's theorem, the posterior distribution of an image is given by

$$P(X|Y) = \frac{\Pi(X)L(Y|X)}{P(Y)}, \qquad (7.8)$$

where $\Pi(X)$ is the prior distribution and $L(Y|X)$ is the transition probability between observation Y and image X. In a Bayesian framework, the segmentation problem can be formulated as the maximization of the posterior probability $P(X|Y)$, which is proportional to

$$\Pi(X)L(Y|X). \qquad (7.9)$$

So the segmentation work is equivalent to finding the correct image X^* such that

$$X^* = \arg\max_X [\Pi(X)L(Y|X)]. \qquad (7.10)$$

We use Y to denote the observation and X to denote the true image. Assuming there is Gaussian noise in the image, then $L(Y|X)$ has the form

$$L(Y|X) = \frac{1}{\sqrt{2\pi\sigma^2}} \exp(-H_{noise}(X,Y)), \qquad (7.11)$$

where

$$H_{noise}(X,Y) = (2\sigma^2)^{-1} \sum_{s\in X} (y_s - x_s)^2, \qquad (7.12)$$

where y_s is the observed datum of pixel s, x_s is the estimated value, and σ is the standard deviation.

We can also define the term $H_{noise}(X,Y)$ as the constraint from the observation [17] and rename it $H_{observation}(X,Y)$.

$$H_{observation}(X,Y) = \vartheta_3 \sum_{s\in X} (y_s - x_s)^2, \qquad (7.13)$$

where ϑ_3 is the weight for the constraint of observation. Using Equations (7.1), (7.9) and (7.11), the posterior probability is

$$P(X|Y) \propto \exp(-H_{posterior}(X,Y)), \qquad (7.14)$$

where

$$H_{posterior}(X,Y) = H_{prior}(X) + H_{observation}(X,Y). \qquad (7.15)$$

Using $H_{posterior}(X,Y)$ instead of $H(X)$ in the energy minimization will lead to a MAP estimation of the object region. The constraint of the observation will compete with the user-defined prior distribution during the minimization process. Therefore the result of the minimization process will still be close enough to the original image observation. Some important image features, such as some irregular edges, will be kept regardless of the prior distribution.

7.4   Energy Minimization

We use the MAP estimator to get an image that fulfills the prior distribution while still being constrained by the observation. The computation of a MAP estimation for a Gibbs field amounts to the minimization of the energy function.

In the minimization process, we use the MRF properties of the image to form a computationally feasible Markov chain. We use strictly positive Markov kernels P in order to make $\Pi(X)L(Y|X)$ invariant during the minimization process. Using the Gibbs sampler developed by Geman and Geman in [10], which is based on the local characteristics of $\Pi(X)L(Y|X)$, for every $I \subset S$, a Markov kernel on $\mathbf{X}$ is defined by

$$P_I(X,Y) = \begin{cases} Z_I^{-1}\exp(-H_{prior}(X_{S\backslash I}Y_I)) & if Y_{S\backslash I} = X_{S\backslash I}, \\ 0 & otherwise, \end{cases} \qquad (7.16)$$

where $Z_I = \sum_{z_I}\exp(-H_{prior}(z_I X_{S\backslash I}))$. These Markov kernels can also be termed as the local characteristics of $\Pi(X)L(Y|X)$. Sampling from $P_I(X,\cdot)$ only changes pixels within I. Therefore these local characteristics can be evaluated in a reasonable time since they depend on a relatively small number of neighbors.

We used two kinds of cooling schedules: the simulated annealing and the ICM method. Simulated annealing can lead to a global solution. However, it is computationally expensive and not time efficient. The ICM can lead to a local minimum that is the closest to the observation. It is fast and easy to code. In a hybrid framework, we can use a deformable model to get the solution out of the local minima later.

7.5   Experiments and Results

The Gibbs prior model has been applied to segment objects in synthetic data and medical images. In the following experiments, a 3×3 clique-based potential system has been used to create the energy function of Gibbs prior models. User-defined start regions have been used to get the initial guess for the probable statistical property of the object. The initial values for the parameters of the Gibbs prior models are estimated based on these statistics.

7.5.1 Phantom Test

In the first experiment, we show the Gibbs prior model's segmentation results on a synthetic phantom image. The image size is 20×20 pixels. The gray values for pixels in the left upper quarter of the image are evenly distributed in the range $[270, 320)$, while all the remaining pixels are evenly distributed in the range $[300, 349)$.

In Figure 7.2(a) we show the test image. We first try the traditional region-based method by setting the weight for the second term of the energy function ϑ_2 to 0. The final result is equal to a piecewise smoothness filter with the boundary threshold set to 15. The result is shown in Figure 7.2(b). The effect of the boundary smoothness factor is shown in Figure 7.2(c). We set ϑ_2 to 1.0, while ϑ_1, the weight for the first term of the energy function is set to 0.2. We set potentials of local characteristics that represent smooth boundaries and homogeneous regions to -5.0, all other local characteristics' potentials are set to 0.0. Finally, for the result shown in Figure 7.2(d), we set both ϑ_1 and ϑ_2 to 1 to show the effectiveness of the Gibbs prior model. In the experiments shown in Figures 7.2(c) and 7.2(d), we use simulated annealing cooling pattern with $10,000$ steps. For experiments shown in Figures 7.2(b), 7.2(c), and 7.2(d), we calculate the initial statistics in a square region whose upper left corner is at $(3, 3)$, lower right corner at $(7, 7)$.

The results of these experiments clearly show that the Gibbs prior model has a better performance than region-based methods which do not integrate the boundary information. The results also show that by changing the weight of the region homogeneity and boundary smoothness, the performance of the Gibbs prior model varies. In most cases, we set both ϑ_1 and ϑ_2 to 1. The setting which has the best performance may be different for other data sets.

The Gibbs prior model is only applied onto the testing image once all parameters are assigned to the default value or estimated by the statistics in the start region. Although the final result is close to known truth, there are still small errors. The Gibbs prior model itself cannot correct these errors if there is no further information included. This is the reason for using the GP-DM hybrid framework in segmentation tasks.

7.5.2 Clinical Data Test

We have also tried our Gibbs prior model segmentation method on clinical images. We here show the segmentation results of applying the Gibbs prior model onto

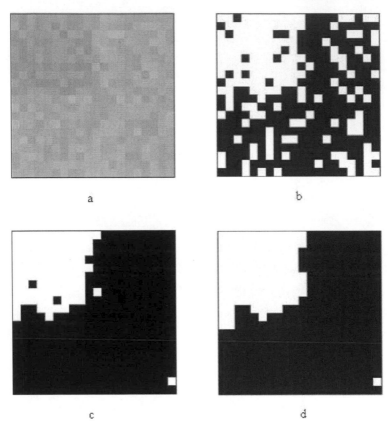

Figure 7.2. The segmentation behavior of the Gibbs prior model: (a) The synthetic testing image; (b) segmentation result when $\vartheta_1 = 1$ and $\vartheta_2 = 0$; (c) segmentation result when $\vartheta_1 = 0.1$ and $\vartheta_2 = 1$; (d) segmentation result when $\vartheta_1 = 1$ and $\vartheta_2 = 1$.

simulated brain data (SBD) provided by A. C. Evans' research group at McGill University ([6, 15, 14, 7]).

The SBD contains simulated brain MRI data based on two anatomical models: normal and multiple sclerosis (MS). For both data sets, full 3D datasets have been simulated using three sequences (T1-, T2-, and proton-density- (PD-) weighted) and a variety of slice thicknesses, noise levels, and levels of intensity non-uniformity. These data are available for viewing in three orthogonal views (transversal, sagittal, and coronal).

There are several separated but related tasks in the field of brain segmentation: the segmentation of cerebral cortex; the segmentation of white matter; the segmentation of brain with/without cerebrospinal fluid (CSF); and segmentation for tumor in brain.

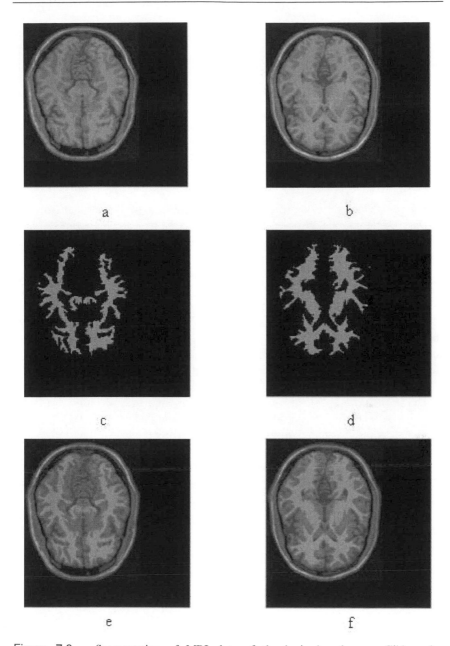

Figure 7.3.   Segmentation of MRI data of the brain based on a Gibbs prior model: (a) and (b) original MRI image from SBD data showing two different slices; (c) and (d) Results of white matter segmentation; (e) and (f) projection of segmentation results to the original images.

Geometrically, the brain is composed of three layers. The white matter is the deepest layer. Outside the white matter, there is a thin, folded sheet of gray matter, which is also called the cerebral cortex. There is also a layer of cerebrospinal fluid between the brain and the skull.

Notice that the surface of cerebral cortex features numerous cortical folding and the white matter has complex surface structures. To model such complicated boundary features, it is necessary for us to use methods that consider the boundary conditions explicitly, such as the Gibbs prior model we used in our segmentation methodology. In Figure 7.3 we show the segmentation result of applying Gibbs prior model to the segmentation of white matter (WM) of brain in two slices in the SBD data set.

The segmentation begins with a user defined seed pixel inside of the object (the WM). The initial statistics is calculated in its 10 by 10 neighborhood. Another 2 seeds have been picked to calculate the statistics of the background. The segmentation process uses a compete strategy to decide the object region and the background. The threshold of the boundary is set to be the variance between the mean value of inside and outside seed pixel neighbors. Both ϑ_1 and ϑ_2 are set to 1. The weights of local characteristics at boundaries are set to -5.0.

In Figure 7.3, 7.3(a) and 7.3(b) are 2 slices from the SBD data set. The data is a normal brain with 3% blurring noise. The image size is 181×181 pixels. 7.3(c) and 7.3(d) show the Gibbs prior segmentation results of those two slices. 7.3(e) and 7.3(f) show the projection of the segmentation result back onto the original image.

The Gibbs prior segmentation process takes about 7 to 8 seconds on a Pentium III 1GHz laptop with 256M memory. The operating system is WinXP. The segmentation result captures most details of the white matter with an error rate under 6%.

We also try the Gibbs prior segmentation method onto other clinical data, such as the lung, the heart and the eyeball. The results are of consistent quality and can be achieved in a very short time.

References

[1] Ballard, D. and C. Brown. *Computer Vision*. Englewood Cliffs, NJ, Prentice Hall, 1982.

[2] Chan, M., Herman, G., and E. Levitan. "A Bayesian approach to PET reconstruction using image-modeling Gibbs prior," *IEEE Transaction on Nuclear Science* 44(3)(June 1997).

[3] Chan, M., Herman, G., and E. Levitan. *Bayesian Image Reconstruction Using Image-Modeling Gibbs Prior*. New York: John Wiley & Sons, Inc, 1998.

[4] Chen, T. and D. Metaxas. "Image segmentation based on the integration of Markov random fields and deformable models," in *Medical Image Computing and Computer-Assisted Intervention* (Proceedings of MICCAI 2000), S. Delp, A. DiGioia, and B. Jaramaz, eds. Springer, LNCS1935(October 2000): 256–265.

[5] Choi, H., Haynor, D., and Y. Kim. "Partial volume tissue classification of multichannel magnetic resonance images - a mixed model," *IEEE Transactions on Medical Imaging* 10(3)(1991): 395–407.

[6] Cocosco C., Kollokian, V., Kwan, R., and A. Evans. "BrainWeb: online interface to a 3D MRI simulated brain database," *NeuroImage*, (Proceedingss of 3rd International Conference on Functional Mapping of the Human Brain, Copenhagen, May 1997). 5(4)(1997): part 2/4.

[7] Collins, D., *et al.* "Design and construction of a realistic digital brain phantom," *IEEE Transactions on Medical Imaging* 17(3)(June 1998): 463–468.

[8] Derin, H. and W. Cole. "Segmentation of textured images using Gibbs random fields," *Computer Vision, Graphics, and Image Processing* 35(1986): 72–98.

[9] Fwu, J. and P. Djuric. "Unsupervised vector image segmentation by a tree structure–ICM algorithm," *IEEE Transactions on Medical Imaging* 15(6)(1996): 871–880.

[10] Geman, S. and D. Geman. "Stochastic relaxation, Gibbs distributions, and the Bayesian restoration of images," *IEEE Transactions Pattern Analysis and Machine Intelligence* 6(1984):721–741.

[11] Hammersley J. and P. Clifford. *Markov fields on finite graphs and lattices.* (Preprint University of California, Berkeley, 2004).

[12] Held, K., *et al.* "Markov random filed segmentation of brain MR images," *IEEE Transactions on Medical Imaging* 16(6)(1997): 878–886.

[13] Hurn, M., Mardia, K., Hainsworth, T., and E. Berry. "Bayesian fused classifcation of medical images," *IEEE Transactions on Medical Imaging* 15(6)(1996): 850–858.

[14] Kwan, R., Evans, A., and G. Pike. "An extensible MRI simulator for postprocessing evaluation," in *Visualization in Biomedical Computing* (Proceedings of VBC'96, Hamburg, Germany 1996), K.H. Hohne, ed. Springer LNCS1131(1996): 135–140.

[15] Kwan, R., Evans, A., and G. Pike. "MRI simulation-based evaluation of image processing and classification methods," *IEEE Transactions on Medical Imaging* 18(11)(Nov 1999): 1085–97.

[16] Lakshmanan, S. and H. Derin. "Simultaneous parameter estimation and segmentation of Gibbs random fields using simulated annealing," *IEEE Transactions on Pattern Analysis and Machine Intelligence* 11(8)(1989): 799–813.

[17] Li, S. *Markov Random Field Modeling in Computer Vision.* Berlin: Springer, 1995.

[18] Liang, Z., MacFall, J., and D. Harrington. "Parameter estimation and tissue segmentation from multispectral MR images," *IEEE Transactions on Medical Imaging* 13(3)(1994): 441–449.

Isosurfaces and Level Sets

Ross Whitaker
University of Utah

8.1   Introduction

8.1.1   Motivation

This chapter describes mechanisms for processing *isosurfaces*. The underlying philosophy is to use isosurfaces as a modeling technology that can serve as an alternative to parameterized models for a variety of important applications in visualization and computer graphics. Level-set methods [17] rely on partial differential equations (PDEs) to model deforming isosurfaces. These methods have applications in a wide range of fields such as visualization [28, 3], scientific computing [6], computer graphics [16, 21], image processing [24, 14] and computer vision [30]. Several nice overviews of the technology and applications are also available [20, 6].

This chapter reviews the mathematics and numerical techniques for describing the geometry of isosurfaces and manipulating their shapes in prescribed ways. Its starts with a basic introduction to the notation and fundamental concepts and then presents the geometry of isosurfaces. We then describe the method of level sets, i.e., moving isosurfaces, and present the mathematical and numerical methods they entail.

8.1.2   Representing Surfaces with Volumes

The *level set* method relies on an implicit representation of a surface, using a scalar function

$$\phi : \underset{x,y,z}{U} \mapsto \mathbb{R}, \tag{8.1}$$

where $U \subset \mathbb{R}^3$ is the domain of the volume (and the *range* of the surface model). Thus, a surface S is

$$S = \{\vec{x} | \phi(\vec{x}) = k\}. \tag{8.2}$$

The choice of k is arbitrary, and ϕ is sometimes called the *embedding*. The surface S is sometimes referred to as an isosurface of ϕ. Notice that surfaces defined in this way divide U into a clear inside and outside; such surfaces are always closed wherever they do not intersect the boundary of the domain. The embedding, ϕ, is represented on a discrete grid, i.e., a volume.

The normal of an isosurface is given by the normalized gradient vector. Typically, we identify a surface normal with a point in the volume domain $\mathcal{D}$. That is,

$$\vec{n}(\vec{x}) = \frac{\nabla \phi(\vec{x})}{|\nabla \phi(\vec{x})|} \quad \text{where } \tilde{x} \in D. \tag{8.3}$$

The convention regarding the direction of this vector is arbitrary; the negative of the normalized gradient magnitude is also normal to the isosurface. The gradient vector points toward that side of the isosurface which has greater values (i.e., brighter). When rendering, the convention is to use *outward pointing* normals, and the sign of the gradient must be adjusted accordingly. However, for most applications any consistent choice of normal vector will suffice. On a discrete grid, one must also decide how to approximate the gradient vector (i.e., first partial derivatives). In many cases central differences will suffice. However, in the presence of noise, especially when volume rendering, it is sometimes helpful to compute first derivatives using some smoothing filter (e.g., convolution with a Gaussian [12, 22]). When using the normal vector to solve certain kinds of partial differential equations, it is sometimes necessary to approximate the gradient vector with discrete, one-sided differences, as discussed in successive sections.

Note that a single volume contains families of nested isosurfaces, arranged like the layers of an onion, as in Figure 8.1. We specify the normal to an isosurface as a function of the position within the volume. That is, $\vec{n}(\vec{x})$ is the normal of the (single) isosurface that passes through the point $\vec{x}$. The k value associated with that isosurface is $\phi(\vec{x})$.

The curvature of the isosurface can be computed from the first-order and second-order structure of the embedding, ϕ. All of the isosurface shape information is contained in a field of normals given by $\vec{n}(\vec{x})$. The 3×3 matrix of derivatives of this vector,

$$N = -[\vec{n}_x \ \vec{n}_y \ \vec{n}_z.] \tag{8.4}$$

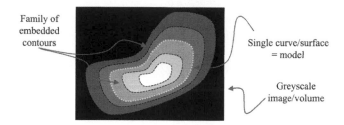

Family of embedded contours

Single curve/surface = model

Greyscale image/volume

Figure 8.1. A scalar field (shown here in 2D) contains a nested family of isosurfaces, each with a difference greyscale value.

The projection of this derivative onto the tangent plane of the isosurface gives the shape matrix, β. Let P denote the normal projection operator, which is defined as

$$P = \vec{n} \otimes \vec{n} = \frac{1}{||\nabla\phi||^2} \begin{pmatrix} \phi_x^2 & \phi_x\phi_y & \phi_x\phi_z \\ \phi_y\phi_x & \phi_y^2 & \phi_y\phi_z \\ \phi_z\phi_x & \phi_z\phi_y & \phi_z^2 \end{pmatrix}. \tag{8.5}$$

The tangential projection operator is $I - P$, and thus the shape matrix is

$$\beta = NT = TD^2\phi T, \tag{8.6}$$

where $D^2\phi$ is the Hessian of ϕ. The shape matrix β has three real eigenvalues which are

$$e_1 = k_1, e_2 = k_2, e_3 = 0. \tag{8.7}$$

The corresonding eigenvectors are the principal directions (in the tangent plane) and the normal, respectively.

The *mean curvature* is the mean of the two principal curvatures, which is one half of the trace of β, which is equal to the trace of N. The total curvature, also called the deviation from flatness [11], D, is the root sum of squares of the two principal curvatures, which is the Euclidean norm of the matrix β.

Notice, these measures exist at every point in U, and at each point they describe the geometry of the particular isosurface that passes through that point. All of these quantities can be computed on a discrete volume using finite differences, as described in successive sections.

8.2 Deformable Surfaces

This section begins with mathematics for describing surface deformations on parametric models. The result is an evolution equation for a surface. Each of the terms in this evolution equation can be re-expressed in a way that is independent of the parameterization. Finally, the evolution equation for a parametric surface gives rise to an evolution equation (differential equation) on a volume, which encodes the shape of that surface as a level set.

8.2.1   Surface Deformation

A regular surface $S \subset \mathbb{R}^3$ is a collection of points in 3D that can be be represented *locally* as a continuous function. In geometric modeling a surface is typically represented as a two-parameter object in a three-dimensional space, i.e., a surface is a local mapping $\vec{S}$:

$$\vec{S} : \underset{r}{V} \times \underset{s}{V} \mapsto \underset{x,y,z}{\mathbb{R}^3}, \tag{8.8}$$

where $V \times V \subset \mathbb{R}^2$. A deformable surface exhibits some motion over time. Thus $\vec{S} = \vec{S}(r,s,t)$, where $t \in \mathbb{R}^+$. We assume second-order continuous, orientable surfaces; therefore at every point on the surface (and in time) there is a surface normal $\vec{N} = \vec{N}(r,s,t)$. We use S_t to refer to the entire set of points on the surface.

Local deformations of $\vec{S}$ can be described by an evolution equation, i.e., a differential equation on $\vec{S}$ that incorporates the position of the surface, local and global shape properties, and responses to other forcing functions. That is,

$$\frac{\partial \vec{S}}{\partial t} = \vec{G}\left(\vec{S}, \vec{S}_r, \vec{S}_s, \vec{S}_{rr}, \vec{S}_{rs}, \vec{S}_{ss}, \dots\right), \tag{8.9}$$

where the subscripts represent partial derivatives with respect to those parameters. The evolution of $\vec{S}$ can be described by a sum of terms that depends on both the geometry of $\vec{S}$ and the influence of other functions or data.

There are a variety of differential expressions that can be combined for different applications. For instance, the model could move in response to some directional "forcing" function [7, 23], $\vec{F} : U \mapsto \mathbb{R}^3$. Alternatively, the surface could expand and contract with a spatially-varying speed, $G : \mathbb{R}^3 \mapsto \mathbb{R}$. The surface motion could also depend on the surface shape, e.g., curvature. There are myriad terms that depend on both the differential geometry of the surface and outside forces or functions to control the evolution of a surface.

The method of level sets, proposed by Osher and Sethian [17] and described extensively in [20, 6], provides the mathematical and numerical mechanisms for computing surface deformations as time-varying iso-values of ϕ by solving a partial differential equation on the 3D grid. That is, the level-set formulation provides a set of numerical methods that describe how to manipulate the greyscale values in a volume, so that the isosurfaces of ϕ move in a prescribed manner (shown in Figure 8.2).

We denote the movement of a point $\vec{x}$ on the deforming surface as $\vec{v}$, and we assume that this motion can be expressed in terms of the position of $\vec{x} \in U$ and the geometry of the surface at that point. In this case, there are generally two options for representing such surface movements implicitly:

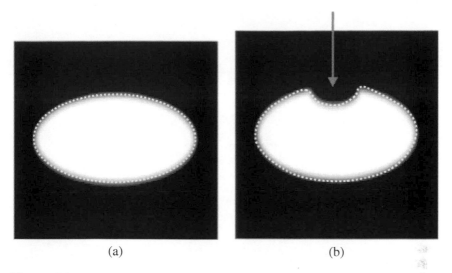

(a)   (b)

Figure 8.2. Level-set models represent curves and surfaces implicitly using greyscale images: (a) an ellipse is represented as the level set of an image; (b) to change the shape we modify the greyscale values of the image.

Static: A single, static $\phi(\vec{x})$ contains a family of level sets corresponding to surfaces at different times t (as in Figure 8.3). That is,

$$\phi(\vec{x}(t)) = k(t) \implies \nabla\phi(\vec{x}) \cdot \vec{v} = \frac{dk(t)}{dt}. \tag{8.10}$$

To solve this static method requires constructing a ϕ that satisfies Equation (8.10). This is a boundary value problem, which can be solved somewhat efficiently starting with a single surface using the fast marching method of Sethian [20]. This representation has some significant limitations, however, because (by definition) a surface cannot pass back over itself over time, i.e., motions must be strictly monotonic: inward or outward.

Dynamic: The approach is to use a one-parameter *family* of embeddings, i.e., $\phi(\vec{x},t)$ changes over time, $\vec{x}$ remains on the k level set of ϕ as it moves, and k remains constant. The behavior of ϕ is obtained by setting the total derivative of $\phi(\vec{x}(t),t) = k$ to zero. Thus,

$$\phi(\vec{x}(t),t) = k \implies \frac{\partial\phi}{\partial t} = -\nabla\phi \cdot \vec{v}. \tag{8.11}$$

This approach can accommodate models that move forward and backward and cross back over their own paths (over time). However, to solve this

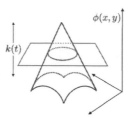

Figure 8.3. In the case of the static level-set approach, Equation (8.10), a single, static embedding, ϕ, captures the motion of a model in its various level values.

requires solving the initial value problem (using finite forward differences) on $\phi(\vec{x},t)$, a potentially large computational burden. The remainder of this discussion focuses on the dynamic case, because of its flexibility.

All surface movements depend on position and geometry, and the level-set geometry is expressed in terms of the differential structure of ϕ. Therefore the dynamic formulation from Equation (8.11) gives a general form of the partial differential equation on ϕ:

$$\frac{\partial \phi}{\partial t} = -\nabla \phi \cdot \vec{v} = -\nabla \phi \cdot \vec{F}(\vec{x}, D\phi, D^2\phi, \ldots), \qquad (8.12)$$

where $D^n\phi$ is the set of order-n derivatives of ϕ evaluated at $\vec{x}$. Because this relationship applies to every level set of ϕ, i.e., all values of k, this equation can be applied to all of ϕ, and therefore the movements of *all* the level-set surfaces embedded in ϕ can be calculated from Equation 8.12.

The level-set representation has a number of practical and theoretical advantages over conventional surface models, especially in the context of deformation and segmentation. First, level-set models are topologically flexible, they can easily represent complicated surface shapes that can, in turn, form holes, split to form multiple objects, or merge with other objects to form a single structure. This topological flexibility can be visualized (for curves) by imagining the model as the boundary of a lake whose shoreline has a fixed altitude (as in Figure 8.4). In order to change the shape of the shoreline, one must modify the height of the surrounding terrain (which is ϕ). As the topography of the terrain changes, new hills for or hills join or split, the topology of the shoreline can change. That is, the model can form new holes, pieces can split off, or separate pieces can join.

Level-set models can also incorporate many (millions) of degrees of freedom, and therefore they can accommodate complex shapes. Indeed, the shapes formed by the level sets of ϕ are restricted only by the resolution of the sampling. Thus, there is no need to reparameterize the model as it undergoes significant deformations.

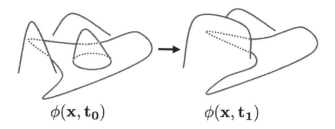

$$\phi(\mathbf{x}, \mathbf{t_0}) \qquad\qquad \phi(\mathbf{x}, \mathbf{t_1})$$

Figure 8.4. If the model is a level-set curve (represented as the shoreline of a lake) it can change as the embedding (terrain surrounding the lake) changes.

Such level-set methods are well documented in the literature [17, 19] for applications such as computational physics [18], image processing [2, 5], computer vision [9, 14], medical image analysis [24, 14], and 3D reconstruction [26, 27]. For instance, in computational physics level-set methods are a powerful tool for modeling moving interfaces between different materials (see Osher and Fedkiw [18] for a nice overview of recent results). Examples are water-air and water-oil boundaries. In such cases, level-set methods can be used to compute deformations that minimize surface area while preserving volumes for materials that split and merge in arbitrary ways. The method can be extended to multiple, non-overlapping objects.

Level-set methods have also been shown to be effective in extracting surface structures from biological and medical data. For instance Malladi *et al.* [14] propose a method in which the level-sets form an expanding or contracting contour which tends to "cling" to interesting features in 2D angiograms. At the same time the contour is also influenced by its own curvature, and therefore remains smooth. Whitaker *et al.* [24, 25] have shown that level sets can be used to simulate conventional deformable surface models, and demonstrated this by extracting skin and tumors from *thick-sliced* (e.g., clinical) MR data, and by reconstructing a fetal face from 3D ultrasound. A variety of authors [8, 31, 5, 13] have presented variations on the method and presented results for 2D and 3D data. Sethian [20] gives several examples of level-set curves and surfaces for segmenting CT and MR data.

8.3 Numerical Methods

By taking the strategy of embedding surface models in volumes, we have converted equations that describe the movement of surface points to nonlinear, partial differential equations defined on a volume, which is generally a rectilinear grid. The expression $\phi^n_{i,j,k}$ refers to the nth time step at position i, j, k, which has an associated value in the 3D domain of the continuous volume $\phi(x_i, y_j, z_k)$.

The discretization of these equations raises two important issues. First is the availability of accurate, stable numerical schemes for solving these equations. Second is the problem of computational complexity and the fact that we have converted a *surface* problem to a *volume* problem, increasing the dimensionality of the domain over which the evolution equations must be solved.

Various level-set terms can be combined, based on the needs of the application, to create a partial differential equation on $\phi(\vec{x},t)$. The solutions to these equations are computed using finite differences. Along the time axis solutions are obtained using finite *forward* differences, beginning with an initial model (i.e., volume) and stepping sequentially through a series of discrete times steps (which are denoted as superscripts on $\phi_{i,j,k}$). Thus the update equation is

$$\phi_{i,j,k}^{n+1} = \phi_{i,j,k}^{n} + \Delta t \Delta \phi_{i,j,k}^{n}, \tag{8.13}$$

The term $\Delta \phi_{i,j,k}^{n}$ is a discrete approximation to $\partial \phi / \partial t$, which is a sum of terms that depend on derivatives of ϕ. These terms must, in turn, be approximated using finite differences on the volume grid.

8.3.1   Up-Wind Schemes

The terms that drive the deformation of the surface fall into two basic categories: the first-order terms and the second-order terms. The first-order terms describe a moving wave front with a space-varying velocity or speed. Equations of this form cannot be solved with a simple finite forward difference scheme. Such schemes tend to overshoot, and they are unstable. To address this issue Osher and Sethian [17] have proposed an *up-wind* scheme. The up-wind method relies on a one-sided derivative that looks in the up-wind direction of the moving wave front, and thereby avoids the over-shooting associated with finite forward differences.

To describe the numerical implementation, we use the notation for finite difference operators introduced in Chapter 4. Second-order terms are computed using the *tightest-fitting* central difference operators. For example,

$$\delta_{xx}\phi_{i,j,k} \equiv \left(\phi_{i+1,j,k} + \phi_{i-1,j,k} - 2\phi_{i,j,k}\right)/h^2, \tag{8.14}$$

$$\delta_{zz}\phi_{i,j,k} \equiv \left(\phi_{i,j,k+1} + \phi_{i,j,k-1} - 2\phi_{i,j,k}\right)/h^2, \quad \text{and} \tag{8.15}$$

$$\delta_{xy}\phi_{i,j,k} \equiv \delta_x \delta_y \phi_{i,j,k}. \tag{8.16}$$

The discrete approximation to the first-order terms $\delta \phi$ are computed using the up-wind proposed by Osher and Sethian [17]. This strategy avoids overshooting by approximating the gradient of ϕ using one-sided differences in the direction that is up-wind of the moving level set, thereby ensuring that no *new* contours are created in the process of updating $\phi_{i,j,k}^{n}$ (as depicted in Figure 8.5). The scheme is separable along each axis (i.e., x, y, and z).

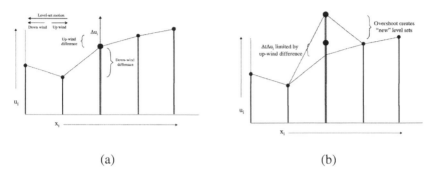

(a)   (b)

Figure 8.5. The up-wind numerical scheme uses one-sided derivatives to prevent over-shooting and the creation of new level sets: (a) a level set embedding with a finite difference Δu_i to be applied at point x_i; (b) using the up-wind limit prevents overshoot. If the down-wind difference is used, the resulting underdamped Δu_i can create a new level set.

Consider the vector-valued advection term $\vec{F}(\vec{x})$. If we use superscripts to denote the vector components, i.e.,

$$\vec{F}(x,y,z) = (F^{(x)}(x,y,z), F^{(y)}(x,y,z), F^{(z)}(x,y,z)), \tag{8.17}$$

the up-wind calculation for a grid point $\phi^n_{i,j,k}$ is

$$\vec{F}(x_i,y_i,z_i) \cdot \nabla\phi(x_i,y_j,z_k,t) \approx \sum_{q\in\{x,y,z\}} F^{(q)}(x_i,y_i,z_i) \begin{cases} \delta^+_q \phi^n_{i,j,k} & F^{(q)}(x_i,y_i,z_i) > 0, \\ \delta^i_q \phi^n_{i,j,k} & F^{(q)}(x_i,y_i,z_i) < 0. \end{cases} \tag{8.18}$$

The time steps are limited: the fastest moving wave front can move only one grid unit per iteration. That is,

$$\Delta t_{\vec{F}} \leq \frac{1}{\sum_{q\in\{x,y,z\}} \sup_{i,j,k\in X}\{|\nabla F^{(q)}(x_i,y_j,z_k)|\}}. \tag{8.19}$$

For speed terms of the form $G(\vec{x})\vec{n}(\vec{x})$ the direction of the moving surface depends on the normal, and therefore the same up-wind strategy is applied in a slightly different form:

$$G(x_i,y_j,z_k)|\nabla\phi(x_i,y_j,z_k,t)| \approx$$
$$\sum_{q\in\{x,y,z\}} G(x_i,y_i,z_i) \begin{cases} \max^2(\delta^+_q \phi^n_{i,j,k},0) & + \\ \min^2(\delta^-_q \phi^n_{i,j,k},0) & G(x_i,y_i,z_i) > 0, \\ \min^2(\delta^+_q \phi^n_{i,j,k},0) & + \\ \max^2(\delta^-_q \phi^n_{i,j,k},0) & G(q)(x_i,y_i,z_i) < 0. \end{cases} \tag{8.20}$$

The time steps are, again, limited by the fastest moving wave front:

$$\Delta t_G \leq \frac{1}{3\sup_{i,j,k \in X}\{|\nabla G(x_i,y_j,z_k)|\}}. \tag{8.21}$$

To compute the approximation update to the second-order terms, one typically uses central differences. Thus, the mean curvature is approximated as:

$$
\begin{aligned}
H^n_{i,j,k} \quad = \quad & \frac{1}{2}\left(\left(\delta_x\phi^n_{i,j,k}\right)^2 + \left(\delta_y\phi^n_{i,j,k}\right)^2 + \left(\delta_z\phi^n_{i,j,k}\right)^2\right)^{-1} \\
& \left[\left(\left(\delta_y\phi^n_{i,j,k}\right)^2 + \left(\delta_z\phi^n_{i,j,k}\right)^2\right)\delta_{xx}\phi^n_{i,j,k}\right. \\
& + \left(\left(\delta_z\phi^n_{i,j,k}\right)^2 + \left(\delta_x\phi^n_{i,j,k}\right)^2\right)\delta_{yy}\phi^n_{i,j,k} \\
& + \left(\left(\delta_x\phi^n_{i,j,k}\right)^2 + \left(\delta_y\phi^n_{i,j,k}\right)^2\right)\delta_{zz}\phi^n_{i,j,k} \\
& -2\delta_x\phi^n_{i,j,k}\delta_y\phi^n_{i,j,k}\delta_{xy}\phi^n_{i,j,k} - 2\delta_y\phi^n_{i,j,k}\delta_z\phi^n_{i,j,k}\delta_{yz}\phi^n_{i,j,k} \\
& \left. -2\delta_z\phi^n_{i,j,k}\delta_x\phi^n_{i,j,k}\delta_{zx}\phi^n_{i,j,k}\right].
\end{aligned}
\tag{8.22}
$$

Such curvature terms can also be computed by calculating the normals at staggered grid locations and then taking finite differences of the normals, as described in [29]. In some cases this is advantageous, but the details are beyond the scope of this chapter.

The time steps are limited, for stability, to

$$\Delta t_H \leq \frac{1}{6}. \tag{8.23}$$

When combining terms, the maximum time steps for each term is scaled by one over the weighting coefficient for that term.

8.3.2 Narrow-Band Methods

If one is interested in only *a single level set*, the formulation described previously is not efficient. This is because solutions are usually computed over the entire domain of ϕ. The solutions $\phi(x,y,z,t)$ describe the evolution of an embedded family of contours. While this dense family of solutions might be advantageous for certain applications, there are other applications that require only a single surface model. In such applications the calculation of solutions over a dense field is an unnecessary computational burden, and the presence of contour families can be a nuisance because further processing might be required to extract the level set that is of interest.

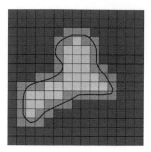

Figure 8.6. Check A level curve of a 2D scalar field passes through a finite set of cells. Only those grid points nearest to the level curve are relevant to the evolution of that curve.

Fortunately, the evolution of a single level set, $\phi(\vec{x}, t) = k$, is not affected by the choice of embedding. The evolution of the level sets is such that they evolve independently (to within the error introduced by the discrete grid). Furthermore, the evolution of ϕ is important only in the vicinity of that level set. Thus, one should perform calculations for the evolution of ϕ only in a neighborhood of the surface $S = \{\vec{x} | \phi(\vec{x}) = k\}$. In the discrete setting, there is a particular subset of grid points whose values control a particular level set (see Figure 8.6). Of course, as the surface moves, that subset of grid points must change to account for its new position.

Adalsteinson and Sethian [1] propose a *narrow-band* approach which follows this line of reasoning. The narrow band technique constructs an embedding of the evolving curve or surface via a signed distance transform. The distance transform is truncated, i.e., computed over a finite width of only m points that lie within a specified distance to the level set. The remaining points are set to constant values to indicate that they do not lie within the narrow band, or *tube* as they call it. The evolution of the surface (they demonstrate it for curves in the plane) is computed by calculating the evolution of ϕ only on the set of grid points that are within a fixed distance to the initial level set, i.e., within the narrow band. When the evolving level set approaches the edge of the band (see Figure 8.7), they calculate a new distance transform and a new embedding, and they repeat the process. This algorithm relies on the fact that the embedding is not a critical aspect of the evolution of the level set. That is, the embedding can be transformed or recomputed at any point in time, so long as such a transformation does not change the position of the kth level set, and the evolution will be unaffected by this change in the embedding.

Despite the improvements in computation time, the narrow-band approach is not optimal for several reasons. First it requires a band of significant width ($m = 12$ in the examples of [1]) where one would like to have a band that is only as wide as necessary to calculate the derivatives of u near the level set (e.g.,

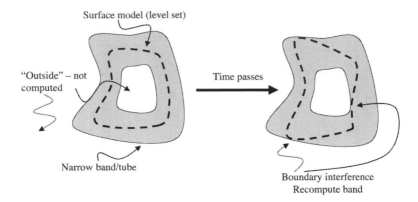

Figure 8.7. The narrow band scheme limits computation to the vicinity of the specific level set. As the level-set moves near the edge of the band the process is stopped and the band recomputed.

$m = 2$). The wider band is necessary because the narrow-band algorithm trades off two competing computational costs. One is the cost of stopping the evolution and computing the position of the curve and distance transform (to sub-cell accuracy) and determining the domain of the band. The other is the cost of computing the evolution process over the entire band. The narrow-band method also requires additional techniques, such as smoothing, to maintain the stability at the boundaries of the band, where some grid points are undergoing the evolution and nearby neighbors are static.

8.3.3   The Sparse-Field Method

The basic premise of the narrow band algorithm is that computing the distance transform is so costly that it cannot be done at every iteration of the evolution process. The sparse-field method [27] uses an approximation to the distance transform that makes it feasible to recompute the neighborhood of the level-set model at each time step. Computation of the evolution equation is computed on a band of grid points that is only one point wide. The embedding is extended from the active points to a neighborhood around those points that is precisely the width needed at each time. This extension is done via a fast distance transform approximation.

   This approach has several advantages. First, the algorithm does precisely the number of calculations needed to compute the next position of the level curve. It does not require explicitly recalculating the positions of level sets and their distance transforms. Because the number of points being computed is so small, it is feasible to use a linked list to keep track of them. Thus, at each iteration the algorithm visits only those points adjacent to the k-level curve. For large 3D data sets, the very process of incrementing a counter and checking the status of all of the grid points is prohibitive.

The sparse-field algorithm is analogous to a locomotive engine that lays down tracks before it and picks them up from behind. In this way the number of computations increases with the surface area of the model rather than the resolution of the embedding. Also, the sparse-field approach identifies a single level set with a specific set of points whose values control the position of that level set. This allows one to compute external forces to an accuracy that is better than the grid spacing of the model, resulting in a modeling system that is more accurate for various kinds of "model fitting" applications.

The sparse-field algorithm takes advantage of the fact that a k-level surface, S, of a discrete image $\phi_{i,j,k}$ (of any dimension) has a set of cells through which it passes, as shown in Figure 8.6. The set of grid points adjacent to the level set is called the *active set*, and the individual elements of this set are called *active points*. As a first-order approximation, the distance of the level set from the center of any active point is proportional to the value of ϕ divided by the gradient magnitude at that point. Because all of the derivatives (up to second order) in this approach are computed using nearest neighbor differences, only the active points and their neighbors are relevant to the evolution of the level set at any particular time in the evolution process. The strategy is to compute the evolution given by Equation 8.12 on the active set and then update the neighborhood around the active set using a fast distance transform. Because active points must be adjacent to the level-set model, their positions lie within a fixed distance to the model. Therefore the values of ϕ for locations in the active set must lie within a certain range. When active-point values move out of this *active range* they are no longer adjacent to the model. They must be removed from the set and other grid points, those whose values are moving into the active range, must be added to take their place. The precise ordering and execution of these operations is important to the operation of the algorithm.

The values of the points in the active set can be updated using the up-wind scheme for first-order terms and central differences for the mean-curvature flow, as described in the previous sections. In order to maintain stability, one must update the neighborhoods of active grid points in a way that allows grid points to enter and leave the active set without those changes in status affecting their values. Grid points should be removed from the active set when they are no longer the nearest grid point to the zero crossing. Because the embedding ϕ is a discrete approximation to the distance transform of the model, the distance of a particular grid point, $x_m = (i, j, k)$, to the level set is given by the value of ϕ at that grid point. If the distance between grid points is defined to be unity, then we should remove a point from the active set when the value of ϕ at that point no longer lies in the interval $[-\frac{1}{2}, \frac{1}{2}]$ (see Figure 8.8). If the neighbors of that point maintain their distance of 1, then those neighbors will move into the active range just as x_m is ready to be removed.

There are two operations that are significant to the evolution of the active set. First, the values of ϕ at active points change from one iteration to the next. Second, as the values of active points pass out of the active range they are removed from the active set and other, neighboring grid points are added to the active set to take their place. In [27] the author gives some formal definitions of active sets and the operations that affect them, which show that active sets will always form a boundary between positive and negative regions in the image, even as control of the level set passes from one set of active points to another.

Because grid points that are near the active set are kept at a fixed value difference from the active points, active points serve to control the behavior of non-active grid points to which they are adjacent. The neighborhoods of the active set are defined in *layers*, $L_{+1}, \ldots, L_{+N}$ and $L_{-1}, \ldots, L_{-N}$, where the i indicates the distance (city-block distance) from the nearest active grid point, and negative numbers are used for the outside layers. For notational convenience the active set is denoted L_0.

The number of layers should coincide with the size of the footprint or neighborhood used to calculate derivatives. In this way, the inside and outside grid points undergo no changes in their values that affect or distort the evolution of the zero set. Most of the level-set work relies on surface normals and curvature, which require only second-order derivatives of ϕ. Second-order derivatives are calculated using a $3 \times 3 \times 3$ kernel (city-block distance 2 to the corners). Therefore only five layers are necessary (two inside layers, two outside layers, and the active set). These layers are denoted L_1, L_2, L_{-1}, L_{-2}, and L_0.

The active set has grid point values in the range $[-\frac{1}{2}, \frac{1}{2}]$. The values of the grid points in each neighborhood layer are kept 1 unit from the next layer closest to the active set (as in Figure 8.8). Thus the values of layer L_i fall in the interval $[i - \frac{1}{2}, i + \frac{1}{2}]$. For $2N + 1$ layers, the values of the grid points that are totally inside and outside are $N + \frac{1}{2}$ and $-N - \frac{1}{2}$, respectively. The procedure for updating the image and the active set based on surface movements is as follows:

1. For each active grid point, $x_m = (i, j, k)$, do the following:

   (a) Calculate the local geometry of the level set.

   (b) Compute the net change of ϕ_{x_m}, based on the internal and external forces, using some stable (e.g., up-wind) numerical scheme where necessary.

2. For each active grid point x_j add the change to the grid point value and decide if the new value $\phi_{x_m}^{n+1}$ falls outside the $[-\frac{1}{2}, \frac{1}{2}]$ interval. If so, put x_m on lists of grid points that are changing status, called the *status list*; S_1 or S_{-1}, for $\phi_{x_m}^{n+1} > 1$ or $\phi_{x_m}^{n+1} < -1$, respectively.

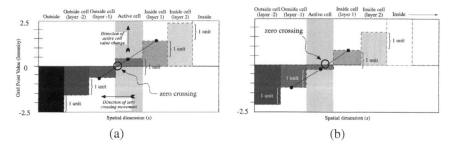

Figure 8.8. The status of grid points and their values at two different points in time show that as the zero crossing moves, *activity* is passed one grid point to another: (a) a 1D embedding shows how the active cell may have a value change that creates motion of the level-set front in an adjacent cell; (b) the resulting level set showing the zero crossing displaced to the left.

3. Visit the grid points in the layers L_i in the order $i = \pm 1, \ldots, \pm N$, and update the grid point values based on the values (by adding or subtracting one unit) of the next inner layer, $L_{i\mp 1}$. If more than one $L_{i\mp 1}$ neighbor exists then use the neighbor that indicates a level curve closest to that grid point, i.e., use the maximum for the outside layers and minimum for the inside layers. If a grid point in layer L_i has no $L_{i\mp 1}$ neighbors, then it gets demoted to $L_{i\pm 1}$, the next level away from the active set.

4. For each status list $S_{\pm 1}, S_{\pm 2}, \ldots, S_{\pm N}$ do the following:

    (a) For each element x_j on the status list S_i, remove x_j from the list $L_{i\mp 1}$, and add it to the L_i list, or, in the case of $i = \pm(N+1)$, remove it from all lists.

    (b) Add all $L_{i\mp 1}$ neighbors to the $S_{i\pm 1}$ list.

This algorithm can be implemented efficiently using linked-list data structures combined with arrays to store the values of the grid points and their states as shown in Figure 8.9. This requires only those grid points whose values are changing, the active points and their neighbors, to be visited at each time step. The computation time grows as m^{n-1}, where m is the number of grid points along one dimension of ϕ (sometimes called the resolution of the discrete sampling). Computation time for dense-field approach increases as m^n. The m^{n-1} growth in computation time for the sparse-field models is consistent with conventional (parameterized) models, for which computation times increase with the resolution of the surface, rather than the volume in which it resides.

Another important aspect of the performance of the sparse-field algorithm is the larger time steps that are possible. The time steps are limited by the speed of

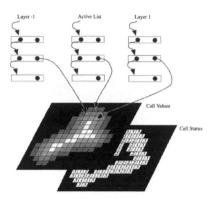

Figure 8.9. Linked-list data structures provide efficient access to those grid points with values and status that must be updated.

the "fastest" moving level curve, i.e., the maximum of the force function. Because the sparse-field method calculates the movement of level sets over a subset of the image, time steps are bounded from below by those of the dense-field case, i.e.,

$$\sup_{x \in \mathcal{A} \subset X} (g(x)) \leq \sup_{x \in X}(g(x)), \tag{8.24}$$

where $g(x)$ is the space varying speed function and $\mathcal{A}$ is the active set.

Results from previous work [27] have demonstrated several important aspects of the sparse-field algorithm. First, the manipulations of the active set and surrounding layers allow the active set to "track" the deformable surface as it moves. The active set always divides the inside and outside of the objects it describes (i.e., it stays closed). Empirical results show significant increases in performance relative to both the computation of full domain and the narrow-band method, as proposed in the literature. Empirical results also show that the sparse-field method is about as accurate as both the full, discrete solution, and the narrow-band method. Finally, because the method positions level sets to sub-voxel accuracy it avoids aliasing problems and is more accurate than these other methods when it comes to *fitting* level-set models to zero sets of scalar or vector-valued functions. This sub-voxel accuracy is an important aspect of the implementation, and will significantly impact the quality of the results for the applications that follow.

8.4 Applications

There are a variety of applications of level sets to problems in image processing, vision, graphics, computational physics, and visualization. The following sections present a pair of applications from the literature that ellucidate some of the technical aspects of level sets.

8.4.1   Segmentation

This section describes a mechanism by which level-set models can be used to segment volumes. It starts with a description of the speed function and the associated parameters and then shows results on MRI datasets.

Level-set models, like the *snakes* technology that preceded them, can lock onto edges using second-derivative information from the input data. In this way, they can be used to find Canny-type edges [4] or Marr-Hildreth [15] edges in a region of interest (specificied by the initialization), while also introducing a smoothness term. If we let the model move *uphill* on the gradient magnitude of the input, this gives

$$\phi_t = \alpha \nabla |\nabla I| + \beta I \nabla \phi |H, \qquad (8.25)$$

where I is the input datum. Even with the appropriate level of smoothing for ∇I this strategy requires the initial model to be near edges in the input image in order for the second derivatives of those edges to pull the surface model in the correct direction. This can be alleviated if we add another speed term that encourages the model to flow into regions with greyscale values that lie within a specified intensity range. This gives

$$\phi_t = \alpha |\nabla \phi| H + \beta \nabla |\nabla I| \cdot \nabla \phi + \gamma |\nabla \phi| (\varepsilon - |I - T|), \qquad (8.26)$$

where T controls the brightness of the region to be segmented and ε controls the range of greyscale values around T that could be considered inside the object. Thus when the model lies on a voxel with a greyscale level between $T - \varepsilon$ and $T + \varepsilon$, the model expands and otherwise it contracts. This speed term is gradual, and thus the effects of the D diminish as the model approaches the boundaries of regions whose greyscale levels lie within the $T \pm \varepsilon$ range. This second speed term is a simple approximation to a one-dimensional statistical classifier, which assumes a single density (with noise) for the regions of interest. However, it allows very poor initializations to deform a long way in order to fill regions of interest.

With this combined edge- and region-based scheme, a user must specify four free parameters, T, ε, α, and γ, *as well as* an initialization.

If a user were to initialize a model in a volume and use only the region-based term, i.e., $\alpha = \beta = 0$, the results would be the same as a simple flood-fill over the region bounded by the upper and lower thresholds. However, the inclusion of the curvature and edge terms alleviate the critical *leaking* problem that arises when using flood-filling as a segmentation technique. The leaking effect is particularly acute in 3D segmentations and is easily demonstrated on a brain tumor dataset,

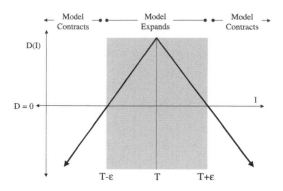

Figure 8.10. A speed function based on image intensity causes the model to expand over regions with greyscale values within the specified range and contract otherwise.

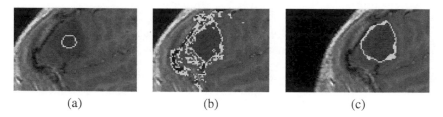

(a)   (b)   (c)

Figure 8.11. Showing one slice of a MRI volume: (a) The spherical initialization; (b) A model expands to fill the tumor but leaks through gaps and expands into other anatomy; (c) The same scenario with a degree of curvature prevents unwanted leaking. The level set isosurface is shown in white.

as shown in Figure 8.11. Figure 8.12 shows the results of using this strategy to segment the cortex, white matter, and a brain tumor from an MRI dataset.

8.4.2 Surface Processing

The proliferation of high-resolution volume and surface measurement technologies has generated an interest in producing a set of tools for processing surfaces which mimic the wide range of tools available for processing images or volumes. One strategy for generalizing image processing algorithms to surfaces is via the *surface normal vectors*. Thus a smooth surface is one that has smoothly varying normals. Penalty functions on the surface normals typically give rise to fourth-order partial differential equations (PDEs). Tasdizen et al. [21] proposed processing level sets in this way using a two step approach: (i) operating on the normal map of a surface while letting the surface shape to lag the normals, and (ii) manipulating the surface to fit the processed normals. Iterating this two-step process, we

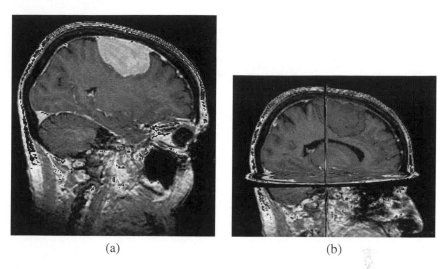

(a)   (b)

Figure 8.12. (See Plate VIII) (a) A 2D slice from an MRI image; (b) 3D rendering of the level-set tumor segmentation.

can efficiently implement fourth-order flows by solving a set of coupled second-order PDEs.

For general fourth-order surface flows, both of these steps involve solving second-order PDEs. The first second-order PDE is used for minimizing a penalty function on the normals. The other second-order PDE minimizes the discrepancy between the modified normals and the surface; in other words, it refits the surface to the normals. Figure 8.13 shows this three step process graphically in 2D: shapes give rise to normal maps, which, when filtered, give rise to new normal maps, which finally give rise to new shapes.

The first step is to formulate the total curvature of a surface from its normal map. Using this, we can derive the variational PDEs on the normal map that minimize functions of total curvature. When using implicit representations, one must account for the fact that derivatives of functions defined on the surface are computed by projecting their 3D derivatives onto the surface tangent plane. The shape matrix for an implicit surface is the gradient of the 3D normal field projected onto the tangent plane [10]: $\nabla \mathbf{N} \mathbf{P}_T$ where $\mathbf{P}_T$ is the projection operator.

The Euclidean norm of the shape matrix is the sum of squared principal curvatures, i.e., total curvature,

$$\kappa^2 = ||(\nabla \mathbf{N}) \mathbf{P}_T||^2. \tag{8.27}$$

We can use Equation (8.27) to define an energy of the normal map,

$$\mathcal{G}_\mathbf{N} = \int_U G(||(\nabla \mathbf{N}) \mathbf{P}_T||^2) \, d\mathbf{x}. \tag{8.28}$$

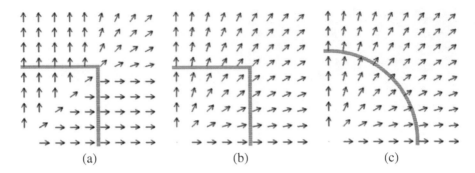

Figure 8.13. Shown here in 2D, the process begins with a shape and constructs a normal map from the distance transform (a), modifies the normal map according to a PDE derived from a penalty function (b), and re-fits the shape to the normal map (c).

The first variation of this energy with respect to the normals is a second order PDE. It is crucial to observe that, even though the projection operator $\mathbf{P}_T$ is a function of ϕ, it is independent of $\mathbf{N}$ because we fix ϕ as we process $\mathbf{N}$.

By choosing appropriate energy functions G, we can obtain isotropic and anisotropic diffusion processes on the normals. However, the final goal is to process the surface, which requires deforming ϕ. Therefore, the next step is to relate the deformation of the level sets of ϕ to the evolution of $\mathbf{N}$. Suppose that we are given the normal map $\mathbf{N}$ to some set of surfaces, but not necessarily level sets of ϕ, as is the case if we filter $\mathbf{N}$ and let ϕ lag. We can manipulate ϕ so that it fits the normal field $\mathbf{N}$ by minimizing a penalty function that quantifies the discrepancy between the gradient vectors of ϕ and the target normal map. That penalty function is

$$\mathcal{D}_\phi = \int_U D(\phi)\ d\mathbf{x}, \quad \text{where } D(\phi) = \sqrt{\nabla\phi \cdot \nabla\phi} - \nabla\phi \cdot \mathbf{N}. \qquad (8.29)$$

The first variation of this penalty function with respect to ϕ is

$$\frac{d\mathcal{D}}{d\phi} = -\nabla \cdot \left[\frac{\nabla\phi}{||\nabla\phi||} - \mathbf{N} \right] = -\left[H^\phi - H^{\mathbf{N}} \right], \qquad (8.30)$$

where H^ϕ is the mean curvature of the level set surface and $H^{\mathbf{N}}$ is half the divergence of the normal map. Then, the gradient descent PDE that minimizes Equation (8.29) is the $d\phi/dt = -||\nabla\phi||d\mathcal{D}/d\phi$ level set, which is embedded in ϕ. According to Equation (8.30), the surface moves as the difference between its own mean curvature and that of the normal field. We have derived a gradient descent for the normal map based on a certain class of penalty functions that use the total curvature. This process is denoted in Figure 8.14 as the $dG/\mathbf{N}$ loop. The

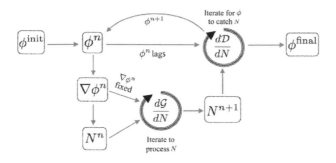

Figure 8.14. Flow chart of surface filtering strategy.

surface refitting to the normal map is formulated as a gradient descent in Equation (8.30). This process is the $d\mathcal{D}/d\phi$ loop in Figure 8.14. The overall algorithm shown in Figure 8.14 repeats these two steps to minimize the penalty functions in terms of the surface.

The normal map energy minimization allows us to experiment with various forms of G in Equation (8.28) that give rise to different classes of penalty functions. The choice of $G(\kappa^2) = \kappa^2$ leads to an isotropic diffusion. This choice works well for smoothing surfaces and eliminating noise, but it also deforms or removes important features. This type of smoothing is called isotropic because it corresponds to solving the heat equation on the normal map with a constant, scalar conduction coefficient, which is the same as Gaussian smoothing, for images. The problem of preserving features while smoothing away noise has been studied extensively in computer vision. Anisotropic diffusion introduced by Perona and Malik has been very successful in dealing with this problem in a wide range of images. A generalization of Perona and Malik diffusion to surfaces is achieved by

$$G(\kappa^2) = 2\mu^2 \left(1 - e^{-\frac{\kappa^2}{2\mu^2}} \right), \tag{8.31}$$

where μ is a positive, free parameter that controls the level of feature preservation. Figure 8.15(a) illustrates an example of the skin surface, which was extracted, via isosurfacing, from an MRI data set. Notice that the roughness of the skin is noise, an artifact of the measurement process. Isotropic diffusion, shown in Figure 8.15(b), is marginally effective for denoising the head surface. Notice that the sharp edges around the eyes, nose, lips and ears are lost in this process. The differences between anisotropic diffusion and isotropic diffusion can be observed clearly in Figure 8.15(c). The two processes produce similar smoothing results in those areas of the surface which are somewhat featureless, such as the forehead

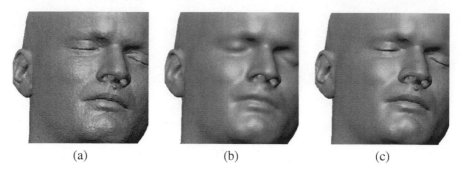

| (a) | (b) | (c) |

Figure 8.15. Processing results on the MRI head model: (a) original isosurface; (b) isotropic diffusion; and (c) anisotropic diffusion. The small protrusion under the nose is a physical marker used for registration.

and the cheeks. However, these results show significant differences exist around the lips and the eyes. The creases in these areas, which have been eliminated by isotropic diffusion, are preserved by the anisotropic process.

8.5 Summary

Volumes provide a powerful tool for modeling deformable surfaces, especially when dealing with measured data. With measured data, the shape, topology, and complexity of the surface are dictated by the application rather than the user. Implicit deformable surfaces, implemented as level sets, provide a natural mechanism for processing such data in a manner that relieves the user of having to decide on an underlying parameterization. This technology easily handles the many degrees of freedom that are important to capturing the fine detail of measured data. Furthermore, the level set approach provides a powerful mechanism for constructing *geometric flows*, which results in output which depends only on the shape of input (and the resolution) and does not produce artifacts which are tied to an arbitrary, intermediate parameterization.

References

[1] Adalsteinson, D. and J. A. Sethian. "A fast level set method for propogating interfaces," *Journal of Computational Physics* 118(1995): 269–277.

[2] Alvarez, L. and J.-M. Morel. "A morphological approach to multiscale analysis: From principles to equations," in *Geometry-Driven Diffusion in Computer Vision* B. ter Haar Romeny, ed. Kluwer Academic Publishers, 1994. 229–254.

[3] Breen, D. and R. Whitaker. "A level-set approach to 3d shape meta-
    morphosis," *IEEE Transactions on Visualization and Computer Graphics*
    7(2)(2001): 173–192.

[4] Canny, J. "A computational approach to edge detection," *IEEE Transactions
    on Pattern Analysis and Machine Intelligence* 8(6)(1986): 679–698.

[5] Caselles, V., Kimmel, R., and G. Sapiro. "Geodesic active contours," in *Fifth
    International Conference on Computer Vision* (Proceedings of ICCV 1995)
    Los Alamitos: IEEE Computer Society Press, 1995. 694–699.

[6] Fedkiw, R. and S. Osher. *Level Set Methods and Dynamic Implicit Surfaces.*
    Berlin: Springer, 2002.

[7] Kass, M., Witkin, A., and D. Terzopoulos. "Snakes: Active contour models,"
    International Journal of Computer Vision 1(1987): 321–331.

[8] Kichenassamy, S., Kumar, A., Olver, P., Tannenbaum, A., and A. Yezzi.
    "Gradient flows and geometric active contour models," in *Fifth International
    Conference on Computer Vision* (Proceedings of ICCV 1995) Los Alamitos:
    IEEE Computer Society Press, 1995. 810–815.

[9] Kimmmel, R. and A. Bruckstein. "Shape offsets via level sets," *Computer
    Aided Design* 25(5)(1993): 154–162.

[10] Kindlmann, G., Whitaker, R., Tasdizen, T., and T. Möller. "Curvature-Based
     Transfer Functions for Direct Volume Rendering: Methods and Applica-
     tions," in *Proceedings of IEEE Visualization 2002* (October 2003): 513–520.

[11] Koenderink, J. "Solid shape," *IEEE Computer* 20(Jaunary 1987): 10–19.

[12] Lindeberg, T. "A scale-space for discrete signals," *IEEE Transactions on
     Pattern Analysis and Machine Intelligence* 12(3)(1990): 234–245.

[13] Lorigo, L., Faugeraus, O., Grimson, W., Keriven, R., and R. Kikinis. "Seg-
     mentation of bone in clinical knee MRI using texture-based geodesic active
     contours," in *Medical Image Computing and Computer-Assisted Interven-
     tion* (Proceedings of MICCAI 1998), W. Wells, A. Colchester, and S. Delp,
     eds. Springer, LNCS1496(1998): 1195–1204.

[14] Malladi, R., Sethian, J., and B. Vemuri. "Shape modeling with front pro-
     pogation: A level set approach," *IEEE Transactions on Pattern Analysis and
     Machine Intelligence* 17(2)(1995): 158–175.

[15] Marr, D. and E. Hildreth. "Theory of edge detection," *Proc. Royal Society
     of London* B207(1980): 187–217.

[16] Museth, K., Breen, D., Zhukov, L., and R. Whitaker. "Level-set segmentation from multiple non-uniform volume datasets," in *Proceedings of IEEE Visualization 2002* (October 2002): 179–186.

[17] Osher, S. and J. Sethian. "Fronts propogating with curvature-dependent speed: Algorithms based on Hamilton-Jacobi formulations," *Journal of Computational Physics* 79(1988): 12–49.

[18] Osher, S. and R. Fedkiw. "Level set methods: An overview and some recent results," Tech. Rep. 00–08, (UCLA Center for Applied Mathematics, Department of Mathematics, University of California, Los Angeles, 2000).

[19] Sethian, J. *Level Set Methods: Evolving Interfaces in Geometry, Fluid Mechanics, Computer Vision, and Material Sciences.* Cambridge: Cambridge University Press, 1996.

[20] Sethian, J. *Level Set Methods and Fast Marching Methods Evolving Interfaces in Computational Geometry, Fluid Mechanics, Computer Vision, and Materials Science.* Cambridge: Cambridge University Press, 1999.

[21] Tasdizen, T., Whitaker, R., Burchard, P., and S. Osher. "Geometric surface processing via normal maps," To appear: *ACM Transactions on Graphics.*

[22] ter Haar Romeny, B., Florack, L., Koenderink, J., and M. Viergever. "Scale space: its natural operators and differential invariants," in *Medical Image Computing and Computer-Assisted Intervention* (Proceedings of MICCAI 1991), A. Colchester and D. Hawkes, eds. Springer, LNCS511(1991): 239–255.

[23] Terzopoulos, D. and K. Fleischer. "Deformable models," *The Visual Computer* 4(December 1988): 306–331.

[24] Whitaker, R. "Volumetric deformable models: Active blobs," in *Visualization In Biomedical Computing 1994* (Proceedings of VBC94, Rochester, Minnesota 1994), R. A. Robb, ed. SPIE 2359(1994): 122–134.

[25] Whitaker, R. "Algorithms for implicit deformable models," in *Fifth International Conference on Computer Vision* (Proceedings of ICCV 1995) Los Alamitos: IEEE Computer Society Press, 1995. 822–827.

[26] Whitaker, R. and D. Breen. "Level-set models for the deformation of solid objects," in *The Third International Workshop on Implicit Surfaces*, (Eurographics, 1998): 19–35.

[27] Whitaker, R. "A level-set approach to 3D reconstruction from range data," *International Journal of Computer Vision* 29(3)(October 1998): 203–231.

[28] Whitaker, R. "Reducing aliasing artifacts in isosurfaces of binary volumes," in *Proceedings of IEEE Symp. on Volume Visualization and Graphics* (October 2000): 23–32,

[29] Whitaker, R. and X. Xue. "Variable-conductance, level-set curvature for image denoising," in *IEEE International Conference on Image Processing*, (Proceedings of ICIP 2001) Los Alamitos: IEEE Computer Society Press, 2001. 142–145.

[30] Whitaker, R. and V. Elangovan. "A direct approach to estimating surfaces in tomographic data," *Medical Image Analysis* 6(2002): 235–249.

[31] Yezzi, A., Kichenassamy, S., Kumar, A., Olver, P., and A. Tannenbaum. "A geometric snake model for segmentation of medical imagery," *IEEE Transactions on Medical Imaging* 16(2)(April 1997): 199–209.

Deformable Models

Dimitris N. Metaxas
Rutgers University

Ting Chen
Rutgers University

9.1   Introduction

In the past decade, deformable models have been widely used in medical image segmentation tasks. In ITK (the Insight Toolkit), there are several C++ classes that can help the user to construct deformable models. There are also some filters in the toolkit that can be used to fit the deformable model to various organ shapes.

Deformable models are dynamic models whose 2D or 3D shape evolves under the influence of external and internal forces. The process of surface evolution corresponds to the minimization of an energy function [17]. The energy function is composed of three components: 1) the internal energy which helps to keep the continuity and regularity of the contour/surface, 2) the image energy which pushes the deformable model to edge features in the original image, and 3) the user-defined energy that can speed up the deformation and/or push the model out of local minima.

There are several advantages in using deformable models for segmentation. By using deformable models, we can use the internal energy constraint (which includes global or local prior information about the object) to define the object's boundary even if there is not enough gradient information in the boundary region. The deformable model is robust to noise and has good performance for images with a low signal-to-noise ratio (SNR). By using the parameterized deformable model, it is easy to achieve sub-pixel accuracy in segmentation applications.

Finally deformable models are easy to interact with based on user defined forces ([7], [4] and [5]).

In the following chapter, we will first present some of the previous work related to deformable models. Then we will describe the basic theory of deformable models. Finally, we will show some experimental results of 2D/3D segmentation using the Insight Toolkit. This will help illustrate how to use ITK to build 2D/3D deformable models for segmentation.

9.2   Previous Work

9.2.1   Snakes

Active contour models (snakes) developed by M. Kass *et al.* [7] is an interactive segmentation methodology based on energy minimization. It can be regarded as a special case of the multi-dimensional deformable model theory [17] for 2D segmentation.

An active contour is a deformable contour defined by a set of points with the initial position vector $\mathbf{p}(s)$ and the displacement vector $\mathbf{d}(s)$, where s is the arc-length. The motion of the contour is governed by the minimization of the following energy function:

$$E = \int E_{internal}(\mathbf{d}) + E_{image}(\mathbf{p} + \mathbf{d}) + E_{user}(\mathbf{p} + \mathbf{d}) ds. \qquad (9.1)$$

The internal energy $E_{internal}$ controls the stretching and bending of the snake. The E_{image} term is a function of short-range image operators (gradient magnitude or gradient vector flow for segmentation). It is used to attract contour points to appropriate image features. E_{user} is the energy derived from the position of the user-controlled cursor. It is used to guide interactively the contour away from the local minimum.

9.2.2   Fourier Parameterized Models

Worring et al. [19] proposed the Fourier parameterized model and its application in segmentation. The Fourier parameterized model approaches the segmentation (or edge finding) as an optimization problem. In an image $I(x,y)$, let $\mathbf{v}(t;\mathbf{p})$ be the deformable boundary template, where $t \in [0, 2\pi)$ and $\mathbf{p}$ is the parameter vector. The segmentation procedure is to find the parameter vector $\mathbf{p}_{opt}$ which will fit the model to the object of interest in an optimal way. To achieve the optimal parameter vector, Worring et al. [19] defined a global objective function $H(\mathbf{p})$ which is the line integral over the template $\mathbf{v}(t;\mathbf{p})$ of the local objective function $h(t; \mathbf{p})$:

$$H(\mathbf{p}) = \int_t h(t; \mathbf{p}) \|\mathbf{v}'(t)\| dt, \tag{9.2}$$

where $h(t; \mathbf{p})$ is derived from image information, e.g. image gradient. To locate the edge feature of the object of interest, we let $h(t; \mathbf{p}) = \nabla I(x, y)$. The global objective function, Equation (9.2), will then have the following form:

$$H(\mathbf{p}) = \int_t \|\nabla I(\mathbf{v}(t; \mathbf{p}))\| \|\mathbf{v}'(t)\| dt \tag{9.3}$$

The global objective function $H(\mathbf{p})$ is evaluated at N discrete points on the curve to approximate the optimal solution for Equation (9.3).

The Fourier parameterized deformable boundary template is built based on the fact that for closed curves the values of the x- and y-coordinates are periodic. In [19], Fourier coefficients are used to parameterize the model as a series of orthonormal basis functions. High-indexed basis functions are associated with high frequencies. We truncate the function series to represent a relatively smooth object surface. By doing so, the Fourier parameterized model can describe a wide variety of smooth surfaces using a small number of parameters.

It should be noted that in [19], a Gaussian smoothed gradient vector field is used to create an external force for the deformable model instead of the gradient magnitude. In a Gaussian smoothed gradient vector field, the edge feature has a larger attraction range so that the deformable model does not have to be initialized near the object boundary and has a bigger chance of getting out of the local minimum.

9.2.3 3D Deformable Models

3D deformable models used in [17, 10, 14, 127] are defined as surfaces that deform under the influence of internal and external forces. There are two major classes of 3D deformable models: explicit deformable models ([17, 10, 15, 14]) and implicit deformable models ([9, 131, 21]). Although these two classes of models are different at the implementation level, the underlying principles are similar.

Explicit deformable models construct the object surfaces explicitly based on parametric formulations during the deformation process. The user can control the deformation process by changing the values of parameters or using user-defined external forces. Moreover, explicit deformable models are easy to represent and faster in implementation. However, the adaptation of model topology is difficult for explicit deformable models. It is not easy to split or merge models during model deformation.

Implicit deformable models in [9] represent the curves and surfaces implicitly as a level set [13] of higher-dimensional scalar function. Implicit deformable models can have topological changes during the model deformation. However, the description of implicit deformable models is more complicated and the deformation process takes more time.

There are two different formulations for explicit deformable models: the energy-minimization formulation [7, 19] and the dynamic-force formulation [4, 5]. The energy-minimization formulation has the advantage that its solution satisfies a minimum principle; while the dynamic force formulation provides the flexibility of applying different types of external forces onto the deformable model.

The energy minimization formulation can be represented briefly as follows. A deformable model can be regarded as a curve or surface that moves through the spatial image domain and minimizes an energy function. The potential energy function has smaller values at features of interest, such as the object boundary. The deformation process minimizes the model's energy until it reaches a minimum. Both the snakes [7] and the Fourier parameterized model [19] use the energy minimization formulation.

In dynamic force formulation models, external forces are applied onto the model surface. The external forces can be potential forces derived from the gradient, non-potential forces such as balloon force or the combination of both. External forces push the model to the features of interest in the image. The deformation stops when these external forces equilibrate or vanish.

9.3 Deformable Model Theories

9.3.1 Geometry of the Deformable Model

Given the material coordinates $\mathbf{u} = (u, v, w)$, the position of a point on the model is defined as a 3D vector $\mathbf{x}$:

$$\mathbf{x}(\mathbf{u},t) = (x_1(\mathbf{u},t), x_2(\mathbf{u},t), x_3(\mathbf{u},t))^T. \tag{9.4}$$

We set up a model-centered reference frame to express these positions as

$$\mathbf{x} = \mathbf{c} + \mathbf{R}\mathbf{p}, \tag{9.5}$$

where $\mathbf{c}(t)$ is the origin of the reference frame, $\mathbf{p} = (\mathbf{u},t)$ denotes the position of the point relative to the model frame, and $\mathbf{R}$ is the rotation matrix that transfers those relative points' positions to the real 3D coordinates.

Given a reference shape $\mathbf{s}$ (the superquadric ellipsoid) and the displacement $\mathbf{d}$ (the local deformation), the position of a point $\mathbf{p}$ on the model is defined by

$$\mathbf{p} = \mathbf{s} + \mathbf{d}. \tag{9.6}$$

We define the 3D reference surface $\mathbf{s}$ as

$$\mathbf{s} = \mathbf{T}(\mathbf{e}(\mathbf{u}; a_0, a_1, \ldots); b_0, b_1, \ldots) = \mathbf{T}(\mathbf{e}; \mathbf{b}), \tag{9.7}$$

where global parameters a_i define the geometric primitive $\mathbf{e}$. We use additional global parameters b_i to control the tapering and bending deformation of $\mathbf{e}$.

We define the vector of global deformation parameters:

$$\mathbf{q}_s = (a_0, a_1, \ldots, b_0, b_1, \ldots)^T. \tag{9.8}$$

The Jacobian matrix that defines the internal forces during global deformation will be calculated by

$$\mathbf{J} = \frac{\partial \mathbf{s}}{\partial \mathbf{q}_s}. \tag{9.9}$$

For example, for a superquadric ellipsoid, $\mathbf{e}$ has the following form:

$$\mathbf{e} = a_0 w \begin{bmatrix} a_1 C_u^{\varepsilon_1} C_v^{\varepsilon_2} \\ a_2 C_u^{\varepsilon_1} S_v^{\varepsilon_2} \\ a_3 S_u^{\varepsilon_1} \end{bmatrix}, \tag{9.10}$$

where $-\pi/2 \le u \le \pi/2$, $-\pi \le v \le \pi$, $0 \le w \le 1$, $0 \le a_1, a_2, a_3 \le 1$, $0 \le a_0$, $0 \le \varepsilon_1, \varepsilon_2$, $S_u^{\varepsilon} = \text{sgn}(\sin(u))|\sin(u)|^{\varepsilon}$, $C_u^{\varepsilon} = \text{sgn}(\cos(u))|\cos(u)|^{\varepsilon}$, and similarly for $S_v \varepsilon$ and C_v^{ε}. $a_i \ldots$ defines the aspect ratio of the deformable model, while ε_i defines the "squareness" of the model.

9.3.2 Dynamics of the Deformable Model

Motion Equation

The deformable model dynamics can be described by the first order Lagrangian mechanics

$$\dot{\mathbf{d}} + \mathbf{K}\mathbf{d} = \mathbf{f}_{ext}, \tag{9.11}$$

where $\dot{\mathbf{d}} = \frac{\partial d}{\partial t}$, $\mathbf{f}_{ext}$ are the external forces and $\mathbf{K}$ is the stiffness matrix. The external forces here can be the gradient-derived force, the balloon force or the combination of both. The model deformation is updated using an Euler equation

$$\mathbf{d}_{new} = \dot{\mathbf{d}} \cdot \Delta t + \mathbf{d}_{old} \tag{9.12}$$

where Δt is the step.

In the next section, we will demonstrate how to derive the stiffness matrix $\mathbf{K}$ in Equation (9.11) from a local deformation strain energy.

Stiffness Matrix

We impose a $\mathbf{C}^0$-continuous loaded membrane deformation strain energy ε_m to keep the continuity of the model surface. It is defined as

$$\varepsilon_m(\mathbf{d}) = \int \omega_{10} \left(\frac{\partial \mathbf{d}}{\partial u}\right)^2 + \omega_{01} \left(\frac{\partial \mathbf{d}}{\partial v}\right)^2 + \omega_{00} \mathbf{d}^2 du, \tag{9.13}$$

where ω_{00} controls the local magnitude of the deformation and ω_{01}, ω_{10} control the local variation of the deformation in the u, v directions, respectively. For simplicity, we assume $\omega_{01} = \omega_{10}$.

The stiffness matrix is derived from the strain energy ε_m. In accordance with the theory of elasticity, the local deformation strain energy can also be expressed as

$$\varepsilon = \int \sigma^T \varepsilon du, \tag{9.14}$$

where σ and ε are the stress and strain vectors respectively. The relationship between the strain vector ε and the local deformation $\mathbf{d}$ is defined as

$$\varepsilon = \partial \mathbf{d}, \tag{9.15}$$

where ∂ is a differential operator derived from the local deformation strain energy, i.e.,

$$\partial = \begin{bmatrix} \frac{\partial}{\partial u} & 0 & 0 \\ 0 & \frac{\partial}{\partial v} & 0 \\ 0 & 0 & 1 \end{bmatrix}. \tag{9.16}$$

We now consider the local displacement $\mathbf{d}$, which can be approximated in terms of a finite number of basis functions as

$$\mathbf{d} = \mathbf{S} \mathbf{q}_d, \tag{9.17}$$

where $\mathbf{S}$ is the basis matrix whose elements are basis functions and $\mathbf{q}_d$ is the vector of local degrees of freedom. Substituting Equation (9.17) into Equation (9.15), we get

$$\varepsilon = \partial \mathbf{S} \mathbf{q}_d. \tag{9.18}$$

According to the theory of elasticity,

$$\sigma = \mathbf{D}\varepsilon, \tag{9.19}$$

where $\mathbf{D}$ is a symmetric matrix derived from local deformations. In our case,

$$\mathbf{D} = \begin{bmatrix} \omega_{10} & 0 & 0 \\ 0 & \omega_{01} & 0 \\ 0 & 0 & \omega_{00} \end{bmatrix}. \tag{9.20}$$

By using the results of Equations (9.15), (9.18), and (9.19), we rewrite Equation (9.14) in the following form:

$$\varepsilon = \int \mathbf{q}_d^T (\partial \mathbf{S})^T \mathbf{D} \partial \mathbf{S} \mathbf{q}_d du. \tag{9.21}$$

Since $\mathbf{q}_d$ depends only on time, we can rewrite Equation (9.21) as

$$\varepsilon = \mathbf{q}_d^T \left(\int (\partial \mathbf{S})^T \mathbf{D} \partial \mathbf{S} du \right) \mathbf{q}_d, \tag{9.22}$$

or in the compact form,

$$\varepsilon = \mathbf{q}_d^T \mathbf{K} \mathbf{q}_d, \tag{9.23}$$

where

$$\mathbf{K} = \int (\partial \mathbf{S})^T \mathbf{D} \partial \mathbf{S} du. \tag{9.24}$$

$\mathbf{K}$ is the stiffness matrix we saw in Equation (9.11). However, in order to use Equation (9.11) to calculate the model deformation, we also need to define external forces. In the next section we will introduce external forces that can push the deformable model out of local minima and fit to features of interest.

External Forces

The external forces that have been applied onto the deformable model can be classified into two categories based on their source: the image-based forces and the non-image-based forces.

We can use non-image-based external forces such as balloon force to fit contours and surfaces to the estimated object boundary. The balloon force was proposed by Cohen [4]. In [4], the deformable model starts as a small circle (2D) or sphere (3D) inside the object. The balloon force can help to push the model closer to the boundary by adding a constant inflation force inside the deformable model. The balloon force will push nodes on the deformable surface outward in the direction of their normal vertices. Therefore the balloon force-driven deformable model will expand like a balloon being filled with gas. However, for objects with complex surface structures, the deformable model may leak out of the boundary when using the balloon force. One solution to this problem is to use region-based segmentation methods. The result of the region-based segmentation method can be used as a rough binary mask of the object. The magnitude of the balloon force decreases as the deformable model gets close to the estimated object boundaries. We can define the balloon force to be effective only inside this binary mask. Therefore when the deformable model expands beyond the binary mask, the balloon force does not push it outwards anymore.

When the deformable model is close enough to the features of interest, we use image-based forces to lead the model to these features. Image-based forces are derived from image information, e.g., the gradient information. They can attract the deformable model to the boundary of the object or other features of interest in the image. We introduce two widely used image-based forces here: the second order derivative gradient vector and the gradient vector flow.

The second order derivative gradient vector can lead the deformable model to edges in the images. It is defined as

$$f_G(x,y,z) = -\nabla P(x,y,z) = -\nabla(w_e|\nabla[G_\sigma(x,y,z) * I(x,y,z)]|)^2 \qquad (9.25)$$

where $I(x,y,z)$ is the original image, w_e is a positive weighting parameter, $G_\sigma(x,y,z)$ is a 3D Gaussian function with standard deviation σ, ∇ is the gradient operator, and $*$ is the convolution operator. The Gaussian filter is used to blur the original image in order to remove small noisy regions and expand the effective range of the gradient-derived force.

During the fitting process, we calculate the dot product of the second order gradient vector and the normal vector at every node on the deformable surface. It will yield a positive value if the model node locates inside the edge feature and a negative value if the model node locates outside. The magnitude of the dot product result will be used as the magnitude of the force vector at the node. The direction of the force vector will be the same as the normal vector.

The gradient vector flow proposed by Xu and Prince [20] has a large attraction range and good performance in leading the deformable model into concavities on the object surface. According to experimental results, it has better performance than operators which use only the gradient magnitude. However, the creation of the gradient vector flow field takes a long time for 3D image volumes. At this stage, we only use the gradient vector flow in 2D segmentation tasks.

9.3.3   Fitting Deformable Models to the Data

Marching Cubes Method

We cannot use the super-quadric ellipsoid model to segment 3D objects with complex surface structures. The ellipsoid model cannot fit easily into the concavities on the object surface so the final segmentation result may be oversmoothed. One way to deal with such objects is to use the *Marching Cubes* method [8, 11]. By using the marching cubes method, we can construct a 3D deformable mesh which is close to the surface of the object. This mesh has no difficulty in fitting into concavities on the object's surface.

To construct the deformable mesh, we first use region-based methods such as the Gibbs prior model to create a rough binary mask of the object. We then use the marching cubes method to create a mesh which is very close to the surface of this 3D mask. Given a binary mask that is not too far from the real surface of the object, the created mesh should be in the effective range of the gradient force. We can use this mesh as the initial location of the deformable surface and apply gradient-derived forces onto it from the beginning of the fitting process.

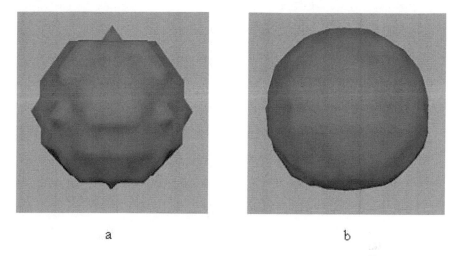

a   b

Figure 9.1. (a) The initial deformable mesh created based on the binary mask of a sphere (the radius is 8 pixels) using marching cube method; (b) the deformable mesh after 100 fitting steps. It is clear that using the marching cubes method to initialize the deformable model will lead to sub-pixel accuracy in final segmentation.

There are several advantages in using the marching cubes method for 3D deformable mesh construction. The deformable surface created by the marching cubes method is close to the object surface we want to segment. We can assume that the global displacement is small enough to be neglected during the deformation. Therefore we only need to consider the local displacement during the fitting process. The deformable mesh created by the marching cubes method is composed of sub-pixel triangular elements. Therefore the segmentation result can also achieve sub-pixel accuracy.

Fitting Process

The fitting processes for 2D and 3D segmentation are different.

2D Models: In the first stage of the 2D deformable model fitting, the deformable model will begin as a closed curve inside the object. It will be pushed to the estimated boundary under the influence of the balloon forces. We use the result of a region-based method to define the effective range for the balloon forces. When the model comes close to the estimated boundaries, the magnitude of balloon forces will gradually decrease to zero. On the other hand, the magnitude of gradient-derived forces will gradually increase as the model is getting close to the estimated boundary. In the second stage of the deformable model fitting process, the deformable surface will be led to the edge feature by using gradient-derived forces.

3D Models: For 3D segmentation, we also need a rough estimation of the location of the object, which can be the segmentation result of a region-based method such as segmentation methodologies using Markov random field models [3]. Then a 3D deformable surface will be created near the estimated object surface using the marching cubes method. The combination of the balloon force and the gradient-based force will then be applied onto the deformable model to improve the segmentation result. We can change the weight for the balloon force and the weight for the gradient-based force to control the behavior of the model deformation.

9.3.4   Finite Element Implementation

Local deformations of deformable models are computed based on the use of the finite element method. The 3D deformable model surface is divided into many small elements. To compute the displacement vector at each node, we need to compute the displacement of each element using Equation (9.11). The Lagrange equations of motion are discretized, and we need to compute the stiffness matrix of every element on the deformable surface. Then we use the summation of nodal displacement vectors in each element as the final nodal displacement vector.

For example, in the case of calculating the deformation of deformable meshes created by using the marching cubes method, the deformable surface is discretized into triangular elements. All these elements are C^0-continuous and shape functions are linear within each element. We can use coordinate transformation to transfer all the elements into a triangular element in local coordinates (ξ, η), as shown in Figure 9.2. Here we introduce how to calculate the stiffness matrix for the triangular element in Figure 9.2. The stiffness matrices for other kinds of triangular elements can be calculated in a similar way. We use shape functions to interpolate the displacement within the element. The nodal shape functions for the element shown in Figure 9.2 are

$$N_1(\xi, \eta) = 1 - \xi - \eta,$$

$$N_2(\xi, \eta) = \xi, \tag{9.26}$$

$$N_3(\xi, \eta) = \eta,$$

where $\sum N_i = 1$ everywhere in the element, $N_i = 1$ at node i and $N_i = 0$ on the edge formed by nodes other than i. The relation between the material coordinates (u, v) and the local coordinates (ξ, η) is

$$\xi = \frac{1}{a}(u - u_1), \tag{9.27}$$

$$\eta = \frac{1}{b}(v - v_1). \tag{9.28}$$

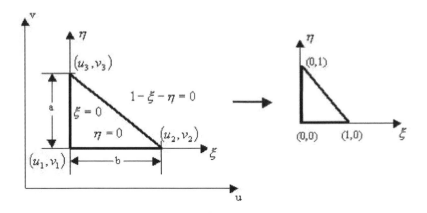

Figure 9.2. Transfer an arbitrary triangular element in (u, v) into a triangular element in (ξ, η)

We compute the derivatives of the nodal shape functions and get the following results:

$$\frac{\partial N_1}{\partial u} = -\frac{1}{a} \quad \frac{\partial N_1}{\partial v} = -\frac{1}{b},$$

$$\frac{\partial N_2}{\partial u} = \frac{1}{a} \quad \frac{\partial N_2}{\partial v} = 0, \tag{9.29}$$

$$\frac{\partial N_3}{\partial u} = 0 \quad \frac{\partial N_3}{\partial v} = \frac{1}{b}.$$

In the case of a triangular element, $\partial \mathbf{S}$ has the following form:

$$\partial \mathbf{S} = \begin{bmatrix} \dfrac{\partial N_1}{\partial u} & \dfrac{\partial N_2}{\partial u} & \dfrac{\partial N_3}{\partial u} \\ \dfrac{\partial N_1}{\partial v} & \dfrac{\partial N_2}{\partial v} & \dfrac{\partial N_3}{\partial v} \\ N_1 & N_2 & N_3 \end{bmatrix} = \begin{bmatrix} -\dfrac{1}{a} & \dfrac{1}{a} & 0 \\ -\dfrac{1}{b} & 0 & \dfrac{1}{b} \\ 1-\xi-\eta & \xi & \eta \end{bmatrix} \tag{9.30}$$

If we denote $\Gamma = \partial \mathbf{S}$, we can rewrite Equation (9.24) as follows:

$$\mathbf{K}_i = \int_{E_i} \Gamma^T \mathbf{D} \Gamma \, du, \tag{9.31}$$

where E_i is the ith element and $\mathbf{K}_i$ is the stiffness matrix for the ith element. We assume that $\omega_{01} = \omega_1$, $\omega_{10} = \omega_1$, and $\omega_{00} = \omega_0$. The elements of the local stiffness matrix $\mathbf{K}$ are

$$\mathbf{K}_{11} = \frac{1}{2}\left(\frac{a}{b} + \frac{b}{a}\right)\omega_1 + \frac{ab}{12}\omega_0,$$

$$\mathbf{K}_{12} = \mathbf{K}_{21} = -\frac{b}{2a}\omega_1 + \frac{ab}{24}\omega_0,$$

$$\mathbf{K}_{13} = \mathbf{K}_{31} = -\frac{a}{2b}\omega_1 + \frac{ab}{24}\omega_0, \tag{9.32}$$

$$\mathbf{K}_{22} = \frac{b}{2a}\omega_1 + \frac{ab}{12}\omega_0,$$

$$\mathbf{K}_{23} = \mathbf{K}_{32} = \frac{ab}{24}\omega_0,$$

$$\mathbf{K}_{33} = \frac{a}{2b}\omega_1 + \frac{ab}{12}\omega_0.$$

The internal force will then be calculated using the stiffness matrix $\mathbf{K}$. We also assume that the nodal range force only has effects on elements containing the node. Now we can compute the nodal displacement vectors within every element using Equation (9.11).

9.4   Experiments, Results, and Applications in ITK

We can use deformable models to segment medical images. ITK provides class to construct 2D curves and 3D superquadric ellipsoids. These curves and ellipsoids can be placed inside the object to be segmented as initial locations of deformable models. Balloon forces can then be applied to deformable surfaces to make them expand.

Alternately, a deformable mesh can be created using the marching cubes method based on a binary mask of the object. Two look-up tables have been read into the memory before the construction of the mesh in order to speed up the process. After the mesh has been created, we can apply gradient forces to the deformable surface.

We use a series of filters to create the second derivative gradient field including Gaussian derivative filters to measure a regularized gradient and gradient magnitude. The result is the second order derivative gradient of the original image blurred by a gaussian filter. The output of these filters can also be used as

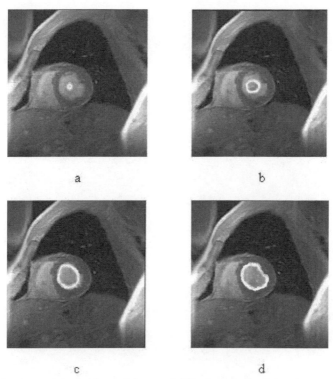

Figure 9.3. (a) A deformable model starts inside the left ventricle; (b) the deformable model expand under the effect of balloon force; (c) the deformable model begins to deform under the effect of gradient flow when close enough to the boundary; (d) final fitting result. [Data courtesy of Dr. Leon Axel, New York University]

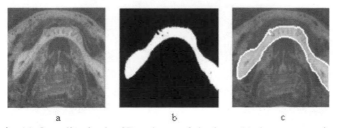

Figure 9.4. (a) One slice in the 3D volume of the jaw; (b) the segmentation result of the Voronoi diagram method and deformable model (see [6]); (c) the projection of the segmentation result onto the original image. [Data from the National Library of Medicine, Visible Human Project]

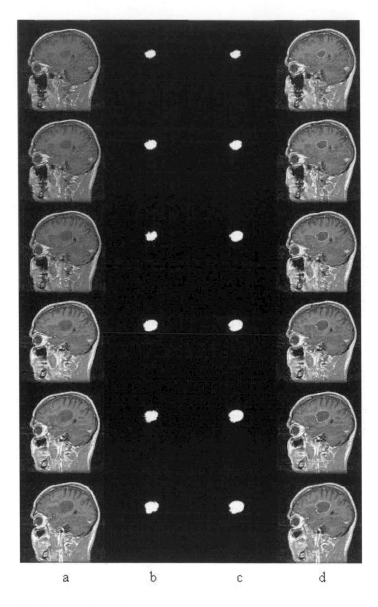

a   b   c   d

Figure 9.5. Column (a) six slices from the tumor data volume; (b) the Gibbs prior segmentation results on these six slices; (c) the deformable model segmentation results on
these slices; (d) the projection of the segmentation results onto the original image. [Data
courtesy of Dr. Peter Ratiu, Harvard Brigham and Women's Hospital.]

the input of the gradient vector flow filter, also provided in ITK. External forces will push the deformable surface to the features of interest in 3D volumes or 2D images.

We use the following three experiments to show how to use these classes. In the first experiment, we use a deformable model to segment the left ventricle. This is a 2D segmentation task. We initialize the model as a deformable curve inside the object using the a superquadric ellipsoid. The closed curve expands under the effect of a balloon force. We use a region estimation provided by a segmentation methodology using the Gibbs prior model to define the effective range of the balloon force. When the surface of the deformable model is close enough to the boundary, the magnitude of the balloon force decreases, while the magnitude for the gradient flow increases. The curve then deforms to fit to the boundary feature in the image. The final result is shown in Figure 9.3. (The data we used in this experiment are a 256×256 MRI image provided by Professor Leon Axel's group at New York University.)

In the second experiment, we use a 3D deformable model to segment the bone structure of a human jaw. The data is part of the Visible Human Project data set provided by National Library of Medicine. The image size is $256 \times 256 \times 102$ voxels. The jaw is a 3D object with complex surface structures so we create a deformable mesh based on a binary mask created by a Voronoi-based segmentation method (by Dr Celina Imielinska's group at Columbia University [6]). The mesh deforms under the effect of a second order derivative gradient flow. The segmentation process stops after 50 steps. One slice of the result is shown in Figure 9.4.

In the third experiment, we use a deformable model to segment a tumor in the brain. The volume size of the tumor dataset is $256 \times 256 \times 29$ voxels. (It was provided by Dr. Peter Ratiu's group at Harvard University.) This is also a 3D object with complex surface structures. We use a region-based segmentation method using the Gibbs prior model to create the 3D binary mask and then create a deformable mesh. Gradient derived external forces are applied to attract the deformable mesh to the surface of the tumor. The experiment was done on a PIII 1GHz pc. The operating system is WinXP. The segmentation process takes about thirteen minutes. Several slices of the final 3D segmentation result are shown in Figure 9.5.

References

[1] Caselles, V., Kimmel, R., and G. Sapiro. "Geodesic active contours," in *Fifth International Conference on Computer Vision* (Proceedings of ICCV 1995) Los Alamitos: IEEE Computer Society Press, 1995. 694–699.

[2] Chan, T. and L. Vese. "Active contours without edges," *IEEE Transactions on Image Processing* 10(2)(2001): 266–277.

[3] Chen, T. and D. Metaxas. "Image segmentation based on the integration of Markov random fields and deformable models," in *Medical Image Computing and Computer-Assisted Intervention* (Proceedings of MICCAI 2000), S. Delp, A. DiGioia, and B. Jaramaz, eds. Springer, LNCS1935(October 2000): 256–265.

[4] Cohen, L. "On active contour models and balloons," *Computer Vision, Graphics, and Image Processing: Image Understanding* 53(2):211–218:1991.

[5] Cohen, L. and I. Cohen. "Finite element methods for active contour models and balloons for 2D and 3D images," *IEEE Transactions on Pattern Analysis and Machine Intelligence* 15(11)(1993): 1131–1147.

[6] Imielinska, C., Metaxas, D., Udupa, J., Jin, J., and T. Chen. "Hybrid Segmentation of Anatomical Data," in *Medical Image Computing and Computer-Assisted Intervention* (Proceedings of MICCAI 2001), W. Niessen and M. Viergever, eds. Springer, LNCS2208(October 2001): 1048–1057.

[7] Kass, M., Witkin, A., and D. Terzopoulos. "Snakes: Active contour models," *International Journal of Computer Vision* 1(4)(1987): 321–331.

[8] Lorensen, W. and H. Cline. "Marching cubes: a high resolution 3D surface construction algorithm," *Computer Graphics* 21(4)(1987): 163–168,

[9] Malladi, R., Sethian, J., and B. Vemuri. "Shape modeling with front propogation: A level set approach," *IEEE Transactions on Pattern Analysis and Machine Intelligence* 17(2)(1995): 158–175.

[10] McInerney, T. and D. Terzopoulos. "A dynamic finite element surface model for segmentation and tracking in multidimensional medical images with application to cardiac 4D image analysis" *Computerized Medical Imaging and Graphics* 19(1)(1995): 69–83.

[11] Montani, C., Scateni, S., and R. Scopigno, "Discretized marching cubes," in *Proceedings of IEEE Visualization 1994* IEEE Computer Society Press (October 1994): 281–287.

[12] Park, J., Metaxas, D., and L. Axel. "Volumetric deformable models with parameter functions: a new approach to the 3D motion analysis of the LV from MRI-SPAMM," in *Fifth International Conference on Computer Vision* (Proceedings of ICCV 1995) Los Alamitos: IEEE Computer Society Press, 1995. 700–705.

[13] Sethian, J. *Level Set Methods: Evolving Interfaces in Geometry, Fluid Mechanics, Computer Vision, and Material Sciences.* Cambridge: Cambridge University Press, 1996.

[14] Staib, L. and J. Duncan. "Deformable fourier models for surface finding in 3D images," in *Visualization In Biomedical Computing 1992* (Proceedings of VBC92, Chapel Hill, NC 1992), R. A. Robb, ed. SPIE 1808(1992): 90–104.

[15] Staib, L. and J. Duncan. "Boundary finding with parametrically deformable models," *IEEE Transaction on Pattern Analysis and Machine Intelligence* 14(11)(1992): 1061–1075.

[16] Terzopoulos, D. "On matching deformable models to images, " Topical Meeting on Machine Vision, *Technical Digest Series*, 12(1987): 160–167.

[17] Terzopoulos, D. "Regularization of Inverse Visual Problems Involving Discontinuities," *IEEE Transaction on Pattern Analysis and Machine Intelligence* 8(4)(1986): 413–424.

[18] Whitaker, R. "Volumetric deformable models: Active blobs," in *Visualization In Biomedical Computing 1994* (Proceedings of VBC94, Rochester, Minnesota 1994), R. A. Robb, ed. SPIE 2359(1994): 122–134.

[19] Worring, M., Smeulders, A., Staib, L., and J. Duncan. "Parameterized feasible boundaries in gradient vector fields," *Computer Vision and Image Understanding* 63(1)(1996): 135–144.

[20] Xu, C. and J. Prince. "Snakes, shapes, and gradient vector flow," *IEEE Transactions on Image Processing* 7(1998): 359–369.

[21] Xu, C. and J. Prince. "Generalized Gradient Vector Flow External Forces for Active Contours," *Signal Processing* 71(2)(December 1998): 131–139.

[22] Yezzi, A., Kichenassamy, S., Kumar, A., Olver, P., and A. Tannenbaum. "A geometric snake model for segmentation of medical imagery," *IEEE Transactions on Medical Imaging* 16(2)(April 1997): 199–209.

Part Three

Registration

Medical Image Registration: Concepts and Implementation

Lydia Ng
Insightful, Inc.

Luis Ibanez
Kitware, Inc.

10.1   Introduction

Registration is the process of finding the spatial transform that maps points from one image to the corresponding points in another image. Medical image registration has many clinical and research applications [17, 13, 7, 9]. For example, repeated image acquisition of a subject is often used to obtain time series information that captures disease development, treatment progress and contrast bolus propagation. Although gross changes in the serial images can be detected by a visual comparison of the images at different time points, image registration enables the detection of subtle changes by eliminating the effect of patient placement and motion artifacts. Once the serial images have been aligned, subtraction can be used for visualization and quantification.

Registration can also be a valuable tool for correlating information obtained from different imaging modalities. For example, magnetic resonance (MR) images have good soft tissue discrimination for lesion identification, while CT images provides bone localization useful for surgical guidance. On the other hand, PET (positron emission tomography) and SPECT (single photon emission computed tomography) images provide functional information that can be used to locate abnormalities such as tumors. Example of multi- (or intra-) modality applications include the study of brain tumors (PET/MR) and radiation treatment planning (PET/planning x-ray CT).

Mean images and atlases are useful to assess abnormalities objectively and quantitatively, for example, in the study of the relationship between brain function and structure. In this scenario, registration algorithms are used to determine the complex differences in structure between subjects and the resulting deformation field is used to encode patterns of anatomic variability within a population.

The purpose of this chapter is to provide a practical "how-to" guide for performing medical image registration. In particular, the chapter begins with an introduction to basic image registration followed by a presentation of a generic and extensible software framework for solving registration problems.

10.2   Image Registration Concepts

10.2.1   Registration Criteria

In the literature, many criteria have been used as the basis for aligning two images. Generally these criteria can be *landmark-*, *segmentation-* or *intensity-*based [17, 7]. Landmark-based registration uses salient features selected by the user. These features are usually points but can also be lines or more complex structures such as corners or crossings. Since the number of identified features is sparse compared to the image content, landmark-based methods are fast to compute. The major drawback is that the method requires user interaction to locate the landmarks. Human location of landmarks suffers from a lack of consistency and reproducibility, therefore undermining the precision of the final registration.

Segmentation-based methods attempt either rigidly or deformably to align the binary structure (curves, surfaces or volumes) obtained by segmentation. The segmented structure of one image can be aligned either to a segmented structure on the second image or to the whole unsegmented second image. In the latter case, the criteria typically require that the boundary of the binary structure matches to "edges" in the second images. Due to the reduction of information, segmentation-based methods are faster than methods using full image content. On the other hand, one of the drawbacks of segmentation-based methods is that the performance of the registration relies on the accuracy of the segmentation pre-processing step.

Intensity-based methods operate directly on the image intensity. They are more flexible than landmark-based or segmentation-based methods as they use all of the available information without previous reduction of data either by the user or by a segmentation algorithm and are typically automatic. However, using full image content is computationally very expensive especially for 3D images and hence may not be suited to time-constrained applications. In practice, it is common to use a multi-resolution approach both to speed up the computational time and to increase the capture range of the algorithm.

In landmark-based and segmentation-based methods, the procedure typically involves minimizing the distance between physical points. In contrast, intensity-based methods involve minimizing a cost function that measures the similarity between the image intensity of corresponding points of the two images. When the two images to be registered have been acquired using the same imaging modality, measuring the similarity is fairly straightforward. When the images arise from different imaging modalities, the intensity mapping between homologous points is more complex. In the literature, popular approaches include correlation ratios and the information theoretic measure of mutual information [4, 26].

10.2.2   Spatial Transformation

A registration problem is also classified by the type of the spatial transformation used to map points from the space of one image to the space of the second image. A transform is typically defined by a set of parameters. The goal of registration is to find the optimal parameters with respect to the registration criterion used. Typically, the more parameters or degrees of freedom a transformation has, the more difficult it is to solve the optimization problem.

For 2D/2D or 3D/3D registration problems, the spatial transformation can be rigid, affine or deformable. In a rigid transformation, only rotations and translations are allowed. Affine transformations allow skew and scaling in addition to rotation and translation. Deformable transformations define free-form mappings and are typically used with a regularization constraint to limit the allowable solution space.

Rigid body transformation can be used when there are no changes in the shape of the structure being imaged. A common application is for images of the human head, where the skull prevents non-rigid deformation. Affine transform is typically used in rigid body problems where the imaging scaling factors are unknown.

Rigid body techniques are currently well-validated. Deformable registration techniques are more complex and, as a result, difficult to validate; however, deformable registration has been expanding in clinical applications including inter-subject brain registration and dynamic contrast breast MRI registration. The topic of deformable registration will be covered in detail in the next chapter.

10.2.3   Parameter Optimization

The registration process can be viewed as an optimization problem, where the registration criterion is the cost function to be minimized over the search space spanned by the spatial transformation parameters. Starting from an initial set of parameters, the optimization procedure iteratively searches for a solution by evaluating the cost function at different positions in the search space. Optimization algorithms can be divided broadly into those that require the use of derivatives and those which do not. Typically, if the derivative of the cost function to be optimized is available, the former type should be used. Detailed discussion of optimization algorithms will not be presented in this chapter. The reader is referred to various monographs on optimization methods such as [22, 21] for more information.

10.2.4   Image Interpolation

When mapping points from one image space to another, the mapped point will generally fall on a non-grid position in the target image. In these cases, to obtain the image intensity at the position, an image interpolation method is needed. In the context of intensity-based image registration, the actual interpolation method used affects the smoothness of the optimization search space and since these interpolations are preformed thousands of times in a single optimization cycle, the selection of an interpolation scheme must trade off simplicity of computation versus the ease of optimization. In the literature, the popular methods include linear and B-spline interpolation.

10.3   A Generic Software Framework for Image to Image Registration

Registration problems involving images from different modalities and/or different anatomical structures have their own unique requirements. Monolithic registration algorithms in general can not satisfy these often potentially competing requirements. Moreover, techniques that have more flexibility (e.g., affine versus rigid or multi-modality versus mono-modality) typically come at the price of a higher computational burden and a more complex optimization search space. For example, when dealing with images of the human head, it is more efficient to limit the transform to rigid body motions instead of an affine transform where the extra degrees of freedom are not required and unnecessarily complicate the optimization process. Even when solving one particular registration problem, the most efficient approach may be to apply several techniques, starting with simple and inexpensive methods to perform the gross alignment and then moving to more complex methods to refine the registration.

What would be useful in practice is to have access to a suite of different registration methods to allow a user to choose the right tool for a particular problem or to compare and contrast between techniques. Literature on registration typically highlights differences between methods, however, registration techniques also have vast commonality. In the implementation of software for registration,

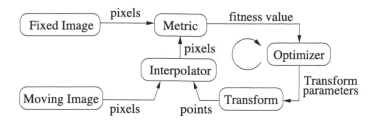

Figure 10.1. The basic components of the registration framework are two input images, a transform, a metric, an interpolator, and an optimizer.

good software engineering practices focus on identifying and isolating common functionalities. Factorization of registration techniques into common functional components simplifies software design, implementation and code maintenance by facilitating encapsulation, modulariy and code reuse [1, 2, 3, 8].

In this section, we will describe the design of a generic software framework for performing image-to-image intensity-based registration. A generic image-to-image registration method can be functionally decomposed into image, similarity cost function, spatial transformation, optimization and interpolation components. For particular registration methods, the role of each component can be accomplished by different techniques. Thus, in theory a combinatorial variety of possible registration methods exists. This structural characteristic suggests the use of the *strategy* software design pattern where different algorithms are encapsulated separately but can be used interchangeably [1, 8].

Figure 10.1 shows the components of the image-to-image registration framework and their interconnections. The input data to the registration process are two images: one is defined as the *fixed* image, $F(x)$, and the other defined as the *moving* image, $M(x)$. The goal of registration is to find the spatial mapping that will bring the moving image into alignment with the fixed or target image. The *transform* component, $T(p)$, is used to map points between the fixed and moving images for a given set of transform parameters, p. The *metric* component, $S(p|F,M,T)$, represents the similarity measure of how well the fixed image is matched by a transformed moving image. This measure forms the quantitative criterion to be optimized by the *optimizer* over the search space defined by the parameters of the transform.

10.3.1 Transforms

Overview

The final goal of registration is to obtain a mapping from the coordinate system of one image into the coordinate system of another image. This mapping is termed as a *transform*. In principle any algorithm capable of finding for every point of the fixed image space an associated point in the moving image space can be considered to be a satisfactory transform. In general, a transform may not necessarily be invertible, although this may be a desirable property in some applications. There are a large variety of possible choices for the transform. The most common are the geometric transformations associated with rigid movements and those associated with conservation of rectilinear features.

A transform $T(p)$ is typically defined by a set of transform parameters, p. A point x of the fixed image coordinate system S_f will be mapped into the point x' of the moving image coordinate system S_m by the expression

$$x' = T(x|p). \tag{10.1}$$

The whole image space is mapped according to:

$$T : \mathbb{R}^n \mapsto \mathbb{R}^m \text{ such that } T(S_f) = S_m, \tag{10.2}$$

where n and m are, respectively, the dimension of the fixed image and moving image spaces.

As an example, let's consider the case of a rigid transform in 2D space. The transform T will be completely defined by the three parameters $p = \{t_x, t_y, \theta\}$. The parameters t_x and t_y define a translation in 2D space, while the parameter θ defines a rotation angle in 2D. Given a point x in the coordinates system of the fixed image S_f, its homologous point x' can be found in the coordinate system of the moving image S_m by applying the transform:

$$x' = \begin{bmatrix} x' \\ y' \end{bmatrix} = \begin{bmatrix} \cos\theta & -\sin\theta \\ \sin\theta & \cos\theta \end{bmatrix} \cdot \begin{bmatrix} x \\ y \end{bmatrix} + \begin{bmatrix} t_x \\ t_y \end{bmatrix}, \tag{10.3}$$

hence

$$x' = T(x|p) = T(x, y|t_x, t_y, \theta). \tag{10.4}$$

Mapping from the Image Discrete Grid

When performing image registration and image resampling it is very important to keep in mind that in addition to the transform mapping one coordinate system into the other, there are two implicit transforms mapping the discrete grid of each image into its space coordinate system. Overlooking this fact is a common source of confusion in practical image registration applications. Figure 10.2 illustrates the sequence of transformations required to map the integer coordinates of a pixel from an image grid into the integer coordinates of its homologous pixel in another image grid. Given the integer coordinates of a pixel in the discrete grid of the fixed image, its associated position in the coordinate system S_f can be computed by a transform involving the image origin, the pixel spacing and any specification of

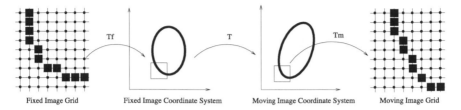

| Fixed Image Grid | Fixed Image Coordinate System | Moving Image Coordinate System | Moving Image Grid |

Figure 10.2. The mapping of pixels from the fixed image into the moving image involves three different transforms: the mapping T_f from the fixed image grid to the fixed image space, the mapping T from the fixed image space to the moving image space and the mapping T_m from the moving image space to the discrete grid of the moving image.

the image orientation. The definition of this mapping starts with the fact that the image grid is composed of samples. From an engineering point of view, these samples are the result of computations performed over a number of physical measures. When we say that the image $I(x)$ has a pixel (i,j) with value $I(i,j)$ we are describing the fact that this value is in some way representative of a certain region around the location of the grid node (i,j). Following this reasoning, the nodes on the image grid represented by small filled circles on the left side of Figure 10.2 are the integer positions (i,j). Each one of these nodes is the center of a pixel.

Filled rectangles represent the area covered by a pixel. Coverage of a pixel is an open choice. In theory, the shape of the pixel coverage should be deduced from the acquisition and reconstruction process of the particular image modality. For example, if the image was acquired through an optical microscope, the pixel will be given by the point spread function (PSF) of the optical system, which defines the effective resolution of the image. If, on the other hand, the image was acquired with a MRI scanner, the parameters of the numerical method used for the reconstruction will determine the area of the space for which the value $I(i,j)$ can be considered a good representative of the continuous field $I(x)$.

The shape of the pixel defines also the optimal kernel function that could be used for recovering the continuous field $I(x)$ by interpolation from the discrete image grid $I(i,j)$. In practice, performance considerations take precedence over the mathematical rigor of image reconstruction, leading to the use of simplified pixel shapes associated with linear and B-spline interpolators. Unfortunately, it is common to find software implementations of image processing that represent pixels as rectangular regions and associate their coordinates to the corner of one of those regions. These representations do not honor the physical reality of the image acquisition process nor the fundamental principles of digital signal processing. Formally, the values of a digital image in the nodes $I(i,j)$ are associated to Dirac deltas $\delta(x)$ as illustrated in Figure 10.3. The evaluation of the continuous image field $I(x)$ requires a process of interpolation from the discrete values $I(i,j)$.

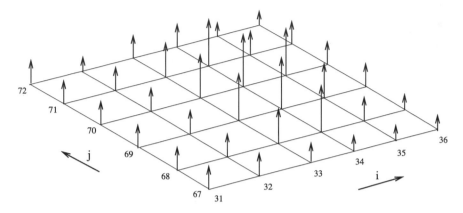

Figure 10.3. Dirac deltas representing the pixel values in a digital image.

The mapping from image grid coordinates into physical space coordinates is computed by taking into account the image elements illustrated in Figure 10.4. The grid point on the corner of the image has a physical space position associated with it. This position is known as the *origin* of the image. In other words, the origin of the image in physical space is the coordinates of the pixel center in position $(i = 0, j = 0)$. In the case illustrated in Figure 10.4, the image origin is located on the physical point $O = (60.0\text{mm}, 70.0\text{mm})$. The physical distance between two consecutive points in the image discrete grid is known as *pixel spacing* and is

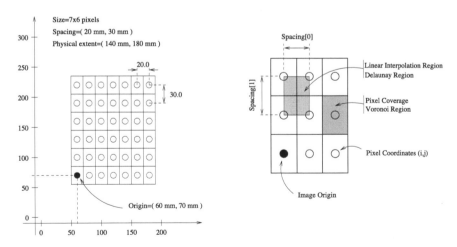

Figure 10.4. Elements defining the mapping from the discrete grid into the physical space. Image origin, pixel spacing, and the notion of pixel center are illustrated.

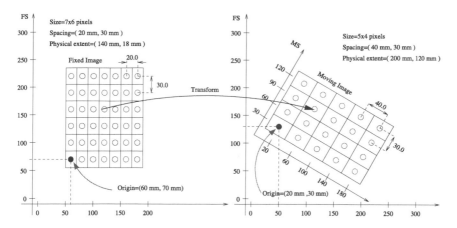

Figure 10.5. A transform maps points from the coordinate system of the fixed image into the coordinate system of the moving image. The main image characteristics involved in coordinate system transformation are shown in this figure.

specified along each dimension. In the example of Figure 10.4, the image spacing is 20.0mm along the x-axis and 30.0mm along the y-axis. These elements lead to the following simple transform between the grid coordinates and the physical space:

$$ x' = \begin{bmatrix} x' \\ y' \end{bmatrix} = \begin{bmatrix} D_x & 0 \\ 0 & D_y \end{bmatrix} \cdot \begin{bmatrix} i \\ j \end{bmatrix} + \begin{bmatrix} O_x \\ O_y \end{bmatrix}, \tag{10.5} $$

where $O = (O_x, O_y)$ are the origin coordinates and $D = (D_x, D_y)$ are the values of pixel spacing. The spacing values act as scaling factors while the origin coordinates act as a translation factor.

Figure 10.5 illustrates the relationship between the coordinate system of the moving image S_m and the coordinate system of the fixed image S_f. The left side of the figure illustrates the case of a fixed image 7×6 pixels in size, with pixel spacing $(20.0\text{mm}, 30.0\text{mm})$. The physical extent covered by the image is then $(140.0\text{mm}, 180.0\text{mm})$. The origin of the image in this coordinate system is in the position $(60.0\text{mm}, 70.0\text{mm})$. The right side of the figure presents the case of a moving image whose coordinate system S_m is rotated, translated, and scaled with respect to the fixed-image coordinate system S_f. The moving image is 5×4 pixels in size, with a pixel spacing $(40.0\text{mm}, 30.0\text{mm})$ for a total physical extent of $(200\text{mm}, 120\text{mm})$. The moving image has its origin in the position $(20\text{mm}, 30\text{mm})$ in S_m. The role of the transform T in the registration process is to convert coordinates from the fixed-image coodinate system S_f into the moving-image coordinate system S_m.

In the framework of medical image applications, the transforms T_f and T_m

account for most of the burden of converting coordinates between the fixed image and moving image spaces. It is then, of fundamental importance to ensure that the image parameters of *origin* and *spacing* are correctly carried over from the acquisition process into the registration process.

Progressive Registration

From the point of view of the registration framework, a transform is a black box capable of receiving a point in the fixed-image coordinate system and computing its homologous point in the moving-image coordinate system. The transform is generally defined by a set of parameters. These parameters constitute the dimensions of the parametric space used by the optimizer as a search space. The more complex the transform is, the larger its number of parameters will be and the harder the optimization problem.

It is a common practice to perform registration as a series of intermediate stages, starting with fairly simple transforms and using progressively more complex ones. This approach improves at the same time as the capture radius and reduces the computational burden of the overall registration process. The motivation and results for this approach are similar to those of the multi-resolution approach, since a registration is obtained by starting with a fairly simple registration problem and progressively adding details. It is not unusual to use different types of transforms at each level of the pyramid in a multi-resolution process. Details on the application of multi-resolution methods are presented in Section 10.4.1.

Transforms and Metric Derivatives

The optimizer searches for better transform parameters by exploring how their values impact the computation of the image similarity metric. The goal of registration is to find the set of transform parameters that optimizes the value of the image similarity metric. In a generic software framework, the communication between the optimizer and the transform is decoupled by passing the parameters in the neutral representation such as an array of floating point numbers. In this way, the optimizer does not have to be aware of the existence of the transform or any of its implementation details and vice versa, the transform does not require knowledge of the optimizer definition and implementation.

Some of the most common optimizers used for registration require cost function derivatives information, that is, the derivatives of the image similarity metric with respect to the parameters of the transform. This computation can be performed in some cases by simple finite differences. However the finite difference approach incurs a significant computational load, especially when the transform has a large number of parameters. It is possible to improve performance by performing the derivative computation analytically using the differential chain-rule, as illustrated in the following equation:

$$\frac{\partial S(p|F,M,T)}{\partial p_i} = \sum_j \frac{\partial S(p|F,M,T)}{\partial x'_j} \cdot \frac{\partial x'_j}{\partial p_i}, \tag{10.6}$$

where the matrix $[\partial x'_j/\partial p_i]$ is called here the *Jacobian* of the transformation. Note that the term Jacobian is ambiguous in this case. Given a transform T that maps a point $x = x_1, x_2, ..., x_N$ into another point $x' = x'_1, x'_2, ..., x'_N$ it is customary to call *Jacobian* the matrix [15]

$$J = \begin{bmatrix} \frac{\partial x'_1}{\partial x_1} & \frac{\partial x'_1}{\partial x_2} & \cdots & \frac{\partial x'_1}{\partial x_n} \\ \frac{\partial x'_2}{\partial x_1} & \frac{\partial x'_2}{\partial x_2} & \cdots & \frac{\partial x'_2}{\partial x_n} \\ \vdots & \vdots & \ddots & \vdots \\ \frac{\partial x'_n}{\partial x_1} & \frac{\partial x'_n}{\partial x_2} & \cdots & \frac{\partial x'_n}{\partial x_n} \end{bmatrix}. \tag{10.7}$$

The determinant $|J|$ of the Jacobian matrix above is often loosely called the *Jacobian of the transformation*. For the purpose of registration, we are rather interested in the following Jacobian:

$$J = \begin{bmatrix} \frac{\partial x'_1}{\partial p_1} & \frac{\partial x'_1}{\partial p_2} & \cdots & \frac{\partial x'_1}{\partial p_m} \\ \frac{\partial x'_2}{\partial p_1} & \frac{\partial x'_2}{\partial p_2} & \cdots & \frac{\partial x'_2}{\partial p_m} \\ \vdots & \vdots & \ddots & \vdots \\ \frac{\partial x'_n}{\partial p_1} & \frac{\partial x'_n}{\partial p_2} & \cdots & \frac{\partial x'_n}{\partial p_m} \end{bmatrix}, \tag{10.8}$$

where the elements p_i are the parameters p of the transformation. This matrix indicates how much the components of a mapped point will change with respect to a variation on the transform parameters.

Note that, strictly speaking, the transformation mapping is a function of the input point coordinates and the transformation parameters

$$x'_j = T(x_1, x_2, ..., x_N, p_1, p_2, ..., p_m). \tag{10.9}$$

It is then equally valid to compute derivatives with respect to the coordinates $\{x_i\}$ considering $\{p_j\}$ to be fixed parameters or to compute derivatives with respect to the parameters $\{p_j\}$ considering the input coordinates $\{x_i\}$ to be fixed. For the purpose of computing the derivative of the image similarity metric, the Jacobian matrix required for the chain-rule is the one presented in Equation 10.8.

This approach for computing the derivatives of the similarity metric increases considerably the performance of the registration process, since the first term of Equation 10.6 is simply dependent on the gradient of the moving image and the

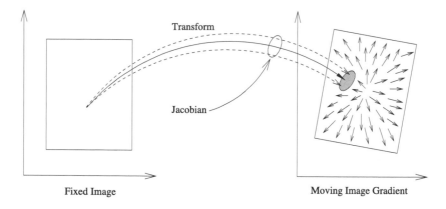

Figure 10.6. Relationship between the variations of the mapped point in the moving image space, as evaluated by the transform Jacobian, and the gradient of image intensities in the moving image. The combination of the Jacobian and the image gradient allow evaluation of the variation of one pixel contribution to the total image metric, and henceforth cumulate values for computing the image metric derivative.

values of the fixed image. These values remain constant during the registration process. The Jacobian J on the other hand, has to be evaluated for every new set of transform parameters.

Figure 10.6 illustrates the interaction between the transform Jacobian and the gradient of the moving image during the computation of image metric derivatives. When mapping a point from the fixed-image space into the moving-image space, the Jacobian indicates how the point resulting from the mapping would move in the moving-image space as a function of variations in the transform parameters. This is represented in the figure by the gray ellipse in the moving-image space. This spatial variation combined with the vectors of the moving-image gradient allow us to estimate how the intensity in the moving image at a mapped point will vary as a function of the transform parameters. The values of such intensity variations are then in the computation of the metric derivative.

In the simple case of a similarity metric whose computation requires visits to all the pixels in the image, the numerical complexity of computing the metric derivatives by finite diferences is given by

$$N_t \times 2N_p \times N \times C_m, \tag{10.10}$$

where N_t is the number of iterations performed by the optimizer, N_p is the number of parameters in the transform, N is the total number of pixels in the image, and C_m is the complexity of computing the contribution of a pixel to the total metric value. When the factorization as described in Equation 10.6 is used, the numerical

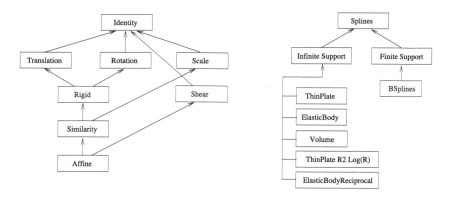

Figure 10.7. Taxonomy of spatial transforms. Each transfrom has an arrow pointing to the transforms it can represent as a particular case.

complexity becomes

$$C_g + N_t \times N \times (C_d + C_j),$$ (10.11)

where C_g is the complexity of computing the moving-image gradient, C_d is the complexity of computing the contribution of one pixel to the derivative of the metric in terms of the moving-image gradient and C_j is the complexity of computing the transform Jacobian in a single point. The complexity C_j is dependent on the number of transform parameters N_p, while the complexity C_d is dependent on the image dimension. From Equation 10.11 it can be observed that the savings in computation becomes more significant when transforms with large number of parameters are used.

Transforms Taxonomy

A large variety of spatial transforms can be considered for registration. The most commonly used can be classified in a taxonomy like the one illustrated in Figure 10.7. The classification is arranged in such a way that the transform in the top of the hierarchy is less general than the ones below. The identity transform is naturally placed at the top of the hierarchy. Transforms down in the hierarchy can be set to behave as the identity transform by selecting a particular combination of their parameters. The arrows in the diagram point from one transform to another that can be considered to be one of its particular cases [6].

Identity Transform.   The identity transform simply maps every point to itself. It has no parameters, and it is only used in the context of software frameworks for validation and testing purposes. Most transforms can be configured to behave as an identity transform by selecting a particular set of parameters. This fact is commonly used as a first test for verifying the correct implementation of a registration method. Performing the registration of an image with itself should produce the set of parameters that makes the transform equivalent to the identity.

One of the fundamental characteristics defining a transform is its *fixed set* [18, 28, 5]. The fixed set of a transform is the set of points that remain unchanged by the transformation. In the case of the identity transform, the fixed set is the entire space.

Following the identity transform in the diagram we find the transforms representing pure translation, pure rotation and pure scaling.

Translation Transform.   Pure translation is in general represented by a vector V indicating the displacement to be applied to every point of the input space in order to map it into the output space:

$$
x' = \begin{bmatrix} x'_1 \\ x'_2 \\ \vdots \\ x'_N \end{bmatrix} = \begin{bmatrix} x_1 \\ x_2 \\ \vdots \\ x_N \end{bmatrix} + \begin{bmatrix} V_1 \\ V_2 \\ \vdots \\ V_N \end{bmatrix}. \tag{10.12}
$$

A translation transform has N parameters, where N is the image dimension. The parameters are naturally the same components of the V vector. Translation transforms are easily managed in the context of image registration. They are in general the first transform to be considered, especially when a progressive registration approach is being used. Since translations are represented by a vector, they are compatible with most common optimizers which have also been devised for exploring vector spaces [16].

The fixed set of the translation transform is an empty set since no point remains unchanged during the translation. However, a translation can be approximated locally to any degree of precision by a scale transform or a rotation transform whose origin is placed arbitrarily far from the region in which we are interested. In the case of medical images, we are interested in very restricted regions of space. It is then relatively easy to place a fixed point far enough from the image in order to represent a translation by using a rotation or a scale change.

This fact is illustrated in Figure 10.8. The left side of the figure shows how a translation on the image could be generated by a rotation around a fixed point located far away. The resulting translation will not be a pure translation since the object will be rotated by the same angle used in the rotation transform. However, the rotation angle can be reduced to an arbitrarily small value by moving the fixed point farther away. The translation vector is the tangent to the arc of the rotation transform. The right side of the figure illustrates how a translation can be locally approximated by a scaling transform using a fixed point located far away from the region of interest. In this case the translation vector is parallel to the line joining

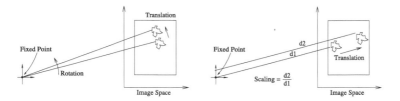

Figure 10.8. Translation can be approximated locally by scaling or rotations using a fixed point located far away from the region of interest.

the region of interest with the fixed point of the scaling transform. The translation is not perfect because the final object will have a different scale, however, the scaling factor can be set arbitrarily close to 1.0 by placing the fixed point farther away.

For image registration applications, it is very important to understand the interplay between the translation component and the center of rotation and scaling for the purposes of initializing the optimization process and also for intepreting the final results.

Scaling Transform.   The scaling transform has a false appearance of simplicity. Despite the fact of being represented with few parameters, it is fundamentally different from the translation transform because its parameters have non-linear effects on the coordinates of mapped points. There are two very different levels at which the scaling transform can be considered. The first and simpler one is the isotropic scaling in which a single scaling factor is applied to all components of the points' positions. The second and more complex level is the anisotropic scaling in which different scaling factors are considered along each dimension of space [18, 28].

In the case of isotropic scaling, the transform is defined by the equation

$$
\mathbf{x'} = \begin{bmatrix} x'_1 \\ x'_2 \\ \vdots \\ x'_N \end{bmatrix} = D \begin{bmatrix} x_1 \\ x_2 \\ \vdots \\ x_N \end{bmatrix}, \tag{10.13}
$$

where D is the scaling factor. One of the common difficulties with this transform is to overlook the fact that the scaling is performed irradiating from the origin of coordinates. In other words, the fixed point of the transform is the origin of coordinates [18]. Since it is common to place the origin of coordinates in the corner of an image, the effects of the transform are counter-intuitive for human observers who implicitly assume the fixed point of the scaling to be in the center of the image. This produces the impression that both a scaling and a translation

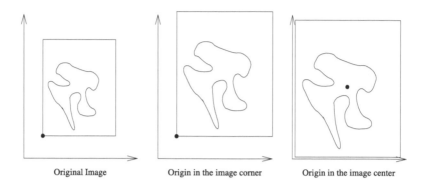

Original Image   Origin in the image corner   Origin in the image center

Figure 10.9. Effect of the position of coordinate system origin in the mapping of the scaling transform.

have been applied simultaneously. We already discussed in the context of the translation transform, how a translation can be applied by a scaling transform and a judicious choice of the scaling fixed point.

Image Center. The selection of the origin of coordinates has dramatic effects in the process of image registration. The intuitive picture of scaling and rotation operations is that they should be applied using the center of the image as the origin of coordinates. That is, when we say that we want to rotate an image by 30 degrees we usually mean to use the center of the image as the fixed point and rotate 30 degrees around this center of coordinates. The different effects of the scaling transform that depend on the position of the coordinate system origin are illustrated in Figure 10.9. The actual coordinate system of the image however, is carried from the image acquisition process. This simple fact leads to the use of a more realistic transform in which the fixed point of the scaling is selected as part of the transformation parameters. In this case, the transform is fully specified by a point C and a scaling factor D. The point C is actually the fixed point of the transformation. The effect of the transform is given by the following equation:

$$
x' = \begin{bmatrix} x_1' \\ x_2' \\ \vdots \\ x_N' \end{bmatrix} = D \begin{bmatrix} x_1 - C_1 \\ x_2 - C_2 \\ \vdots \\ x_N - C_N \end{bmatrix} + \begin{bmatrix} C_1 \\ C_2 \\ \vdots \\ C_N \end{bmatrix}. \tag{10.14}
$$

In this case the transform has $(N+1)$ parameters. The first parameter is the scale factor D, and the remaining N parameters are the N components of the point C coordinates. This configuration of the scaling transform is capable of representing translations and scale changes simultaneously. The translation can be made explicit in the following expression:

$$
\begin{bmatrix} x_1' \\ x_2' \\ \vdots \\ x_N' \end{bmatrix} = D \begin{bmatrix} x_1 \\ x_2 \\ \vdots \\ x_N \end{bmatrix} + (1-D) \begin{bmatrix} C_1 \\ C_2 \\ \vdots \\ C_N \end{bmatrix} = D \begin{bmatrix} x_1 \\ x_2 \\ \vdots \\ x_N \end{bmatrix} + \begin{bmatrix} T_1 \\ T_2 \\ \vdots \\ T_N \end{bmatrix}, \qquad (10.15)
$$

where T stands for the translation induced by the scaling. Any translation can be represented by a judicious choice of the fixed point C. In fact, the fixed point is simply computed as $C = T/(1-D)$. This representation provides poor precision when the scaling factor is close to one, since in this case the fixed point tends to be at infinite distance. In such cases it is a moderate compromise to use an auxiliary translation transform T, so the expression of the total transform becomes

$$x' = D(x - C) + C + T. \qquad (10.16)$$

With this expression, the fixed point C is taken as simply a relocation of the origin of coordinates and the transform parameters used for the optimization are the scaling factor D and the components of the auxiliary translation T.

Optimization.   The optimization of the scaling transform is a sensitive process. The fundamental difficulty originates in the non-linear relationship between the scaling factor and the coordinates of the mapped points. Optimizers designed for vector spaces are not well-suited for managing the scaling parameter [16]. In principle such optimizers should optimize the logarithm of the scaling factor, so that the step length of the optimizers maps uniformly in the parameter space. In practice, however, the scale factors found in medical image registration tend to be close to 1.0 making unecessary the use of the logarithm.

A scale factor update is illustrated in the following equation where a typical gradient descent optimizer is applied. Starting from the scaling value D, the gradient G of the cost function in the parameter space, and a step length λ, the next scale value in the parameter space is computed as

$$D' = D + G \cdot \lambda. \qquad (10.17)$$

In the context of medical image registration, we rarely expect the scaling factor to be far from 1.0. Since any compensation of the image resolution should have to be solved by the T_f and T_m transforms (see Section 10.3.1), the T transform should only deal with small scale corrections. Different instances of human bodies are not extremely different in scale. Unless we attempt to register infants to adults, we rarely will require the use of scaling factors below 0.5 or over 2.0.

Let's consider now the optimization problem of the scaling transform with a specified origin of coordinates, as expressed in Equation 10.14. In this case, the coordinates of the point used as fixed point are linear with respect to the co-ordinates of the mapped point. This means that if an optimizer computes the derivative of the image metric with respect to the parameters of the transform $[D, C_1, C_2, \cdots, C_N]$, we will obtain a vector of $(N+1)$ components like

$$\left[\frac{\partial S}{\partial D}, \quad \frac{\partial S}{\partial C_1}, \quad \cdots, \quad \frac{\partial S}{\partial C_N} \right]. \tag{10.18}$$

When this vector is multiplied by a time step and used for updating the current parameters $[D, C_1, C_2, \cdots, C_N]$, it should be applied as

$$\left[D + \frac{\partial S}{\partial D} \cdot \lambda, \quad C_1 + \frac{\partial S}{\partial C_1} \cdot \lambda, \quad \cdots, \quad , C_N + \frac{\partial S}{\partial C_N} \cdot \lambda \right]. \tag{10.19}$$

Attention should be paid here to the relative dynamic range of the parameters. In particular, the scaling factor is expected to be in the range $[0.5 : 2.0]$ while the fixed point coordinates are expected to be in the range of the image extent in millimeters. It is convenient to normalize these parameters to a common dynamic range in order to create the conditions in which each paramater can exert a similiar influence on the evolution of the optimization process.

Anisotropic Scaling Transform. Anisotropic scaling is a difficult variant of the scaling transform. It uses a different scaling factor for each dimension of the space. In the context of medical images it is useful for correcting minor calibration errors in scanners which result in the pixel spacing parameters being innaccurate. For example, a dataset acquired with pixel spacing of $(1.2\text{mm}, 1.2\text{mm}, 2.0\text{mm})$ may be incorrectly reported by a miscalibrated scanner as having a pixel spacing of $(1.3\text{mm}, 1.3\text{mm}, 1.9\text{mm})$. In this case, an anisotropic scaling transform will allow one to compensate for the incorrect spacings when these data are registered against another dataset for which the pixel spacing is set correctly.

The following equation represents the effect of the anisotropic scaling filter:

$$\mathbf{x}' = \begin{bmatrix} x'_1 \\ x'_2 \\ \vdots \\ x'_N \end{bmatrix} = \begin{bmatrix} D_1 & 0 & \cdots & 0 \\ 0 & D_2 & \cdots & 0 \\ \vdots & \vdots & \ddots & \vdots \\ 0 & 0 & 0 & D_N \end{bmatrix} \cdot \begin{bmatrix} x_1 \\ x_2 \\ \vdots \\ x_N \end{bmatrix}. \tag{10.20}$$

As in the case of isotropic scaling, this transform is sensitive to the choice of the fixed point. In a more general representation the fixed point can be added explicitly to the previous equation as

$$\mathbf{x}' = \begin{bmatrix} x'_1 \\ x'_2 \\ \vdots \\ x'_N \end{bmatrix} = \begin{bmatrix} D_1 & 0 & \cdots & 0 \\ 0 & D_2 & \cdots & 0 \\ \vdots & \vdots & \ddots & \vdots \\ 0 & 0 & 0 & D_N \end{bmatrix} \cdot \begin{bmatrix} x_1 - C_1 \\ x_2 - C_2 \\ \vdots \\ x_N - C_N \end{bmatrix} + \begin{bmatrix} C_1 \\ C_2 \\ \vdots \\ C_N \end{bmatrix}. \tag{10.21}$$

An appropriate choice of the fixed point C could be used to represent an additional translation. In practice the equation above can be used for representing anisotropic scaling and translation.

The parameters of this transform are

$$[D_1, D_2, \cdots, D_N, C_1, C_2, \cdots, C_N]. \tag{10.22}$$

The gradient of the cost function with respect to the parameter will produce the vector

$$\left[\frac{\partial S}{\partial D_1}, \frac{\partial S}{\partial D_2}, \cdots, \frac{\partial S}{\partial D_N}, \frac{\partial S}{\partial C_1}, \frac{\partial S}{\partial C_2}, \cdots, \frac{\partial S}{\partial C_N} \right]. \tag{10.23}$$

A gradient descent optimizer using a step length λ would compute the next transform from the set of updated parameters as given by

$$\left[\frac{\partial S}{\partial D_1} \lambda, \cdots, \frac{\partial S}{\partial D_2} \lambda, \frac{\partial S}{\partial C_1} \lambda, \cdots, \frac{\partial S}{\partial C_N} \lambda \right]. \tag{10.24}$$

Here again, we encounter the conflicting case of representing translations when the scaling factors are very close to one. This scaling factor will force the fixed point to be at a very large distance. In such situations it is a good compromise to use an auxiliary translation transform T, so the expression of the total transform becomes

$$x' = D \cdot (x - C) + C + T, \tag{10.25}$$

and the fixed point C is set only approximatively. The optimizer will use the set of parameters $[D_1, D_2, \cdots, D_N, T_1, T_2, \cdots, T_N]$ as its search space.

Rotation Transform.   The rotation transform shares many of the characteristics of isotropic scaling. In particular, rotations are dependent on the origin of coordinates. The origin will be the fixed point of the transformation. A well-defined rotation transform should include the specification of its fixed point.

We analyze in the following sections the particular cases of rotations in 2D and 3D spaces. Although rotations can be analyzed in higher dimensions in terms of the rotational groups $SO(n)$, they have not been conveniently parameterized [5].

Rotations in 2D.   The classical representation of rotations in 2D space is given by the following expression:

$$x' = \begin{bmatrix} x' \\ y' \end{bmatrix} = \begin{bmatrix} \cos\theta & -\sin\theta \\ \sin\theta & \cos\theta \end{bmatrix} \cdot \begin{bmatrix} x - C_x \\ y - C_y \end{bmatrix} + \begin{bmatrix} C_x \\ C_y \end{bmatrix}, \tag{10.26}$$

where $C = (C_x, C_y)$ is the center of rotation, which is also the fixed point of the transformation. The angle θ defines the rotation around this center C. This representation can recover both translations and rotations. The set of parameters representing the rotation can be selected to be $[\theta, C_x, C_y]$, in which case the angle θ has a non-linear effect on the computation of the mapped point coordinates (x', y').

We can recover the more traditional presentation of the transform as a rotation plus a translation by grouping the terms related to the fixed point:

$$
x' = \begin{bmatrix} x' \\ y' \end{bmatrix} = \begin{bmatrix} \cos\theta & -\sin\theta \\ \sin\theta & \cos\theta \end{bmatrix} \cdot \begin{bmatrix} x \\ y \end{bmatrix}
$$
$$
+ \begin{bmatrix} 1-\cos\theta & \sin\theta \\ -\sin\theta & 1-\cos\theta \end{bmatrix} \cdot \begin{bmatrix} C_x \\ C_y \end{bmatrix}, \quad (10.27)
$$

which is much closer to the typical expression presented in Equation 10.3, where the classical translation T will be given by

$$
T = \begin{bmatrix} T_x \\ T_y \end{bmatrix} = \begin{bmatrix} 1-\cos\theta & \sin\theta \\ -\sin\theta & 1-\cos\theta \end{bmatrix} \cdot \begin{bmatrix} C_x \\ C_y \end{bmatrix}. \quad (10.28)
$$

The position of the fixed point C can be very counter-intuitive, and this is probably the reason why it is more common to find expressions in terms of the translation transform T than in terms of the fixed point C. One of the inconvenient characteristics of the fixed point representation is that when the rotation angle θ is small compared to the magnitude of the translation $|T|$, the fixed point C is located very far away. This makes it difficult to optimize the values of its coordinates (C_x, C_y). If we consider the case of very small θ values, Equation (10.28) can be simplified by using the Taylor's series expansions of $\cos\theta$ and $\sin\theta$ in order to obtain

$$
C = \begin{bmatrix} C_x \\ C_y \end{bmatrix} = \frac{1}{\theta} \begin{bmatrix} -T_y \\ T_x \end{bmatrix}, \quad (10.29)
$$

which simply reminds us of $|T|$ as being the length of the cord subtended by the angle θ in a circle of radius $|C|$ [18, 28].

Rotations with very small angles will result in fixed points being located at large distances making the optimization quite imprecise. In such situations it is a good compromise to use an auxiliary translation transform T, so the expression of the total transform becomes

$$
\begin{bmatrix} x' \\ y' \end{bmatrix} = \begin{bmatrix} \cos\theta & -\sin\theta \\ \sin\theta & \cos\theta \end{bmatrix} \cdot \begin{bmatrix} x \\ y \end{bmatrix}
$$
$$
+ \begin{bmatrix} 1-\cos\theta & \sin\theta \\ -\sin\theta & 1-\cos\theta \end{bmatrix} \cdot \begin{bmatrix} C_x \\ C_y \end{bmatrix} + \begin{bmatrix} T_x \\ T_y \end{bmatrix}, \quad (10.30)
$$

where the fixed point C is set only approximatively and kept unchanged during the optimization. The optimizer will use the set of parameters $[\theta, T_x, T_y]$ as its search space.

Let's now go back to the expression in Equation (10.26) and ignore in the first instance the fixed point C in order to simplify the analysis of how the optimization of the angle θ should be performed. This is equivalent to considering that the center of rotation is the origin of coordinates and henceforth $(C_x, C_y) = (0,0)$, which reduces Equation (10.26) to

$$x' = \begin{bmatrix} x' \\ y' \end{bmatrix} = \begin{bmatrix} \cos\theta & -\sin\theta \\ \sin\theta & \cos\theta \end{bmatrix} \cdot \begin{bmatrix} x \\ y \end{bmatrix}. \tag{10.31}$$

In order to simplify further the notation, it is convenient to represent the coordinates (x, y) in terms of the complex plane:

$$(x, y) = re^{i\phi} = (r\cos\phi, ir\sin\phi) = r\cos\phi + ir\sin\phi \tag{10.32}$$

This is illustrated in the left side of Figure 10.10. The parameters (r, ϕ) are the representation of (x, y) in polar coordinates. The right side of the figure illustrates the effect of applying a rotation by an angle θ. In terms of polar coordinates this rotation can be represented as

$$(x', y') = re^{i(\phi+\theta)} = re^{i\phi}e^{i\theta}. \tag{10.33}$$

When we search for a value of θ that will optimize a cost function S, and we use a gradient descent algorithm, the update stage of the parameter should be presented as

$$\theta' = \theta + \frac{\partial S}{\partial\theta}\lambda, \tag{10.34}$$

where λ is the step length of the gradient descent algorithm. In terms of the rotation, this expression becomes

$$e^{i\theta'} = e^{i\theta}e^{i\frac{\partial S}{\partial\theta}\lambda}, \tag{10.35}$$

which can be presented as

$$e^{i\theta'} = e^{i\theta}\exp(G)^{\lambda}, \tag{10.36}$$

where G is the non-linear incremental factor

$$G = i\frac{\partial S}{\partial\theta}. \tag{10.37}$$

The term $\exp(G)$ represents the variation computed from the gradient of the cost function. λ is again the step length used by the optimizer.

Since a complex factor can represent both rotation and scaling, we can consolidate Equations (10.17) and (10.36) in a single expression by replacing the factor $e^{i\theta}$ with the factor $De^{i\theta}$. This consolidated expression is

$$F' = F\exp(G)^{\lambda}, \tag{10.38}$$

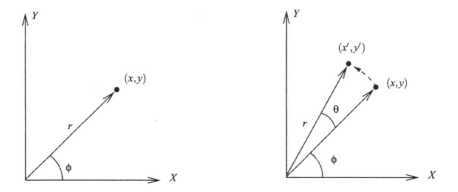

Figure 10.10. Representation of rotation in 2D using the complex plane.

where F, F' and G are given by

$$F = De^{\mathbf{i}\theta},$$
$$F' = D'e^{\mathbf{i}\theta'}, \tag{10.39}$$
$$G = D\frac{\partial M}{\partial D} + \mathbf{i}\frac{\partial M}{\partial \theta}.$$

This generic expression allows the use of a standard gradient descent algorithm for optimizing both scaling and rotation in a single consistent framework. F represents a current rotation and scaling transform; F' represent a variation of the current F transform. F' is computed by the optimizer in order to be used in the next iteration for computing the cost function value. The term G is computed from the partial derivatives of the cost function with respect to the scaling factor and the rotation angle.

The rotation and scaling transform F will be applied to an arbitrary point as

$$(x',y') = Dre^{\mathbf{i}(\phi+\theta)} = re^{\mathbf{i}\phi} \cdot De^{\mathbf{i}\theta}. \tag{10.40}$$

Now that we have unified the notation for scaling and rotation we can reintroduce the translation. This is done by allowing the choice of the fixed point of the rotation and scaling. In order to reintroduce the fixed point we first need to show that given an arbitrary transformation composed of rotation, scaling, and translation, it is always possible to find a fixed point for the final transformations. The procedure for finding the fixed point is illustrated in Figure 10.11. Consider shape Π to be the initial shape in space and shape Π'' to be the result of applying the transformation T on Π. In other words, $\Pi'' = T(\Pi)$. The shape Π has been rotated by an angle θ, scaled by a factor D, and translated in the plane by a vector $\vec{d}$ in order

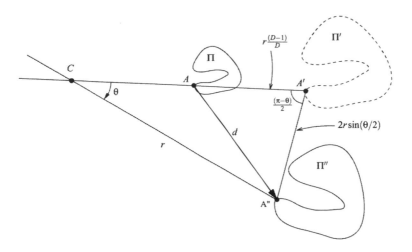

Figure 10.11. Illustration of the procedure for finding the fixed point of a transformation composed of rotation, scaling, and translation.

to produce Π''. We want to show that instead of specifying a translation vector $\vec{d}$, we can always find a fixed point C for the scaling and rotation transforms such that the translation $\vec{d}$ will result as a collateral effect of both the scaling and the rotation.

We assume that the known parameters of the transformation are the scaling factor D, the rotation angle θ, and the translation vector $\vec{d}$. Given these parameters we seek to identify the location of the point C which is the fixed point of the total transformation. We saw in Section 10.3.1 and Figure 10.8 that a scaling transform will induce translations parallel to the lines joining local points with the fixed point. Rotations, on the other hand, induce translations orthogonal to the line joining the local points with the fixed point [6, 18]. In the case of a transformation composed of scaling and rotation, the translation $\vec{d}$ is the vector sum of the collateral translation induced by scaling and the collateral translation induced by the rotation [18, 28].

When applied using the same fixed point, rotation and scaling transforms are orthogonal, which means that the rotation does not result in secondary effects that may look like a scaling, and the scaling does not produce collateral effects that may look like a rotation. This orthogonality also implies that the two transforms commute. That is, applying the rotation followed by the scaling produces the same result as applying the scaling followed by the rotation. Let's note that this is not the case when the rotation and the scaling use different fixed points.

Figure 10.11 presents with a dashed contour the shape Π' which is the result of first applying only the scaling transform to shape Π. That is, $\Pi' = D(\Pi)$. We assume this scaling transform to be applied using a fixed point C. We want now to show that this same fixed point C can be used to define a rotation consistent with the whole transformation. The shape Π'' is the result of applying the rotation transform to the intermediate shape Π', hence

$$\Pi'' = R(\Pi') = R(D(\Pi)) = T(\Pi). \tag{10.41}$$

We select an arbitrary point A in the shape Π as an auxiliary reference point for analyzing the effects of the intermediate transforms. There is nothing particular in the choice of this point, any other point of Π could have been used. The ratio between the line segments CA and CA' is given by the scaling factor D as $CA' = D \cdot CA$. Therefore

$$AA' = CA' - CA = CA(D-1) = CA'\left(1 - \frac{1}{D}\right) = CA'\left(\frac{D-1}{D}\right), \tag{10.42}$$

which is also illustrated in Figure 10.11 where we use the distance r as a synonym for CA' in order to simplify the notation. The distance AA' is the translation induced in the shape Π as a collateral effect of the scaling transform D.

The application of the rotation transform R to shape Π' produces shape Π'' as a result, with the collateral translation indicated by the line segment $A'A''$. The trigonometric relationship between the translation and the rotation angle θ is given by the expression

$$AA'' = 2r\sin(\theta/2). \tag{10.43}$$

Since the angle $A'CA''$ is the rotation angle θ, we know that the angle $A''A'C$ is $(\pi - \theta)/2$.

Using the cosinus theorem we can relate the length of the translation $\vec{d}$ to the two sides AA' and $A'A''$ of the triangle $AA'A''$ and the angle $(\pi - \theta)/2$. Reorganizing the expression resulting from the cosinus theorem we arrive at

$$r^2 = d^2\left[\left(\frac{D-1}{D}\right)^2 + 4\left(\frac{1}{D}\right)\sin^2(\theta/2)\right]^{-1}. \tag{10.44}$$

This expression shows that we can always find a distance r for placing the fixed point C with respect to point A''.

The trivial known cases of pure rotation and pure translation are recovered by setting $D = 1$ or $\theta = 0$;

$$r = \frac{d}{\sin(\theta/2)} \quad \text{or} \quad r = d\left(\frac{D}{D-1}\right). \tag{10.45}$$

Given that we can always find a fixed point C for a combined transform of scaling, rotation, and translation, we continue the analysis of how such a transform can be optimized. The expression of the transform is now

$$P' = T(P) = (P - C)De^{i\theta} + C, \tag{10.46}$$

where P is an arbitrary point being transformed, C is the fixed point of the transformation, D is the scaling factor, and θ is the rotation angle. Note that points P and C are represented here as points on the complex plane, which means that they are in the form $(x + \mathbf{i}y)$ or $re^{\mathbf{i}\theta}$. This equation can be compared with Equation (10.40).

From the optimization process point of view, the main advantage of using the fixed point representation of Equation (10.46) is that it decorrelates both the scaling and the rotation from the translation. That is, a variation of the rotation angle will not result in a collateral translation, since we use the natural fixed point of the transform as the center of rotation. Therefore, when the optimizer uses small variations of the transform parameters for computing the steepest descent direction, it is clear that the variations computed for each parameter are independent of each other.

The parameters of the transform in Equation (10.46) are D, θ, and the coordinates of the fixed point C. Note that the point C must be among the parameters to be optimized because it is through its position that appropriate translations are induced by the transform.

The Jacobian of this transform, as defined in Equation (10.8), can be computed from Equation (10.46) as

$$
\begin{aligned}
\frac{\partial P'}{\partial \theta} &= (P - C)D\mathbf{i}e^{\mathbf{i}\theta}, \\
\frac{\partial P'}{\partial D} &= (P - C)e^{\mathbf{i}\theta}, \\
\frac{\partial P'}{\partial C} &= (1 - De^{\mathbf{i}\theta}),
\end{aligned}
\tag{10.47}
$$

and explicitly expanding all the terms we find the following expressions:

$$
\frac{\partial P'}{\partial \theta} =
\begin{bmatrix} \frac{\partial x'}{\partial \theta} \\ \frac{\partial y'}{\partial \theta} \end{bmatrix} = D
\begin{bmatrix} -\sin\theta & -\cos\theta \\ \cos\theta & -\sin\theta \end{bmatrix}
\begin{bmatrix} x - C_x \\ y - C_y \end{bmatrix},
\tag{10.48}
$$

$$
\frac{\partial P'}{\partial D} =
\begin{bmatrix} \frac{\partial x'}{\partial D} \\ \frac{\partial y'}{\partial D} \end{bmatrix} =
\begin{bmatrix} \cos\theta & -\sin\theta \\ \sin\theta & \cos\theta \end{bmatrix}
\begin{bmatrix} x - C_x \\ y - C_y \end{bmatrix},
\tag{10.49}
$$

$$
\frac{\partial P'}{\partial C} =
\begin{bmatrix} \frac{\partial x'}{\partial C_x} & \frac{\partial x'}{\partial C_y} \\ \frac{\partial y'}{\partial C_x} & \frac{\partial y'}{\partial C_y} \end{bmatrix} =
\begin{bmatrix} 1 - D\cos\theta & D\sin\theta \\ -D\sin\theta & 1 - D\cos\theta \end{bmatrix}.
\tag{10.50}
$$

Arranging these terms in columns we can recover the Jacobian matrix J:

$$
J =
\begin{bmatrix} \frac{\partial x'}{\partial \theta} & \frac{\partial x'}{\partial D} & \frac{\partial x'}{\partial C_x} & \frac{\partial x'}{\partial C_y} \\ \frac{\partial y'}{\partial \theta} & \frac{\partial y'}{\partial D} & \frac{\partial y'}{\partial C_x} & \frac{\partial y'}{\partial C_y} \end{bmatrix}.
\tag{10.51}
$$

This matrix is used for reducing the computational complexity of evaluating cost function derivatives.

Let's analyze now the implications of reintroducing the fixed point C into the expressions found in Equations (10.38) and (10.39) for updating the transform using a transform differential G and an optimizer of step length λ.

The set of transform parameters $[D, \theta, C_x, C_y]$ should be updated with increments given by

$$\left[\ \frac{\partial S}{\partial D} \lambda, \quad \frac{\partial S}{\partial \theta} \lambda, \quad \frac{\partial S}{\partial C_x} \cdot \lambda, \quad \frac{\partial S}{\partial C_y} \cdot \lambda \ \right], \tag{10.52}$$

which leads to the following expressions for the updated parameters of transform T':

$$\left[\ D + \frac{\partial S}{\partial D} \cdot \lambda, \quad \theta + \frac{\partial S}{\partial \theta} \cdot \lambda, \quad C_x + \frac{\partial S}{\partial C_x} \cdot \lambda, \quad C_y + \frac{\partial S}{\partial C_y} \cdot \lambda \ \right]. \tag{10.53}$$

The transform described by Equation (10.46), is called a *rigid* transform when the scaling factor is equal to 1.0, and it is called a *similarity* transform when the scaling factor has any other value. A rigid transform preserves distances between points. That is, if two points A and B are mapped by the transform into homologous points A' and B', the distance AB is equal to the distance $A'B'$. A similarity transform, on the other hand, does not preserve distances because a scaling factor is being applied. It preserves, however, the angles between any two lines.

The fixed point in Equation (10.46) has the disadvantage of being a poor representation in the cases where the scaling factor D is close to one, and the rotation angle θ is close to zero. In such situations it is a good compromise to use an auxiliary translation transform T, so the expression of the total transform becomes

$$x' = (x - C)De^{j\theta} + C + T, \tag{10.54}$$

where the fixed point C is set only approximatively and kept unchanged during the optimization. The optimizer will use the set of parameters $[D, \theta, T_x, T_y]$ as its search space.

Affine Transform.   Following the taxonomy presented in Figure 10.7, the next level of transform complexity is given in the *affine* transform. An affine transform preserves collinearity. That is, if points A, B, and C are collinear in the input space, their homologous points A', B', and C' in the output space will also be collinear.

The generic expression for the affine transform is represented as

$$x' = A \cdot x + T, \tag{10.55}$$

where A is an N-dimensional square matrix and T is an N-dimensional vector, usually called the *translation* vector.

The parameters of the transform are the set of $N \times N$ coefficients of A plus the N components of vector T.

The apparent simplicity of this transform can be misleading. It is useful here to keep in mind that any of the coefficients in the matrix A are as complex as the scaling factor that we analyzed in Section 10.3.1 or as complex as the trigonometric terms of the rotation transform. After all, both the scaling transform and the rotation transform are nothing other than particular cases of the affine transform.

Two fundamental characteristics of the affine transform should then be analyzed here. First, how to find and manage the fixed point of the transformation, and how to update the matrix and vector parameters in the context of a gradient descent algorithm.

Let's start by considering the problem of the fixed point. If a fixed point C exists, it should satisfy the following equation:

$$C = A \cdot C + T, \tag{10.56}$$

which implies

$$C = (I - A)^{-1} \cdot T, \tag{10.57}$$

where I is the $N \times N$ identity matrix. From this expression we can deduce that a fixed point C will exist as long as the matrix $(I - A)$ is invertible. This requires the determinant $|I - A|$ to be non-null. However, being more strict, what the expression is telling us is that in the cases when the determinant tends to zero, the components of the fixed point C tend to infinity.

A typical case is when the matrix A tends to the identity matrix. This is actually equivalent to a *pure translation* transform. We already know that an almost pure translation can be represented with a fixed point tending to infinity and a rotation angle tending to zero. We can then conclude that the fixed point C can always be found, in the worst case scenario as the limit of the determinant $|I - A|$ tending to zero.

We introduce then the fixed point C in the expression of the affine transform. The translation vector T is now equivalent to the collateral effect of applying the A matrix around the fixed point:

$$x' = A \cdot (x - C) + C. \tag{10.58}$$

The transform parameters are now the set of matrix A coefficients and the components of the fixed point C, for a total of $N \times (N + 1)$ parameters.

The translation vector T easily can be computed as

$$T = (I - A) \cdot C. \tag{10.59}$$

Affine Transform Optimizer Update.   We can now analyze the problem of updating parameters using derivatives of the cost function with respect to transform parameters. The matrix A coefficients have the same algebraic structure that the scale in the scaling transform has. We will then compute updates for them using expressions of the form

$$A'_{ij} = A_{ij} + \frac{\partial S}{\partial A_{ij}} \lambda, \tag{10.60}$$

where λ is the step length of the optimizer. The components of the fixed point C, on the other hand, can be updated with the linear expression

$$C'_i = C_i + \frac{\partial S}{\partial C_i} \lambda \tag{10.61}$$

In the cases when the fixed point C is placed at great distances, it is a good compromise to use an auxiliary translation T so that the equation of the affine transform becomes

$$x' = A \cdot (x - C) + C + T, \tag{10.62}$$

where the fixed point C is set only approximatively and kept unchanged during the optimization. The optimizer will use the set of parameters $[A_{ij}, T_k]$ as its search space.

Parameter Scaling. In the affine transform there is also a parameter scaling difference that must be compensated for during the optimization. This scaling difference originates from the fact that typical translations in medical image registration are usually measured in millimeters and therefore they have values in the numeric range $[-1000 : 1000]$. The coefficients of the matrix A, on the other hand, are typically in the range $[-1 : 1]$, and only in extreme cases go into the range $[-2 : 2]$. We already discussed in Section 10.3.1 that the scaling factors found in medical image registration should only compensate for miscallibrations between the image acquisition devices. The relative resolution of the images is already compensated for by the transforms mapping the image grid space into their physical space. The matrix A coefficients are then the result of products of sine and cosine trigonometric functions and scale parameters close to ± 1.0.

Mixing parameters with very different numeric ranges is undesirable in the context of gradient descent optimizers since the parameters with large numeric values take over the ones with small numeric values. As far as the optimizer is concerned, the gradients always look as if the parameters with small numeric values never make any significant contribution to the variation of the cost function.

It is a common practice to introduce a scaling factor in some parameters before they are sent to the optimizer. For example, in an image of size 512×512 pixels and with pixel spacing $(0.2\text{mm}, 0.2\text{mm})$, the full extent of the image will be $(102.4\text{mm}, 102.4\text{mm})$. When performing registration with this image, we could expect translations to be in the range $[-50\text{mm} : 50\text{mm}]$. It is then reasonable to scale the fixed point coordinates by a factor of 100 in order to obtain values in the range $[-0.5 : 0.5]$ which will be comparable with the numeric range of the matrix A coefficients.

Affine Transform Jacobian. As with the previous transforms, the Jacobian of the affine transform may be required for computing the cost function derivatives. From Equation (10.58) we can deduce

$$
\begin{aligned}
\frac{\partial P'_k}{\partial A_{ij}} &= (P_j - C_j) \quad \text{if} \quad i = k, \\
\frac{\partial P'_k}{\partial A_{ij}} &= 0 \qquad\qquad \text{if} \quad i \neq k, \\
\frac{\partial P'_i}{\partial C_j} &= 1 - A_{ij}.
\end{aligned}
\tag{10.63}
$$

Note the Jacobian matrix will have dimensions $N \times N(N+1)$. Along the rows, we will place the derivatives of a specific component of P' with respect to the $N(N+1)$ parameters of the transform. There will be one column per element of the A matrix and one column per component of the fixed point C. The matrix will be composed of $N+1$ matrices, each one being a diagonal matrix of dimension $N \times N$.

Since this Jacobian is in practice quite a sparse matrix, it is not efficient to perform computations using it as a normal matrix. Software implementations of the cost function derivatives could improve performance by avoiding to consider all the null coefficients in the Jacobian matrix. This will reduce operations from $N \times N(N+1)$ to just $N(N+1)$.

One of the advantages of the affine transform is its simple numerical representation. However, the same characteristics make it very hard to interpret since from the values of the matrix A coefficients it is not trivial to determine the major effects of the transform. The large number of parameters in this transform can also be seen as a disadvantatge. In the simple case of a 3D transform we are dealing with an optimization in a parametric space of dimension 12. It is not advisable to undertake a registration using this transform without having first performed a registration with a rigid transform or a similarity transform. As illustrated in Figure 10.7, the affine transform is adding the complexity of shearing transforms to the degrees of freedom already allowed by the similarity transform.

In the cases where the fixed point C is located too far, it is convenient to represent the transform using an auxiliary translation. The components of C will not make part of the transform parameters and will be replaced with the components of the auxiliary translation T.

Quaternions. Quaternions were introduced by William Hamilton in his *Elements of Quaternions* published in 1866 after his death [10]. A considerable number of the current notions in vector calculus were introduced in this text. Curiously, its notation is easier to read than later descriptions of the same body of work. An impressive philosophical content was presented to the reader for justifying the introduction of the new mathematical concepts. Such effort is in general absent in our contemporary literature.

Hamilton introduced the notion of a quaternion as the *quotient* of two vectors. The quaternion was devised to capture the relationships between the two vectors and make it possible to retrieve one of the vectors by operating with the other. In other words, the quaternion could also be seen as an *operator* that, when applied to one of the vectors, will produce the second vector.

Given two vectors $\vec{A}$ and $\vec{B}$, their quotient was given by the quaternion Q:

$$Q = \vec{A}/\vec{B}. \tag{10.64}$$

The application of the quaternion operator Q onto the vector $\vec{B}$ produced the vector $\vec{A}$:

$$\vec{A} = Q \star \vec{B}, \tag{10.65}$$

where the symbol $\star$ stands for the application of the quaternion onto the vector.

Since vectors are relationships betweens points in space, they are not anchored to any particular position. A vector is fully defined with direction and magnitude. Quaternions were then designed to represent the orientation of one vector with respect to another vector as well as the ratio of their magnitudes.

The orientation of one vector relative to another was represented by a *versor* while the ratio of magnitudes was represented by a *tensor*. The term "tensor" was introduced by Hamilton, with the meaning of being an operator intended to modify the length of a vector.

In the particular case of parallel vectors, the tensor was equivalent to the *scalar*. The concept of "scalar" as introduced by Hamilton is quite different from the connotation used in our day according to which a scalar is a *single-valued* quantity as opposed to a *multi-valued* entity like a vector. A scalar was defined by Hamilton as the quotient of two parallel vectors. The scalar operator can then be represented with a single number corresponding to the magnitude ratio of the two parallel vectors in the quotient. The concept of a tensor was also different to what we call tensors now. The tensor in the context of quaternions represented the operation of exerting *tension* over a vector with the intent of changing its length.

A versor was defined as the quotient between two non-parallel vectors of equal length. Since versors represent a orientation change in a vector, they provide a natural representation for rotations in 3D space. Their numerical representation requires only three numbers [10, 12]. Figure 10.12 illustrates the conceptual basis of the definion of scalar and versor operators.

A quaternion is the operator resulting from the composition of a versor operator and a tensor operator. The versor rotates the vector while the tensor changes the vector magnitude as shown in the following equation:

$$\begin{aligned} \vec{A} &= T \star (V \star \vec{B}) \\ &= (T \diamond V) \star \vec{B} \\ &= Q \star \vec{B}, \\ Q &= T \diamond V, \end{aligned} \tag{10.66}$$

Figure 10.12. *Scalar* defined as the quotient of two parallel vectors. *Versor* defined as the quotient of two non-parallel vectors of equal length.

where T is a tensor operator, V is a versor operator, Q is a quaternion, and the symbol $\diamond$ stands for the composition of quaternions. The numerical representation of quaternions require four numbers. This is indeed what originated the term *"quaternion"* from the latin *"quaternio"* deriving from the greek "τετρακτυς" which means *"a set of four"*.

A versor has a direction and a norm associated with it. The direction is parallel to the axis around which the vector $\vec{B}$ should be rotated in order to be mapped into the vector $\vec{A}$. This axis is always orthogonal to the plane defined by the two vectors $\vec{A}$ and $\vec{B}$. A versor also has an angle associated with it. This is naturally the angle between the two vectors for which the versor is a quotient. The norm of the versor is a function of its rotation angle.

Probably the most intuitive representation of a versor is the one that maps its elements onto the surface of a unit sphere. This is illustrated in Figure 10.13. A versor is fully defined by a directed arc traced on the sphere surface. The arc can be easily defined by placing the tails of the two vectors $\vec{A}$ and $\vec{B}$ in the center of the sphere, identifying the plane Π containing both vectors, then tracing the circle which is the intersection of the sphere surface with the plane Π. Along this circle we can trace a directed arc with an angle equal to the angle between the two

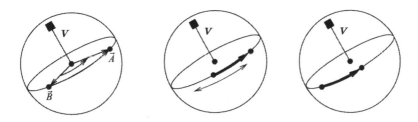

Figure 10.13. Representation of a versor in terms of directed arcs on the surface of a unit sphere. The directed arc can slide freely along the circle in the sphere and still represent the same versor V.

vectors. The arc can be moved freely along the circle without losing its identity, in the same way that vectors can be translated freely in space without losing their identity.

Compositions of versors can be represented very intuitively using directed arcs on the surface of the unit sphere. The arcs representing versors can be combined in the sphere surface just like 2D vectors are combined in the plane. Figure 10.14 illustrates how to compose two versors. At left, the points marked as $\vec{A}$, $\vec{B}$, and $\vec{C}$ indicate the places where unit vectors touch the unit sphere. These vectors $\vec{A}$, $\vec{B}$, and $\vec{C}$ have their tails in the center of the sphere. From this set of vectors we define the following versors V_{CB}, V_{BA}, and V_{CA} as the vector quotients:

$$
\begin{aligned}
V_{CB} &= \vec{B}/\vec{C}, \\
V_{BA} &= \vec{A}/\vec{B}, \\
V_{CA} &= \vec{A}/\vec{C},
\end{aligned}
\tag{10.67}
$$

where the versor V_{CA} is the result of the composition

$$
V_{CA} = V_{BA} \diamond V_{CB}. \tag{10.68}
$$

On the sphere surface the composition is done by sliding the directed arcs in order to make the tail of V_{BA} match the head of V_{CB}. This is always possible because the arcs are placed in major circles of the sphere which are non-coplanar [5, 28]. Once this is done, the tail of V_{CB} is connected to the head of V_{BA} using an arc in a major circle. This arc is the one representing the versor V_{CA} that is the result of the composition.

Versor composition using arcs on the unit sphere is just like composing vectors in the plane, with the single exception that versor composition is non-commutative. The reason why versor composition does not commute is illustrated in the central part of Figure 10.14.

The versors V_{AB} and V_{BC} are composed in this figure in both possible orders. The first order can be seen on the left side of the sphere where the versor

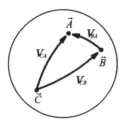 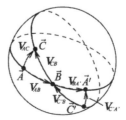 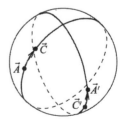

Figure 10.14. Representation of a versor composition using directed arcs in the surface of a unit sphere. Two arcs are arranged head-to-tail in order to obtain the resulting arc.

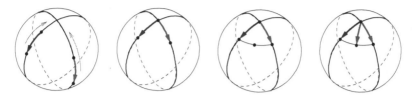

Figure 10.15. Procedure for performing versor *addition* using the directed arc representation. First, arcs slide until their tails meet. Then the middle point of their heads are
located. Finally an arc is traced joining the tails with the heads middle point. This final arc
represents the versor resulting from the addition of the two versors.

V_{AC} is the result of $V_{BC} \diamond V_{AB}$. The second order can be seen on the right side
of the sphere where the versors have been shifted along their respecting arcs. For
clarity, the shifted position of versor V_{AB} has been marked with an auxiliary vector $\vec{A'}$, and the shifted position of versor V_{BC} has been marked with an auxiliary
vector $\vec{C'}$. Thus the versor $V_{C'B}$ is the shifted version of V_{BC} while the versor
$V_{BA'}$ is the shifted version of V_{AB}. The figure shows that the versor $V_{C'A'}$ is
the result of the composition $V_{C'B} \diamond V_{BA'}$. Since the two resulting versors V_{AC}
and $V_{C'A'}$ are in different major circles, it is clear that they cannot be equivalent
and therefore versor composition does not commute.

Optimization of Versors.   The fact that versors do not behave like vectors make
them a poor fit for traditional gradient descent optimizers. One common source of
confusion is the fact that the addition operation has also been defined for versors,
but it does not correspond to the notion of cumulative applications of versors, as
addition does among vectors.

The operation of versor addition is illustrated in Figure 10.15. The directed
arc representation is used in this figure. The first step in the process of adding two
versors is to slide their arcs along their major circles until the arc tails met in the
intersection of the two major circles. Note that this can be done in any one of the
two points where the major circles intersect each other. Then, we trace the major
circle that passes through the heads of both directed arcs, and over this major
circle we identify the middle point between the heads of the two directed arcs.
Finally we trace the arc joining the tails of the two original arcs with the middle
point of their heads. This final arc is the one representing the versor resulting
from the *addition* of the two original versors.

Versor addition is not the operation that we want to perform during optimization. The versor operation that corresponds to the sum among vectors is the versor
composition ⋄. This has an important implication in the process of optimization,
especially when a gradient-descent type of approach is used. The application of
a small variation of the versor should be done by *composing* the current versor
with another versor representing the small variation. The typical update stage of

a gradient descent will be presented in a versor space as

$$V' = dV \diamond V, \qquad (10.69)$$

where V is the current versor, V' is the versor to be used in the next evaluation of the cost function, and dV is a versor increment computed by the gradient descent optimizer. The cost function will be dependent on the versor V and therefore can be writen as $S(V)$. The versor increment will be computed from the derivative of the cost function $S(V)$ with respect to the versor V, and will be weighted by the step length λ. The versor increment is then given by the expression

$$dV = \left[\frac{\partial S(V)}{\partial V} \right]^{\lambda}. \qquad (10.70)$$

The derivative of $S(V)$ with respect to V involves the evaluation of how much S changes with a variation in the angle of V and how much S changes with a variation in the orientation of V.

The weighting by the step length λ is done here using exponentiation. The reason for doing so is that exponentiation is the versor operation equivalent to scaling the angle of the versor. Probably the most intuitive way of introducing the notion of versor exponentiation is by starting with the definition of the *square* of a versor as given by the following equation:

$$V^2 = V \diamond V. \qquad (10.71)$$

That is, the operator V^2 is equivalent to the repeated operation of the operator V. We can also describe it as the operator V^2 being equal to the composition of V with itself. This can easily be extended to the cube operation and in general any integer exponentiation:

$$\begin{aligned} V^3 &= V \diamond V \diamond V, \\ V^n &= V \diamond V \diamond \cdots \diamond V. \end{aligned} \qquad (10.72)$$

The integer root of a versor V can then be defined as the versor R that composed n times with itself results in the original versor. For example the square root

$$\begin{aligned} V^{1/2} &= R \\ V &= R \diamond R \end{aligned} \qquad (10.73)$$

or the cubic root

$$\begin{aligned} V^{1/3} &= R \\ V &= R \diamond R \diamond R. \end{aligned} \qquad (10.74)$$

Figure 10.16 illustrates the effect of this relationship over the directed arcs in the unit sphere. At left in the figure, the versor V is the quotient of vectors $\vec{A}$

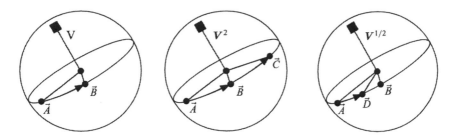

Figure 10.16. Versor exponentiation is equivalent to a linear operation on the versor angle. The square of a versor is a versor with an angle double in size. A square root of a versor is a versor with an angle half in size.

and $\vec{B}$. In the center of the figure, the application of the versor operator V on vector $\vec{A}$ produces vector $\vec{B}$. The application of V on vector $\vec{B}$ produces vector $\vec{C}$. Therefore, the application of the operator V^2 on vector $\vec{A}$ produces vector $\vec{C}$. Which means that the angle of versor V^2 is the double of the one in versor V. The right side of the figure illustrates the effect of the square root on versor V. This results in a versor with an angle half the size of the angle in V, since, in this way, its composition will reconstitute V. We can summarize this discussion with the following expression:

$$\begin{aligned} \Theta(V) &= \theta \\ \Theta(V^n) &= n\theta, \end{aligned} \qquad (10.75)$$

where $\Theta(V)$ is the function that evaluates the angle of a versor V. We can then reinterpret Equation (10.70) saying that the step length λ will multiply the angle of the versor represented by $\partial S(V)/\partial V$. This is consistent with the expression found in Section 10.3.1 for rotations in 2D space, where the increments of the rotation operator $\exp(\theta)$ were also made by exponentiation.

Rigid Transform in 3D.   The versor transform analyzed in the previous section is appropriate for representing rotations around the origin. In order to extend this transform for also representing translations we introduce again the fixed point C into the transform expression:

$$P' = V \star (P - C) + C. \qquad (10.76)$$

The adequate selection of C will allow representation of any translation T:

$$\begin{aligned} P' &= V \star P + [C - V \star C] \\ P' &= V \star P + T \\ T &= C - V \star C. \end{aligned} \qquad (10.77)$$

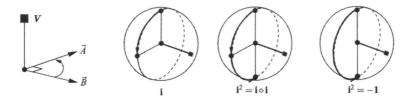

Figure 10.17. A *right* versor is an operator inducing a rotation of $\pi/2$. The square of a right versor is an operator that reverts the direction of a vector, and for this reason is denoted as the *operator* -1.

This new transformation is equivalent to a *rigid* transform in 3D space. We can verify that the transform is defined by six parameters. The three numbers representing the versor and the three components of the fixed point coordinates.

Let's introduce now the numerical expression allowing representation of an arbitrary versor using three numbers. First we introduce the notion of a *right versor*, which is a versor whose angle is a right angle.

Figure 10.17 illustrates the main properties of right versors. The left part of the figure shows that V is a right versor since it is the quotient between two orthogonal vectors of same length. Seen as an operator, V applied on vector $\vec{B}$ will produce the orthogonal vector $\vec{A}$. Next on the figure, the right versor **i** when applied twice, reverts the direction of the vector, which is equivalent to applying an operator -1 on the vector.

When three right versors are arranged in orthogonal directions they satisfy the following relations:

$$\begin{aligned} -\mathbf{i} &= \mathbf{k} \diamond \mathbf{j} \\ -\mathbf{j} &= \mathbf{i} \diamond \mathbf{k} \\ -\mathbf{k} &= \mathbf{j} \diamond \mathbf{i}. \end{aligned} \tag{10.78}$$

The similarity of this expression with the familiar vector product of unit vectors is not an accident. In fact, the concept of *vector product* itself was introduced by Hamilton in his work on quaternions as a particular case of quaternion composition [10, 12]. The reason why the right versors in Equation (10.78) have negative signs is easily explained in Figure 10.18. As with any other versor, the right versors can be represented by directed arcs on the surface of the unit sphere. The three orthogonal versors in the left side of the figure are shown as directed arcs in the middle of the figure. The orientation of the arcs matches the rotation of $\pi/2$ induced by each right versor. The right side of the figure shows how the composition $\mathbf{k} \diamond \mathbf{j}$ is equal to $-\mathbf{i}$. The right versor $\mathbf{j}$ is the quotient $\vec{B}/\vec{C}$. The right versor $\mathbf{k}$ is the quotient $\vec{A}/\vec{B}$, therefore the composition $\mathbf{k} \diamond \mathbf{j}$ is the quotient $\vec{A}/\vec{C}$. The direction of this last quotient is opposed to the one shown in the middle of the figure for versor $\mathbf{i}$. Hence, the composition $\mathbf{k} \diamond \mathbf{j}$ is equal to $-\mathbf{i}$.

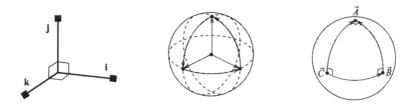

Figure 10.18. Three orthogonal right versors can be used to construct a basis for representing arbitrary versors. The composition of any two out of these three versors is equivalent to the remaining versor.

This set of orthogonal right versors $[\mathbf{i}, \mathbf{j}, \mathbf{k}]$ is also known as the set of *elementary quaternions* [10, 12].

Any right versor $\mathbf{v}$ can be expressed as a linear combination of these elementary versors, following:

$$\mathbf{v} = x\mathbf{i} + y\mathbf{j} + z\mathbf{k}, \tag{10.79}$$

$$x^2 + y^2 + z^2 = 1. \tag{10.80}$$

Any generic versor V can be expressed in terms of the right versor $\mathbf{v}$ parallel to its axis and the rotation angle θ as

$$V = e^{\mathbf{v}\theta}, \tag{10.81}$$

with the trivial cases of the elementary quaternions

$$\begin{aligned} \mathbf{i} &= e^{\mathbf{i}\pi/2} \\ \mathbf{j} &= e^{\mathbf{j}\pi/2} \\ \mathbf{k} &= e^{\mathbf{k}\pi/2}. \end{aligned} \tag{10.82}$$

Equation (10.81) can also be writen as

$$\begin{aligned} V &= \cos\theta + \mathbf{v}\sin\theta \\ V &= \cos\theta + (x\mathbf{i} + y\mathbf{j} + z\mathbf{k})\sin\theta, \end{aligned} \tag{10.83}$$

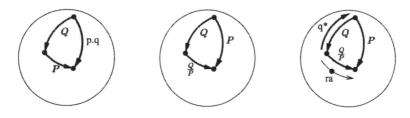

Figure 10.19. Interpolation along the surface of the unit sphere can be implemented using versor operations.

which is probably the most widespread numeric notation for versors. It is common to use the notation of the component arranged in a set as

$$V = (\cos\theta, x\sin\theta, y\sin\theta, z\sin\theta), \tag{10.84}$$

commonly called a *unit quaternion*, even though the correct term is *versor*. This expression has four components but it is constrained by the condition $x^2 + y^2 + z^2 = 1$, which makes it possible to use only three numbers for the representation.

Similarity Transform in 3D.   The expression for a similarity transform in 3D can be obtained by replacing the versor V with a quaternion Q in Equation (10.76) of the rigid 3D transform:

$$x' = Q \star (x - C) + C, \tag{10.85}$$

where x is the point being transformed, C is the fixed point of the transformation, and Q is a generic quaternion. The quaternion in this expression represents both rotation and scaling, while the combination of the quaternion and the fixed point represents an arbitrary translation T, as in

$$\begin{aligned}
x' &= Q \star x + [C - Q \star C] \\
x' &= Q \star x + T \\
T &= [C - Q \star C].
\end{aligned} \tag{10.86}$$

The optimization of this transform with a gradient descent algorithm will require an update of the transform using an expression of the type

$$dQ = \left[\frac{\partial S(Q)}{\partial Q}\right]^{\lambda}, \tag{10.87}$$

where Q represents both rotation and scale changes. Scaling is regulated by the tensor part of the quaternion, while the rotation is controled by the versor part of the quaternion. This means that the step size λ will multiply the angle of the quaternion differential expression

$$\frac{\partial S(Q)}{\partial Q}, \tag{10.88}$$

while its tensor part will be affected by the exponent λ. By writing this equation as the quaternion W, we can better identify the effect of the step length λ on the tensor part T of W as well as in its angle θ:

$$\begin{aligned}
W &= \frac{\partial S(Q)}{\partial Q} = T \cdot e^{v\theta} \\
W^{\lambda} &= T^{\lambda} \cdot e^{v\theta \cdot \lambda},
\end{aligned} \tag{10.89}$$

where T is an incremental scale and θ is an angle differential.

Updates of the fixed point C can be done independently using a linear expression. The independence is derived from the fact that Equation (10.86) decorrelates translations from scaling and rotation. Unfortunately, the decorrelation is only complete when we use the correct C coordinates. During the optimization process this is not the case since the optimizer is actually searching for the optimal fixed point and the optimal quaternion. Decorrelation is only partial under this circumstance.

In the case where the fixed point C is located at a large distance, it is a good compromise to use an auxiliary translation and rewrite Equation (10.86) as

$$x' = Q \star (x - C) + C + T. \tag{10.90}$$

In this case, the fixed point will be approximately set and will remain unchanged during the optimization process. The parameters of the transform will be the set of quaternion components and the components of auxiliary translation T, which will result in a seven-dimensional search space.

Matrix Representation of Similarity Transform. Although the quaternion notation is convenient for mathematical analysis, when it comes to applying a transform to a large number of points, it is more efficient to generate a transformation matrix whose coefficients will produce the same effect as the quaterion operator. In this case we are replacing the quaternion expression

$$x' = Q \star (x - C) + C \tag{10.91}$$

with the matrix operation

$$x' = M \cdot (x - C) + C. \tag{10.92}$$

If the quaternion Q is expressed as

$$Q = w + x\mathbf{i} + y\mathbf{j} + z\mathbf{k}, \tag{10.93}$$

then the equivalent matrix M is given by

$$\begin{bmatrix} \left(w^2 + x^2 - y^2 - z^2\right) & 2\left(xy - zw\right) & 2\left(xz + yw\right) & 0 \\ 2\left(xy + zw\right) & \left(w^2 - x^2 + y^2 - z^2\right) & 2\left(yz - xw\right) & 0 \\ 2\left(xz - yw\right) & 2\left(yz + xw\right) & \left(w^2 - x^2 - y^2 + z^2\right) & 0 \\ 0 & 0 & 0 & \left(w^2 + x^2 + y^2 + z^2\right) \end{bmatrix}$$
$$\tag{10.94}$$

Quaternion composition can also be performed as a matrix operation. Given two quaternions P and Q with components

$$\begin{aligned} P &= w_p + x_p\mathbf{i} + y_p\mathbf{j} + z_p\mathbf{k} \\ Q &= w_q + x_q\mathbf{i} + y_q\mathbf{j} + z_q\mathbf{k}, \end{aligned} \tag{10.95}$$

their composition $R = P \diamond Q$ is given by

$$
R =
\begin{bmatrix}
x_r \\
y_r \\
z_r \\
w_r
\end{bmatrix}
=
\begin{bmatrix}
w_q & z_q & -y_q & x_q \\
-z_q & w_q & x_q & y_q \\
y_q & -x_q & w_q & z_q \\
-x_q & -y_q & -z_q & w_q
\end{bmatrix}
\begin{bmatrix}
x_p \\
y_p \\
z_p \\
w_p
\end{bmatrix}.
\tag{10.96}
$$

10.3.2   Image Interpolators

During the registration process, a metric typically compares the intensity values in the fixed image with the corresponding values in the transformed moving image. When a point is mapped from one image space to another, it will generally be mapped to a non-grid position. Thus, an interpolation method is needed to obtain a intensity value for the mapped point using the information from the neighboring grid positions. Within the registration software framework, the *interpolator* component has two functions: to compute the interpolated intensity value at a requested position and to detect whether or not a requested position lies within the moving-image domain.

In the context of registration, the interpolation method affects the smoothness of the metric space. Since interpolations are evaluated thousands of times in a single optimization cycle, the user has to trade off the efficiency of computation versus the ease of optimization when selecting the interpolation scheme. In this section, we review three of the most popular interpolation methods: nearest-neighbor, linear, and B-spline.

Nearest-Neighbor Interpolation

Nearest-neighbor interpolation simply uses the intensity of the nearest grid position. It is equivalent to assuming that the underlying image is piece-wise constant with jumps midway between grid positions. This interpolation scheme is cheap and does not require floating point calculations.

Linear Interpolation

In linear interpolation, the underlying image is assumed to be piece-wise linear. That is, the intensity varies linearly between grid positions. The interpolated image is spatially continuous but not differentiable. Linear interpolation is computed as a weighted sum of the 2^{n-1} neighbors:

$$
M(\boldsymbol{x}) = \sum_i w_i M(\boldsymbol{x}_i),
\tag{10.97}
$$

where n is the dimension of the image space. The weights are computed from the distance between the requested position and the neighors such that

$$w_i = \prod_k (1 - |(\boldsymbol{x})_k - (\boldsymbol{x}_i)_k|). \tag{10.98}$$

B-Spline Interpolation

In B-spline interpolation, the underlying image is represented using B-spline basis functions. The B-spline coefficients can be computed using a fast recursive filtering technique [24, 25] The intensity at a non-grid position can then be computed by multiplying the B-spline coefficients, $c(\boldsymbol{x}_i)$, with shifted B-spline kernels within the support region of the requested position:

$$M(\boldsymbol{x}) = \sum_i c(\boldsymbol{x}_i)\beta(\boldsymbol{x} - \boldsymbol{x}_j). \tag{10.99}$$

B-spline kernels are separable such that

$$\beta(\boldsymbol{x}) = \prod_k \beta(x_k). \tag{10.100}$$

For spline orders greater than one, both the interpolated image and its derivatives are spatially continuous. The higher the spline order the greater the number of pixels required to compute the interpolated value. Third order B-spline kernels are typically used as they are a good tradeoff between smoothness and computational burden. The equation of a 1D cubic B-spline kernel is given by

$$\beta^{(3)}(x) = \begin{cases} \frac{2}{3} - \frac{1}{2}|x|^2(2 - |x|) & 0 \le x < 1 \\ \frac{1}{6}(2 - |x|)^3 & 1 \le x < 2 \\ 0 & 2 \le x. \end{cases} \tag{10.101}$$

10.3.3 Metrics

Within the registration software framework, the *metric* component is perhaps the most critical. The metric component quantitatively measures how well a transformed moving image "matches" the fixed image. The metric, $S(p|F,M,T)$ can be thought of as a scalar function of the set of transform parameters for a given fixed image, moving image, and transformation type. The metric is used by the optimizer to evaluate the quantitative criterion at various positions in the transform parameter search space. For gradient-based optimization schemes the metric component is also required to provide the derivatives of the measure with respect to each transform parameter, $\frac{\partial S}{\partial p}$.

To compute the measure, the metric component typically samples points within a defined region of the fixed image. For each point $\boldsymbol{x}$, the corresponding moving image point, $x' = T(x|p)$, is obtained using the transform component, then the interpolator component is used to compute the moving image intensity, $M(x')$, at the mapped position.

The selection of the type of metric to use is highly dependent on the registration problem to be solved. For example, some metrics are only suitable for mono-modality registration while others can handle multi-modality comparisions; some metrics have a large capture range while others require initialization close to the optimal position. Unfortunately, there are no clear-cut rules as to how to choose the metric. In this section, we review several of the popularly used similarity metrics.

Mean Squares

When the images to be registered come from the same imaging modality, the image intensity at corresponding points between the two images should be similar. One of the simplest measures of the similarity is the mean squared difference over all the pixels in the images. Mathematically, the measure is defined by

$$S(p|F, M, T) = \frac{1}{N} \sum_i^N [F(x_i) - M(T(x_i, p))]^2, \qquad (10.102)$$

where F is the fixed image intensity function, M is the moving image intensity function, T is the spatial transformation function, x_i is the ith pixel of the fixed image region over which to compute the metric, and N is the total number of pixels in the region. Typically, the value of $M(T(x_i, p))$ needs to be interpolated from the discrete input image. This functionality is provided by the interpolater component. When the mapped position lies outside the domain of the moving image, the contribution due to the pixel pair is discarded.

When a gradient-based optimization scheme is used, the metric component is also required to supply the gradient information. The derivative of the mean squares metric with respect to transform parameters is given by

$$\begin{aligned}
\frac{\partial S}{\partial p} &= \frac{2}{N} \sum_i^N [F(x_i) - M(T(x_i))] \times \frac{\partial M(T(x_i, p))}{\partial p} \\
&= \frac{2}{N} \sum_i^N [F(x_i) - M(T(x_i))] \times \frac{\partial M(y)}{\partial y^T}\bigg|_{y=T(x_i, p)} \frac{\partial T(x_i, p)}{\partial p}.
\end{aligned}$$
$$(10.103)$$

The second term represents the moving image intensity derivative at a mapped location, $T(x_i, p)$. There are several implementation options for this term. The simplest is to compute the derivative for the whole moving image using methods such as central differencing or convolution with a derivative of Gaussian kernel and then use the derivative information at the nearest grid position. Alternatively, a B-spline interpolation scheme can be used where a derivative kernel is used to obtain the interpolated derivative from the B-spline coefficients.

The last term of Equation (10.103) is the *transformation Jacobian* which is a matrix containing the derivatives of a mapped position with respect to the transform parameters. In the software framework, the Jacobian information is supplied by the transform component.

Figure 10.20 illustrates some of the characteristics of the mean squares metric using a 2D synthetized MRI proton density image of the brain [1]. Using the test image as both the fixed and moving image, the metric value was computed for various values of the *x*-axis component of a 2D translation transform.

The mean square metric has a relatively large capture radius with an optimal value of zero. Poor matches between two images result in larger values of the metric. The ripples in the metric plot of Figure 10.20 are artifacts of the linear interpolation method use to compute the moving image intensity values. The peak of the ripples corresponds to integer value translations. In linear interpolation, the intensity value is assumed to vary linearly between integer grid positions. Although the interpolated image is continuous, it is not differentiable.

The plots in Figure 10.21 show the same experiment repeated using cubic B-spline interpolation on the left and nearest neighbor interpolation on the right. It can be observed that the B-spline plot is much smoother. However, this comes at a cost of a six fold increase in computation time. On the other hand, nearest neighbor interpolation is 1.5 times faster than linear interpolation but the resulting metric is discontinuous at whole integer translations. In registration, the cost of interpolation is an important issue since interpolations are computed thousands of times in a single optimization cycle.

The mean squares metric relies on the assumption that intensity representing the same homologous point must be the same in both images. Hence, its use is restricted to images of the same modality. Additionally, any linear change in intensity between the two images will result in a poor similarity score.

Normalized Correlation

This metric computes the pixel-wise cross-correlation between the intensity of images to be registered, normalized by the square root of the autocorrelation of each image. When two images are identical, the measure equals one. Misalignments between the images will result in a measure less than one. This measure is insensitive to a constant multiplicative factor between the images.

---

[1] The synthetized MRI images of the brain were obtained from the BrainWeb project, http://www.bic.mni.mcgill.ca/brainweb

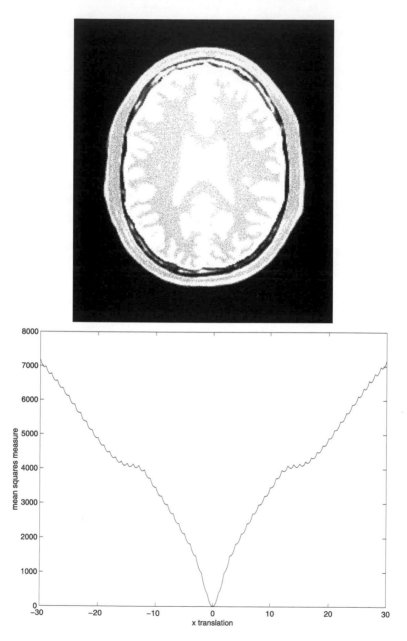

Figure 10.20. Mean squares measure for various *x*-axis translations of a 2D synthetized MRI proton density image of the brain using linear interpolation.

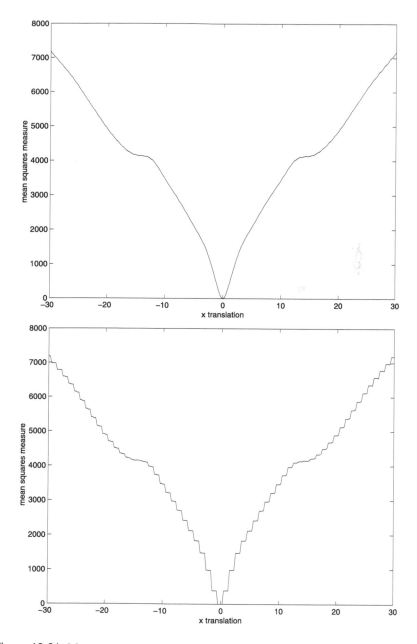

Figure 10.21. Mean squares measure for various *x*-axis translations of the 2D synthetized MRI proton density image of the brain in Figure 10.20 using B-spline interpolation on the top and nearest neighbor interpolation on the bottom.

Mathematically, the measure is given by

$$S(p|F, M, T) = -1 \times \frac{\sum_i^N F(x_i) \cdot M(T(x_i))}{\sqrt{\sum_i^N F^2(x_i) \cdot \sum_i^N M^2(T(x_i))}}. \qquad (10.104)$$

The factor -1 is added so that minimum-seeking generic optimizers can also work with this metric. When a mapped position lies outside the moving image domain, this measure uses the same approach as mean squares where the contribution due to the pixel pair is discarded. The derivative of the normalized correlation metric is given by

$$\frac{\partial S}{\partial p} = -\frac{1}{a} \left[\sum_i^N F(x_i) \cdot \frac{\partial M(T(x_i))}{\partial p} - b \sum_i^N M(T(x_i)) \cdot \frac{\partial M(T(x_i))}{\partial p} \right]$$

$$(10.105)$$

where

$$a = \sqrt{\sum_i^N F^2(x_i) \cdot \sum_i^N M^2(T(x_i))}$$

$$b = \frac{\sum_i^N F(x_i) \cdot M(T(x_i))}{\sum_i^N M^2(T(x_i))}.$$

The derivative of the moving image intensity with respect to the transform parameters required in Equation (10.105) can be computed using the transformation Jacobian as described in Section 10.3.3. The experiment illustrated in Figure 10.20 is repeated with the normalized correlation metric in Figure 10.22 using both linear (top) and fourth order B-spline interpolation (bottom). It can be seen that the shape of the metric curves is similar to that of the mean squares measure with a distinct minimum at zero translation.

Difference Density

For the difference density metric, each pixel's contribution is calculated using $f(d) = \frac{1}{1+(d/\lambda)^2}$ where d is the difference in intensity between the fixed and moving image. The function $f(d)$ is bell-shaped with a maximum of 1 at $d = 0$ and minimum of zero at $d = \pm\infty$. The constant λ controls the rate of drop off as shown in Figure 10.23 and corresponds to the difference in intensity where $f(d)$ has dropped by 50%. The selection of the λ value will depend on the intensity of the two images to be registered.

Mathematically, the difference density metric is given by

$$S(p|F, M, T) = \sum_i^N \frac{1}{1 + \frac{[F(x_i) - M(T(x_i))]^2}{\lambda^2}}. \qquad (10.106)$$

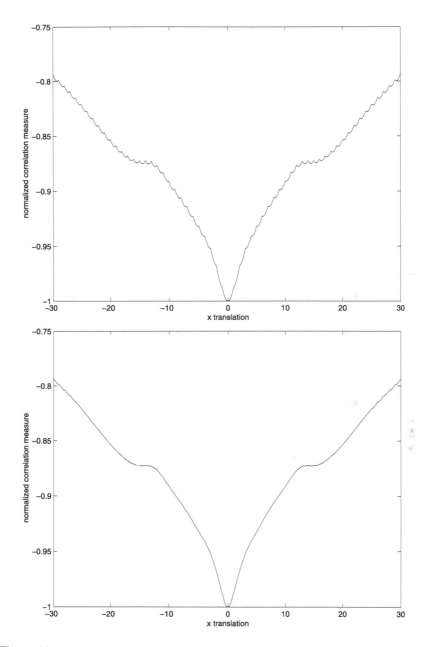

Figure 10.22. Normalized correlation measures for various x-axis translations of the 2D synthetized MRI proton density image of the brain in Figure 10.20 using linear (top) and fourth order B-spline interpolation (bottom).

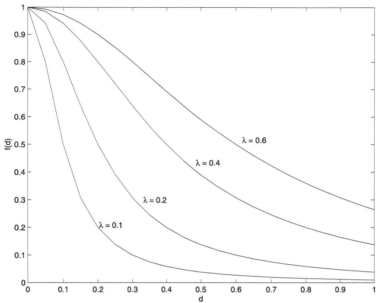

Figure 10.23. Plots of $f(d) = \frac{1}{1+(d/\lambda)^2}$ for various λ values. Function $f(d)$ is used in computing the difference density similarity metric.

The optimal value for the metric is N and poor matches correspond with small measure values. This metric can be interpreted as approximating the probability density function of the difference image and maximizing its value at zero. Since the metric is based on the intensity difference between the two images, it is sensitive to linear differences in intensity.

The derivative of the difference density metric with respect to the transform parameters is

$$\frac{\partial S}{\partial p} = \frac{2}{\lambda^2} \sum_i^N \frac{[F(x_i) - M(T(x_i))]}{\left(1 + \frac{[F(x_i) - M(T(x_i))]^2}{\lambda^2}\right)^2} \times \frac{\partial M(T(x_i))}{\partial p}. \tag{10.107}$$

As previously described in Section 10.3.3, the moving image derivatives can be computed using the transformation Jacobian supplied by the transform component.

The experiment performed in Figure 10.20 is repeated with two different λ values in Figure 10.24. From the plots it can be observed that the profile of this metric exhibits a sharp peak. The bandwidth of the peak is governed by the λ. The plot on the left corresponds to $\lambda = 10$ and with $\lambda = 4$ on the right. The advantage of the sharp peak is that it quickly locks the optimization onto the

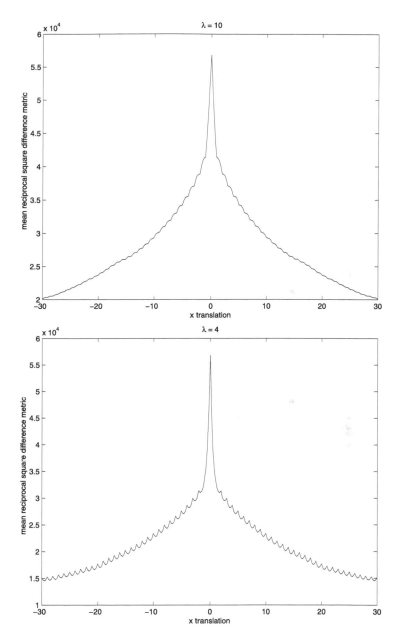

Figure 10.24. Pattern intensity measures for various x-axis translations of the 2D synthetized MRI proton density image of the brain in Figure 10.20 with $\lambda = 10$ (top) and $\lambda = 4$ (bottom).

minima, however, the fact that the metric derivative is very large at the central peak may pose a problem for optimization algorithms that rely on the decrease of the derivatives of the metric as the extrema is reached.

Mutual Information

Mutual information is an information theoretic entity that qualitatively measures how much information is gained about one random variable by the knowledge of another random variable. The use of mutual information for multi-modality registration was independently introduced by two different groups: Viola and Wells [26] and Collignon et al. [4]. In the case of image registration, the image intensity of the fixed and moving images are the random variables. The major advantage of using mutual information is that the actual form of dependency between the two random variables does not have to be specified. Therefore, complex mappings between the two images can be modeled making mutual information well-suited to multi-modality registration.

Mutual information is defined in terms of entropies. Let

$$H(A) \ = \ -\int p_A(a) \log p_A(a) \, da \qquad (10.108)$$

$$H(B) \ = \ -\int p_B(b) \log p_B(b) \, db \qquad (10.109)$$

where p_A and p_B are, respectively, the marginal probability density function for random variables A and B, and $H(A)$ and $H(B)$ are their respective entropies. Additionally, let

$$H(A,B) = -\int p_{AB}(a,b) \log p_{AB}(a,b) \, da \, db \qquad (10.110)$$

where p_{AB} is the joint probability density function and $H(A,B)$ the joint entropy.

If random variables A and B are independent then

$$p_{AB}(a,b) = p_A(a) p_B(b) \qquad (10.111)$$

and

$$H(A,B) = H(A) + H(B). \qquad (10.112)$$

On the other hand, if there are any dependencies then

$$H(A,B) < H(A) + H(B). \qquad (10.113)$$

The difference,

$$I(A,B) = H(A) + H(B) - H(A,B), \qquad (10.114)$$

is called mutual information.

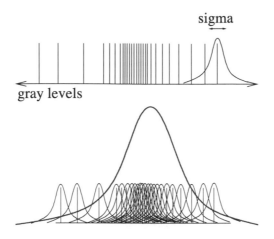

Figure 10.25. In Parzen windowing, a density function is constructed by super-imposing kernel functions (Gaussian function in this example) centered on the intensity samples obtained from the image. In the case, the standard deviation of the Gaussian acts as the smoothing or tuning parameter.

In our image registration framework, random variables A and B correspond to the fixed image intensity, $F(x)$, and moving image intensity, $M(T(x))$. Typically, direct access to the marginal and joint probability densities of the image intensity is not available and hence the densities must be estimated from the image data. To this end, Parzen windowing (also known as kernel density estimators) can be used to estimate the densities. In this method, the image intensity is randomly sampled from the image to form set S. A density function is then constructed by super-positioning a kernel function, $K(s)$, centered on the elements of S along the intensity or grey-level axis as shown in Figure 10.25.

Mathematically, the density estimate is given by

$$P^*(a) = \frac{1}{N_S} \sum_{s_j \in S} K(a - s_j) \qquad (10.115)$$

where N_S is the number of samples in the set S. A variety of functions can be used as the kernel, K, as long as they satisfy the requirements that they are symmetric, have zero means, and integrate to one. For example, Gaussian, box, and B-spline functions all meet these criteria. A smoothing parameter is also used to scale the kernel function. The larger the smoothing parameter, the wider the kernel function, and hence the smoother the density estimate. If the smoothing parameter is too small, the resulting density may be too noisy. On the other hand, if the parameter is too large, features such as modes in the density will get smoothed out. Typically, the optimal value of the smoothing parameter will depend on the data and the number of samples used. Another option is to normalize the image intensity and then use a fixed smoothing parameter to register images of arbitrary magnitude and range.

Using the density estimate, P^*, the entropy integral can be approximated in two different ways: sample mean [26] or summing over discrete bins [19]. In the sample mean method, another set of intensity samples, R, is randomly drawn from the image. The entropy integral is then approximated by

$$H(A) = \frac{1}{N_R} \sum_{r_j \in R} log P^*(r_j) \qquad (10.116)$$

Alternatively, P^* is evaluated at discrete positions or bins uniformly spread within the dynamic range of the images. The entropy is then approximated by summing over the bins. When the chosen kernel function has a finite support interval (e.g., box or B-spline), the probablity distribution function (PDF) estimate calculation can be accelerated by only considering samples which are within the support interval of the position to be evaluated.

Deriving the formula for the mutual information metric derivatives is an involved process where the actual form is dependent on the kernel function and the method used to approximate the entropy integral. In [26], the derivative formulation is presented for Gaussian kernels and sample mean entropy approximation, and in [20], the formulation is presented for cubic B-spline kernels and entropy approximation by summation over bins.

The joint histogram of the images is a useful tool for understanding how mutual information works. Figure 10.26 shows the joint histogram images generated using identical synthetic MRI proton density images of the brain. Each pixel of

Figure 10.26. Examples of joint intensity histograms using identical synthetic MRI proton density images of the brain (top left). In the histogram images, the intensity ranges from white (zero value) to black (the highest value). The middle image is the intensity histogram generated when the two images are aligned. The image on the right is the intensity histogram generated when one of the images is translated horizontally by five pixels. Misalignment has the effect of dispersing the histogram.

Figure 10.27. Examples of joint intensity histograms using the MRI proton density image in Figure 10.26 and the synthetic MRI T1-weighted image on the top left. In the histogram images, the intensity ranges from white (zero value) to black (the highest value). The middle image is the intensity histogram generated when the two images are aligned. The image on the right is the intensity histogram generated when the T1-weighted image is translated horizontally by five pixels. Misalignment has the effect of dispersing the histogram.

the histogram image represents a histogram bin with intensity value ranging from white, representing the zero value, to black, the highest value. The first histogram illustrates the condition where the two images are prefectly aligned. Since the two images are identical the result is a 45° straight line. The second histogram is obtained when one of the image is shifted horizontally by five pixels. Under these conditions, the misalignment causes the histogram values to disperse.

Figure 10.27 shows joint histogram examples when the images arise from different sources. In the figure, the histograms are generated from the MRI proton density image in Figure 10.26 and the MRI T1-weighted image on the left. The first histogram corresponds to when the images are aligned and the second histogram to when the T1-weighted image has been shifted horizontally by five pixels. Since the images are from different pulse sequences, the data are not correlated and we do not get a distinct line as in the aligned case. However, the dispersion effect due to misalignment is still prominent in the multi-modality case. Hence, a possible registration criterion is to seek the transform which will produce a small number of bins with very high values and as many zero-valued bins as possible. This criterion is equivalent to minimizing the joint entropy. Using the joint entropy alone for image registration however is problematic as it favors transformation which cause the images to be as far apart as possible (i.e., minimizing the overlap between the images) [11]. Mutual information overcomes this problem by also trying to maximize the information contributed by each image in the overlap region.

Figures 10.28 and 10.29 illustrate the characteristics of two different implementations of mutual information. The implementation in Figure 10.28 follows the method described in [26] using a Gaussian kernel function for density estimation and the sample mean entropy approximation method. Using the proton

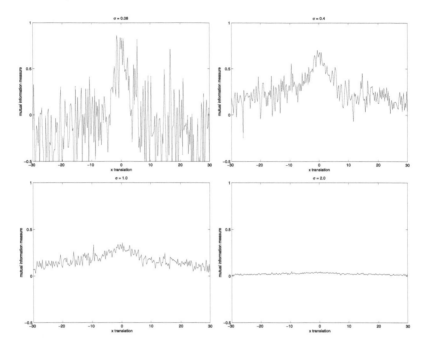

Figure 10.28. Mutual information measures between the proton density and T1-weighted synthetic MRI images of the brain in Figures 10.26 and 10.27 using the method described in [26]. The images were first normalized to zero mean and unit variance, and the measure was computed using different standard deviation values for the Gaussian kernel.

density and T1-weighted images from Figures 10.26 and 10.27, the images were first normalized to have zero mean and unit variance. Mutual information was then evaluated for various x-axis shifts of the T1-weighted image. For each evaluation, 50 random samples were used for density estimation and another 50 samples for entropy calculation. New samples were drawn for each evaluation of the metric for each shift in the T1-weighted image. The experiment was repeated for various standard deviation values for the Gaussian kernel function which acts as the smoothing parameter. It can be observed from the graphs that the smoothing parameter has a dramatic impact on the metric. When the parameter value is small, the resulting plot is very noisy. As the parameter increases, there is a reduction in noise and the maximum corresponding the zero translation becomes more obvious. When the parameter becomes very large, however, the maximum starts to smooth out. In this case, a good choice of parameter value would be between 0.4 and 1.0. It should also be noted that the use of linear interpolation and redrawing of samples for each evaluation of the metric also contributes to the fluctuations in the plots.

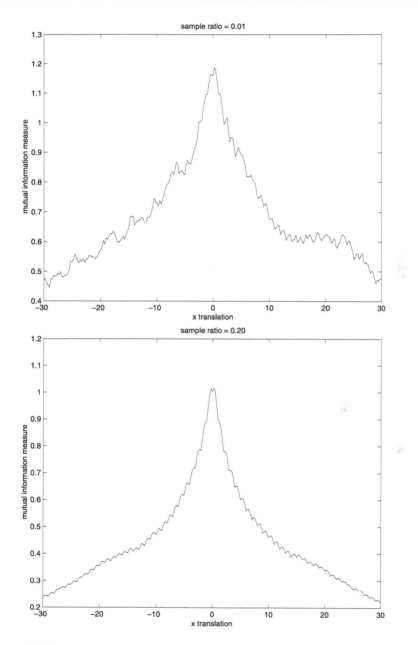

Figure 10.29. Mutual information measures between the proton density and T1-weighted synthetic MRI images of the brain in Figures 10.26 and 10.27 using the method described in [20]. The images were first rescaled to have a minimum of zero and maximum of one and the measure was computed by sampling 1% (top) and 20% (bottom) of the image.

The implementation in Figure 10.29 follows the description in [20]. In this method, a box kernel is used to compute the density for the fixed-image intensity and a cubic B-spline kernel for the moving-image intensity. Since the fixed-image pdf does not contribute to the metric derivatives, it does not need to be smooth, permitting the use of a simple box kernel. On the other hand, a cubic B-spline kernel is used for the moving image to ensure smoothness and stable derivative calculations. One of the major advantages of using a B-spline kernel over a Gaussian kernel is that the B-spline kernel has a finite support region. Thus, to evaluate the density at a given position, only samples within the support region are needed for the calculation. Additionally, by rescaling the image to have a minimum of zero and maximum of one, a fixed B-spline kernel bandwidth can be used to handle image data with arbitrary magnitude and dyanmic range.

The implementation in [20] also differs in the intensity sampling regime. In the former implementation, a small number (50 to 100) samples are used but are redrawn for each metric evaluation. In this method, a large number of samples (1% to 20% of the image, depending on the smoothness of the image) are drawn once and used for the whole registration process. Using one large sample, results in a much smoother metric space. The plots in Figure 10.29 are smoother with a more distinct maximum than in Figure 10.28. The smoothness of the metric space has an impact on the type of optimization algorithms that can be used. The roughness of the metric in first implmentation only allows the use of a simple steepest ascent method in [26], while the relative smoothness of the second implemenation allows the use of the more complex memory-limited Broyden-Fletcher-Goldfarb-Shanno (LBFGS) [21] nonlinear optimization scheme in [20].

10.3.4   Optimization Algorithms

Within the registration framework, the role of the *optimizer* component is to optimize the qualitative measure provided by the *metric* with respect to the parameters of the *transform* component. Starting from an initial set of parameters, the optimization procedures iteratively search for the optimal solution by evaluating the metric at different positions of the transform parameter search space. Optimization algorithms can be broadly divided into those which use derivative information and those which do not. In the registration framework, the metric derivative is also provided by the metric component.

Since transform parameters can have vastly different dynamic ranges, particular attention should be paid in the scale of parameters when solving registratin problems. For example, a 2D rigid transform is specified by a rotation (in radians) and two translation parameters (in mm). A unit change in angle has a much greater impact on an image than a unit change in translation. This difference in scale appears as elongated valleys in the parameter search space causing difficulties for some optimization algorithms. In some cases, it may be neccessary to rescale the parameters to overcome this problem. A simple implementation of rescaling is to divide or multiply the metric gradient by weights chosen to balance the parameters.

Detailed discussion and analysis of optimization algorithms is beyond the scope of this chapter. The reader is referred to various monographs (e.g., [22, 21]) on optimization methods for details.

10.4 Examples

10.4.1 Multi-Modality Multi-Resolution Example

The Problem

In this example, the problem is to register the 3D CT image of the head in Figure 10.30 to the MR T1-weighted image of the same subject in Figure 10.31 The results of this type of registration could be used to aid neurosurgery or radiotherapy planning.

The two images are of different size and resolution: the CT image is of size $512 \times 512 \times 44$ with pixel size of 0.41×0.41mm, the MR-T1 image is $256 \times 256 \times 52$ with pixel size of 0.78×0.78mm. Both images have 3mm slice thickness. There is also a small difference in the field of view with the MR-T1 image extending farther down past the nose than the CT image.

Strategy

To use the registration framework described in Section 10.3 we need to select the underlying algorithm for each of the components in Figure 10.1. In this example, the MR-T1 image is the fixed image and the CT is the moving image. Since the CT and MR-T1 images have very different distributions, the use of simple similarity measures such as mean squares and normalized correlation are not applicable. Mutual information, on the other hand, is well-suited to this problem. In particular we will use the implementation in [20] as our metric component. Since the human skull prevents any non-rigid deformation, we will use a rigid transform. In particular we will use a quaternion to represent the 3D rotation. In this illustrative example, we will also use a tri-linear interpolator and simple steepest descent as the optimizer.

It is important to note that the transformation is rigid with respect to physical coordinates and not image index coordinates. The registration components should be implemented such that communication between the components is with respect to physical coordinates, allowing images of different resolution to be registered without resampling the images to a common isotropic resolution.

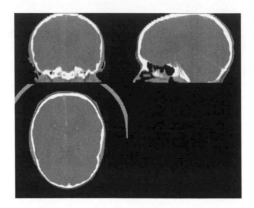

Figure 10.30. 3D CT image of the head used in the multi-modality, multi-resolution registration example in section 10.4.1. The image is of size $512 \times 512 \times 44$ with pixel size of 0.41×0.41mm and 3mm slice thickness. [The images were provided as part of the "Retrospective Image Registration Evaluation" project, National Institutes of Health, Project Number 8R01EB002124-03, Principal Investigator, J. Michael Fitzpatrick, Vanderbilt University, Nashville, TN. See Section 10.4.1 for information.]

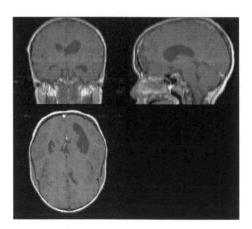

Figure 10.31. 3D MR-T1 image of the head used in the multi-modality, multi-resolution registration example in section 10.4.1. The image is of size $256 \times 256 \times 52$ with pixel size of 0.78×0.78mm and 3mm slice thickness. [The images were provided as part of the "Retrospective Image Registration Evaluation" project, National Institutes of Health, Project Number 8R01EB002124-03, Principal Investigator, J. Michael Fitzpatrick, Vanderbilt University, Nashville, TN. See Section 10.4.1 for information.]

Performing image registration using a multi-resolution strategy helps to improve the computational speed, accuracy, and robustness. In this example, we will perform registration in a coarse-to-fine manner where the transformation computed at one resolution level is used to initialize the registration at the next finer resolution level.

The first step is to generate a pyramid of downsampled images for each of the images to be registered. In this example, we will use four resolution levels. At the coarsest level, the CT image will be downsampled by a factor of eight for each of the in-plane dimensions. Since the slice thickness is large, we will not downsample in the slice direction. At the coarsest level, the image is approximately isotropic and of size $64 \times 64 \times 44$ pixels. The images at the subsequent two levels are formed by downsampling in-plane by factors of four and two. The full resolution image is used at the last level. The coarsest level of the MR-T1 image pyramid is such that the image matches the resolution of the coarsest CT image. That is, downsample by a factor of four in-plane. The next level is obtained by downsampling by a factor of two. The full resolution MR-T1 image is used for the last two levels. Note that before downsampling, the images are first blurred using a Gaussian kernel with variance of 0.5 times the shrink factor. The downsampled CT images are shown in Figure 10.32 with the downsampled MR-T1 images in Figure 10.33.

As described in Section 10.3.1, the location of the center of rotation has a great impact on the performance of the optimization procedure. In practice, the

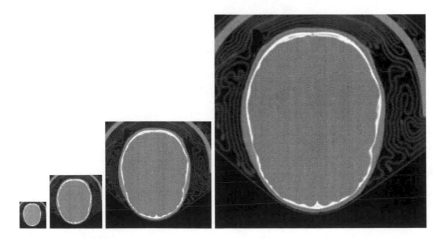

Figure 10.32. Pyramid of downsampled images. The coarsest image (left) is generated by downsampling the original CT by a factor of $\{8, 8, 1\}$. The next two images are obtaineded using downsample factors of $\{4, 4, 1\}$ and $\{2, 2, 1\}$. The full resolution CT image is used at the finest level (right).

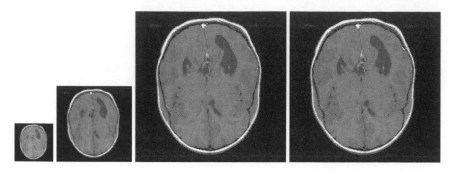

Figure 10.33. Pyramid of downsample images. The coarsest image (left) is generated downsampling the original MR-T1 image by a factor of $\{4,4,1\}$. The next image is obtained by using downsample factors of $\{2,2,1\}$. The full resolution MR-T1 image is used for the last two resolution levels.

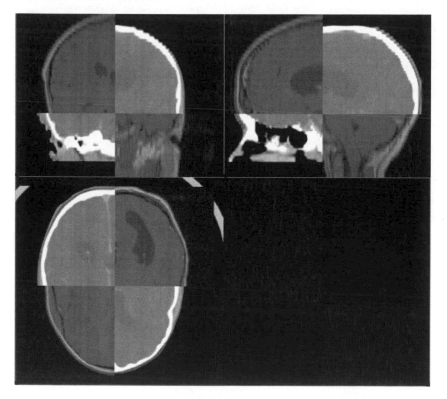

Figure 10.34. (See Plate IX) Registration start with the center of the CT image initially aligned with the center of the MR-T1 image. The CT and MR-T1 images are shown as a checkerboard where each block alternately display data from each dataset.

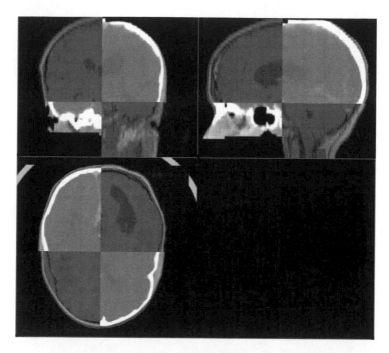

Figure 10.35. (See Plate X) Results of the multi-resolution registration process described in Section 10.4.1. The CT and MR-T1 images are shown as a checkerboard where each block alternately displays data from each dataset.

center of rotation will be closer to the center of the image than at the image origin (center of the first pixel). With this in mind, we will place the center of rotation at the center of the image and start the registration process with the two images aligned at their centers. This initialization is shown in Figure 10.34 where the center of the CT image has been aligned with the center of the MR-T1 image and resampled to the same resolution as the MR-T1 image. The CT image is displayed as a color overlay on top of a gray-scale MR-T1 image. The bright color values represent the bright part of the CT image in Figure 10.30 corresponding to the skull. From the figure, the initial misalignment is quite clear.

In this example, 3D rigid transformation is defined by seven parameters: the first four representing the quaternion and the last three the 3D translation. As described in Section 10.3.4, it is important to address the scale differences between the quaternion and translation parameters. For this example, we scale the metric derivative for the translation parameters by a factor of 4×10^4.

Using the parameters listed in Table 10.1 registration was performed in a coarse-to-fine, manner and the results at each resolution level are shown in Fig-

Level	1	2	3	4
CT Shrink Factors	$8,8,1$	$4,4,1$	$2,2,1$	$1,1,1$
MR-T1 Shrink Factors	$4,4,1$	$2,2,1$	$1,1,1$	$1,1,1$
No. of Spatial Samples	1000	8000	64000	512000
No. of Histogram Bins	50	50	50	50
No. of Iterations	250	250	250	250
Step Size	2×10^{-3}	4×10^{-4}	8×10^{-5}	1.6×10^{-5}

Table 10.1. Parameters used for performing the multi-modality, multi-resolution registration example.

ure 10.36. The two images are mostly aligned as a result of registration at the coarsest level. The registration is further refined in the next two levels. This can be seen as the increasing overlap between the bright region representing the skull on the CT image and the corresponding dark region in the MR-T1 image. The

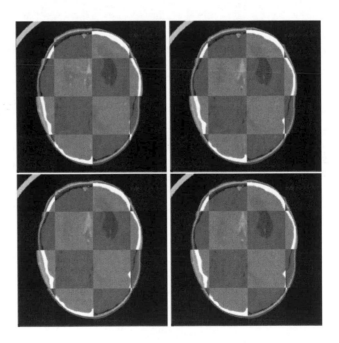

Figure 10.36. (See Plate XI) Results of the multi-resolution registration process in Section 10.4.1 after registration at the first (coarsest) level (top-left), the second level (top-right), the third level (bottom-left) and the fourth (finest) level (bottom-right). The CT and MR-T1 images are shown as a checkerboard where each block alternately displays data from each dataset.

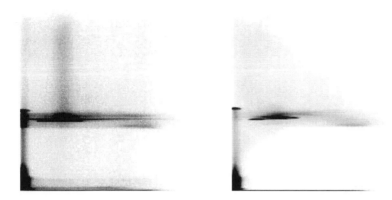

Figure 10.37. Joint intensity histogram of the CT and MR-T1 images of the head before (left) and after (right) registration. In the histogram images, the intensity ranges from white (zero value) to black (highest value). Registration has the effect of concentrating the histogram to fewer non-zero elements.

differences between the last two levels cannot be seen visually. Depending on the application, registration at the highest resolution may not be needed. The final result of the registration is also shown in coronal and sagittal views in Figure 10.35. In the sagittal view the sinus cavity is clearly aligned.

The joint intensity histogram of the CT and MR-T1 images before and after registration are shown in Figure 10.37. The effect of registration is to concentrate the histogram to have fewer non-zero elements and to have a small number of elements with very high values. As described in Section 10.3.3, if the joint histogram is interpreted as the joint probability density function then the effect of registration is equivalent to minimizing the joint entropy.

About the Data

The images used in this example were from the "The Retrospective Image Registration Evaluation Project" (RIRE) conducted by Dr J. Michael Fitzpatrick at Vanderbilt University [27]. For the project, image volumes of three modalities (CT, MR and PET) were taken of patients undergoing neurosurgery at Vanderbilt University Medical Center.

Before imaging, fiducial markers (designed to be visible in all three modalities) were attached to the skull of the patient. These markers were then used to perform point-based registration of CT to MR and PET to MR images. The fiducial markers were then airbrushed out of the images to allow a blind evaluation of retrosepective registration methods.

This project is on-going: participants submit their registration results. Comparisions are then made between the fidicual-based registration (the gold standard) and the submitted results. Registration is scored by assessing its accuracy in ten regions of neurological and/or surgical interest: (1) maximum aperture of fourth ventricle, (2) junction of forth ventricle with aqueduct, (3) right globe, (4) left globe, (5) optic chiasm, (6) apex of left Sylvian fissure, (7) apex of right Sylvian fissure, (8) junction of central sulcus with midline, (9) left occiptal horn, and (10) right occiptal horn.

Accuracy of the registration at each volume of interest is determined by measuring the difference between the registered target position of the submitted method and that of the gold standard. The *target regisration error* (TRE) for each of the ten volumes of interest is published on the RIRE project website[2]. The RIRE project is a valuable resource for researchers in registration to evaluate their registration technique against existing methods in a standarized and systematic manner.

10.4.2   Deformable Registration Example

The Problem

The problem addressed by this example is the registration of the contrast-enhanced breast MRI time series images in Figure 10.38. Dynamic contrast MRI has become a valuable tool for early detection of breast cancer [14]. During the MRI exam, a contrast agent is injected via a catheter. A 3D MRI scan is acquired prior to the injection, followed by a series of scans post-injection. The contrast uptake curves of malignant lesions behave differently to those corresponding to benign lesions and thus characteristics of the uptake curves can be used to discriminate cancerous lesions. A typical exam can take 15 minutes, any patient movement during the course of the exam will result in misalignment of the MRI time series images; as a result accurate registration the MRI time series will faciliate the extraction of the contrast uptake curves.

Strategy

The two challenging factors in registering dynamic contrast breast MRI images are (1) the deformable motion of the breast tissue and (2) the non-uniform intensity change due to the contrast uptake. These factors must be taken into account when choosing components for the registration framework. Following the approach taken in [23], we will model the deformable transform using a regular grid of B-spline control points and use mutual information as the similarity metric. For this example, we will use the mutual information implementation described

---

[2]http://www.vuse.vanderbilt.edu/ image/registration/

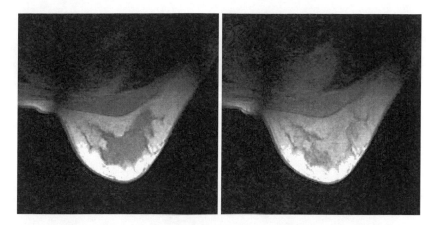

Figure 10.38. 3D contrast-enhanced breast MRI time series images. Each 3D image is of size $192 \times 192 \times 13$ with pixel size of 0.9375×0.9375mm and 8mm slice thickness. Images courtesy of University of Washington.

in [20] using 50×50 histogram bins and sampling 20% of the image to create the joint pdf. We will also use a tri-linear interpolator and a LBFGS [21] optimizer.

The top-left image of Figure 10.39 shows the absolute difference between the images before registration. The large differences result from contrast uptake (in the middle of the breast tissue) and motion artifacts (e.g., at boundary of the breast tissue). It can be observed that the scale of misalignment is small and hence a multi-resolution approach is not needed for this problem.

For this example, we will model the deformation using an $11 \times 11 \times 5$ grid of B-spline control points. It should be noted that for a cubic B-spline, 27 neighboring control points are required to evaluate the deformation at any one position. As a result, the grid should be extended beyond the image domain so that it is always possible to evaluate the deformation at any point inside the image.

The top-right image of Figure 10.39 shows the result of the registration. It can be seen that the registration was able to remove a majority of the motion artifacts while differences due to contrast uptake were left unalterd. For comparision, we repeated the experiment using a different metric and a different transform model. In the bottom-left image, we used a mean squares metric instead of mutual information. It can be observed that for this case, the focus of the registration is to reduce the large difference due to the contrast uptake which is not appropriate to the dynamic contrast MRI application. In the bottom-right image, we used a simple rigid transform instead of the B-spline deformable model. Although some of the motion artifact has been removed, it is clear that a rigid transform is not sufficient to model the deformable motion of the breast tissue.

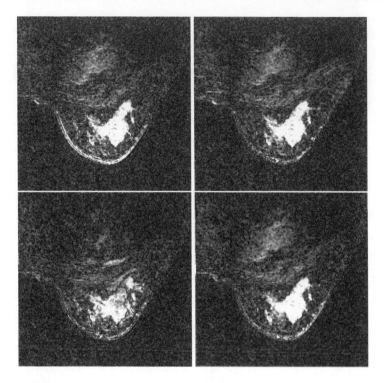

Figure 10.39. Absolute difference between unregistered (top left) and registered contrast-enhanced breast MRI images. The top right image corresponds to deformable registration using mutual information as the basis of registration, the bottom left corresponds to deformable registration using mean squares, and the bottom right, rigid registration using mutual information.

References

[1] Alexandrescu, A. *Modern C++ Design: Generic Programming and Design Patterns Applied.* Professional Computing Series. Addison-Wesley, 2001.

[2] Austern, M. *Generic Programming and the STL:.* Professional Computing Series. Addison-Wesley, 1999.

[3] Booch, G., Jacobson, I., Rumbaugh, J., and J. Rumbaugh. *The Unified Modeling Language User Guide.* Addison-Wesley Pub Co, 1998.

[4] Collignon, A., *et al.* "Automated multimodality image registration based on information theory." in *Proceedings of Information Processing in Medical*

Imaging 1995, Y. Bizais, *et al.*, editor, Dordrecht, The Netherlands: Kluwer Academic Publishers, 1995. 263–274.

[5] Coxeter, H. *Introduction to Geometry*, second edition. New York: John Wiley and Sons, 1969.

[6] Eccles, F. *Transformational Geometry*. Addison Wesley, 1971.

[7] Fitzpatrick, J.M., Hill, D., and C. Maurer. "Image registration." in *Handbook of Medical Imaging, Vol.2: Medical Image Processing and Analysis*, (M. Sonka and J. M. Fitzpatrick, eds.), SPIE Press, July 2000. Chapter 8.

[8] Gamma, E, Helm, R., Johnson, R., and J. Vlissides. *Design Patterns, Elements of Reusable Object-Oriented Software*. Professional Computing Series, Addison-Wesley, 1995.

[9] Hajnal, J., Hawkes, D., and D. Hill. *Medical Image Registration*. CRC Press, 2001.

[10] Hamilton, W. *Elements of Quaternions*. Chelsea Publishing Company, 1969.

[11] Hill, D. and D. Hawkes. "Across-modality registration using intensity-based cost functions," chapter 34, in *Handbook of Medical Imaging: Processing and Analysis*, (I. Bankman, ed.), San Diego: Academic Press, 2000. 537–554.

[12] Joly, C. *A Manual of Quaternions*. MacMillan and Co. Limited, 1905.

[13] Knowlton, R. "Clinical applications of image registration," in *Handbook of Medical Imaging: Processing and Analysis*, chapter 38, (I. Bankman, ed.), San Diego: Academic Press, 2000. 613–622.

[14] Kuhl, C. and H. Schild. "Dynamic image interpretation of MRI of the breast," *Journal of Magnetic Resonance Imaging* 12(6)(2000):965–974.

[15] Lanczos, C. *Space Through the Ages: The Evolution of Geometrical Ideas from Pythagoras to Hilbert and Einstein*. San Diego: Academic Press, 1970.

[16] Luenberger, D. *Optimization by Vector Space Methods*. New York: Wiley Professional Paperback Series. Interscience, April 1997.

[17] Maintz, J. and M. A. Viergever. "A survey of medical image registration," *Medical Image Analysis* 2(1998):1–36.

[18] Martin, G. *Transformation Geometry, An Introduction to Symmetry*. Berlin: Springer Verlag, 1982.

[19] Mattes, D., Haynor, D., Vesselle, H., Lewellen, T., and W. Eubank. "Nonrigid multimodality image registration," in *Proceedings of SPIE: Medical Imaging* 4322(2001): 1609–1620.

[20] Mattes, D., Haynor, D., Vesselle, H., Lewellen, T., and W. Eubank. "PET-CT image registration in the chest using free-form deformations," *IEEE Transactions on Medical Imaging* 22(1)(January 2003):120–128.

[21] Nocedal, J. and S. J. Wright. *Numerical Optimization*. New York: Springer series in operations research. Springer-Verlag, 1999.

[22] Press, W, Flannery, B., Teukolsky, S., and W. Vetterling. *Numerical Recipes in C*, second edition. Cambridge: Cambridge University Press, 1992.

[23] Rueckert, D., Sonoda, L., Hayes, C., Hill, D., Leach, M, and D. Hawkes. "Nonrigid registration using free-form deformations: Application to breast MR images," *IEEE Transactions on Medical Imaging* 18(8)(1999):712–721.

[24] Unser, M., Aldroubi, A., and M. Eden. "B-Spline signal processing: Part I—Theory," *IEEE Transactions on Signal Processing* 41(2)(February 1993):821–833.

[25] Unser, M., Aldroubi, A., and M. Eden. "B-Spline signal processing: Part II—Efficient design and applications," *IEEE Transactions on Signal Processing* 41(2)(February 1993):834–848.

[26] Viola, P. and W. Wells III. "Alignment by maximization of mutual information," *International Journal of Computer Vision* 24(2)(1997):137–154.

[27] West, J., Fitzpatrick, J.M., *et al.* "Comparison and evaluation of retrospective intermodality image registration techniques," in *Proceedings of SPIE: Medical Imaging* SPIE 2710(1996): 332–347.

[28] Yaglom, I. *Geometric Transformations I*. Mathematical Association of America, 1962.

Non-Rigid Image Registration

Brian Avants
Tessa Sundaram
Jeffrey T. Duda
James C. Gee
University of Pennsylvania

Lydia Ng
Insightful, Inc.

11.1   Introduction

Non-rigid (deformable) image registration methods range from fast, basic image processing algorithms to highly advanced physical models that push the boundaries of numerical simulation and computation. Techniques may rely on simple assumptions about the data appearance, statistical relationships between intensities, or detailed anatomical information gained from medical expertise. Algorithms may be engineered for clinical practice, neurosurgery, anatomical and functional research as well as studies of biology, aging, genetics, and disease. Structural-functional research is supported by underlying automated image registration programs [86, 33] that align patient studies within a documented anatomical space [68]. The need for inter-modality non-rigid image registration arises in the clinical context in which patient motion necessitates de-warping of functional images (e.g., PET) acquired over a matter of minutes such that they match rapidly acquired structural CT or MRI data. Furthermore, distortions caused by signal variations or in novel imaging technologies often require non-rigid transformations as a basic image processing operation [48]. Development of atlases for surgical planning and teaching is another valuable application [44, 78]. Deformable image registration for placement of deep brain stimulators has also shown promis-

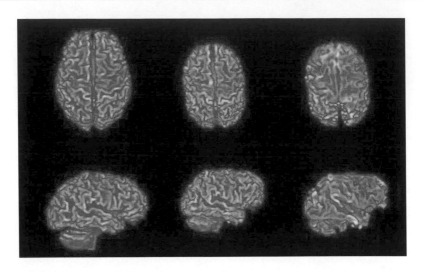

Figure 11.1. (See Plate IV) The gray-white interfaces of a human (left) and a chimp (far right) are registered by their mean curvature [3]. The result is shown in the center column.

ing results [22]. This chapter will attempt to explain some of the theory behind these diverse methods as well as provide illustrations of how this relatively recent field of research may be adapted to different needs [73, 79, 82, 56, 17, 18].

Although non-rigid image registration is an application-driven area with algorithmic innovations largely motivated by clinical and industrial research, the questions that arise in development of these methods constitute many open problems. Current technical research has expanded into determining the statistics of Lie groups [26], surface and volumetric conformal mapping problems [37], novel variations of statistical measures such as mutual information, studies of perception and correspondence [4], and many other interesting areas. Furthermore, questions about anatomical similarity and variations that arise in non-rigid image registration often exceed human understanding of anatomy [49].

Some of these questions have been addressed with volumetric techniques, discussed below. Surface-based algorithms, however, have been developed for sulcal driven registration [74, 20, 73, 84, 52] and used for detailed statistical studies of shape and correlated variables [65, 6, 82, 36]. Surfaces embedded in volumes may also be registered as illustrated in Figure 11.1 where two instances of segmented cortical white matter are put into correspondence. Surface representations are also combined with knowledge of medialness for volumetric analysis of specific anatomy [67, 32]. Many other approaches have also proven to be useful [10, 81, 50], including curve-based matching [61, 24].

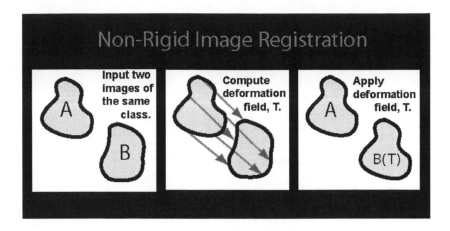

Figure 11.2. This schematic illustrates the process of non-rigid image registration. Note that the vector field is computed in the A to B direction, but B is transformed into the coordinate system of A via the inverse warp.

The more traditional philosophy in non-rigid image registration is to model an image as a continuum (fluid, plastic, elastic, etc.). The medium is then allowed to deform in order to satisfy some *a priori* optimization criterion. For example, one may assume that, given two images, one of the pair should deform in some permissible way until its appearance is similar to the other. This type of transformation will give correspondence between the image domains and also reveals shape differences via the transformation itself. The overall class of problems in medical imaging is illustrated in Figure 11.3.

Tensors derived from deformation fields [1] provide a beautiful means to statistically investigate population data in ways that are not possible with human resources alone. Computational anatomical research has also burgeoned with the availability of large image datasets and the computers and algorithms to process and statistically analyze them. Volumetric techniques are being used in many forms of change detection, including study of age-related and frontotemporal dementia [76, 23], schizophrenia [16], and functional and psychological questions [14]. Non-rigid methods have been used to aid understanding of fetal development [6] and the correlation of specific structures (such as the hippocampus) with specific diseases [63].

Image registration algorithms require that a quantifiable correspondence exists or may be assumed between images to be registered. This is not enough, however, as one is then usually left with an ill-posed problem, in the sense of Hadamard [38]. These problems suffer from being fundamentally under-constrained in that the solution space is full of many indistinguishably good answers. Furthermore, in

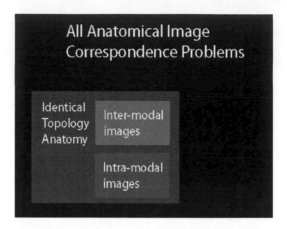

Figure 11.3. A (incomplete) sketch of the anatomical image correspondence problem space. The focus of this chapter is within the domain of topologically similar anatomy. However, research is currently being conducted on methods that incorporate photometric variation, allowing the image to change topology [56].

non-rigid registration, the problem is complicated by the fact that often solutions exist in an infinite dimensional (or at least combinatorially intractable discrete) space.

The standard solution to ill-posedness is Tikhonov regularization [80] . Constraining the energy of the solution's derivatives restricts one to a computable subspace and provable uniqueness, the latter part resolving one of Hadamard's criteria. This strategy is also useful in insuring that solutions are physically meaningful, an additional motivation for using continuum mechanical models [45] in the algorithms. The resulting optimization problem may be formulated with classical variational (or control or Bayesian) theory. No Free Lunch theorems [85] have further clarified the formal need for algorithms which flexibly introduce prior knowledge in real world optimization problems.

This chapter deals mainly with dense high-dimensional image registration with continuum mechanical regularization priors [5, 29, 54, 13]. The goal is to locate a physically admissible transformation that provides correspondence between the images under consideration, a fixed image, I, and a moving image, J. The moving image will ultimately be warped into the reference frame provided by I. However, the transformation is computed from I to J. The Lagrangian view states this as finding correspondence vector field $\vec{u}$ such that,

$$I(\mathbf{X} + \vec{u}) \sim J(\mathbf{x}), \tag{11.1}$$

where $\sim$ denotes that the correspondence approached an extremum with respect to an appropriate measure of equivalence. The Lagrangian view always makes reference to the original coordinates of I at $\mathbf{X}$. Noting that $\mathbf{X} = \mathbf{x} - \vec{u}$, the Eulerian reference frame views the problem as a flow in time such that $\mathbf{x}(t)$. At convergence (time $t = 1$),

$$I(\mathbf{x}(t = 1)) \sim J(\mathbf{x}). \qquad (11.2)$$

We will denote velocity fields in this frame as v. The velocity, v, and displacement fields, $\vec{u}$, are distinguished by the particular physical reference frame that one uses. The Eulerian reference frame of Figure 11.4 in image registration is associated with Christensen's viscous model [12, 9], whereas the Lagrangian frame is associated with elastostatic models that follow Bajcsy's pioneering work [5].

Much work has followed along both of these lines. The static formulation is frequently combined in variational techniques with the finite element method [30, 59, 31, 77, 62] and applied to morphometric study. The fluid framework has led to a series of interesting theoretical publications that explore the connection of the transport equation in fluid frameworks with the diffeomorphism group [56, 55].

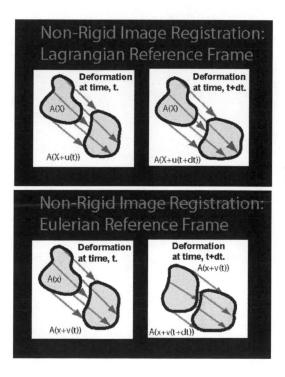

Figure 11.4. The progress of an image registration in the Lagrangian (top) and Eulerian (bottom) reference frames. The Lagrangian frame always refers back to the original configuration of the image, $A(\mathbf{X})$. The Eulerian frame treats the displacements as always occurring from A in its last configuration.

This defines curved anatomical distances and group theoretical shape metrics and statistics.

A goal of this chapter is to put non-rigid image registration in the context of some of the revelant theory that supports it. The connection between variational and differential optimization techniques will also be explored. The regularization requirements that result from the problem class' ill-posedness will lead into a discussion of the physical models that are often used in this role, as well as some pertinent mathematical background that relates to the specific algorithmic techniques for solving the optimization problems that occur. Along the way, we will also introduce data terms as needed, that is, some of the statistical and prior criteria that are used to drive the physical models toward a final equilibrated solution.

11.2   Optical Flow: Fast Mono-Modality Non-Rigid Registration

This section uses one of the oldest problems in the computer vision image correspondence literature, optical flow, to introduce the concepts that are common in non-rigid image registration problems. We will develop an optimization viewpoint on the optical flow problem and then provide an algorithm with which the minimization may be estimated. Aspects of the two dominant formulations for image registration optimization algorithms (differential and variational) are illustrated.

There is significant literature on the subject of optical flow for motion analysis, and here we discuss the most commonly applied gradient-based formulation of the problem. The working (intensity constancy) assumption is that image values are conserved over time:

$$I(x,t) = I(x + v\delta t, t + \delta t), \tag{11.3}$$

where $I(x,t)$ is the image value at position x and time t, v is the velocity (or optical flow) vector, and δt is assumed small. The optical flow constraint, C_{of}, is gained by expanding the right hand side in Taylor series to find, ignoring second order terms,

$$C_{of} = I_x \cdot v + I_t = 0. \tag{11.4}$$

Within the context of registration, the images indicated on the left and right hand sides of Figure 11.3 correspond to the pair that is to be matched, such that $I_t = I(T) - J$, where T is the transformation. According to Equation 11.3 then, an image point at one time instant that is translated to a new position in a later image is assumed to retain its original image value.

This problem is indeed an ill-posed inverse problem. The transformation between time-points is one of a potentially infinite number, each of which would appear the same according to the criterion specified, expressed here as a functional (a non-negative scalar measure associated with bounded domain Ω and operating on v),

$$E_D(v) = \int_\Omega \Phi(C_{of})d\Omega. \tag{11.5}$$

The function $\Phi(x)$ may equal x^2, as in the optical flow error term introduced in Horn and Schunck [40]. Hadamard gave three criterion for a problem to be well-posed:

1. its solution must exist,

2. be unique,

3. and vary continuously with the input data.

The optical-flow problem will violate at least the second constraint. The problem is under-constrained in that the flow must be found throughout Ω, but the optical flow constraint will vanish at many points in the domain. For example, the flow within the overlapping internal regions of a binary image pair will be undetermined, as both terms in C_{of} vanish. Arbitrary solutions in those regions will be indistinguishable.

Tikhonov's solution to this type of problem was to take advantage of the solution's *a priori* physicality. Tikhonov restricted the original ill-posed problem's solution space to a physical subspace and then required a solution to:

1. be known to exist in the restricted space,

2. be unique,

3. and vary continuously with the input data, as long as the variations do not cast the solution from the subspace.

Formally, this led Tikhonov to propose the use of regularizing operators L to stabilize the solutions of problems such as optical flow. Here, the terminology $\|\cdot\|$ is used for the norm of a function and $|\cdot|$ the norm of a vector. A norm with respect to L may also be written $\|\cdot\|_L$. These operators (often linear, differential operators), and norms associated with them, form a stabilizing function, ψ. This may be combined with the original problem to gain a smoothing functional operating on the solution v, such as

$$E(v) = \int \Phi(v)d\Omega + \int \psi(v)d\Omega, \tag{11.6}$$

where $\int \psi(v)d\Omega = \|Lv\|^2$. The above functional makes the original ill-posed problem well-posed in its restricted subspace.

Horn and Schunck [40] incorporated a Tikhonov regularization term, with $L = \nabla^2$, that penalizes the roughness of the velocity field. It can be shown using Green's formulae that that the Laplacian operation ∇^2 gives,

$$\psi(v(\mathbf{x})) = |\nabla v|^2. \tag{11.7}$$

The combination of these two constraints yields a global variational energy to optimize

$$E(v) = \int_\Omega \Phi(C_{of}) + \int |v_x|^2 d\Omega, \tag{11.8}$$

where $\nabla v(\mathbf{x}) = v_x(\mathbf{x})$. The solution, here v, must be found that provides a minimum of the smoothed functional, E. This resolves the lack of constraints previously discussed and allows the use of variational calculus to address the minimization problem. From here, we will assume the domain Ω is bounded with homogeneous boundary conditions at its border.

The standard framework in the variational calculus permits three equivalent views of this minimization problem. The equation above is the first, a global potential energy. The weak equation follows from it and leads to the Euler-Lagrange (E-L) equations [66]. The variational or weak form may be used in the finite element method, while the E-L equations are associated with finite difference techniques.

We will now see how to derive the E-L equations from the variational functional, E. We denote the variational problem as

$$E(v) = \int_\Omega F(\mathbf{x}, v, v_x, \cdots, v_{x^k}) d\Omega, \tag{11.9}$$

where F is a function, here $\Phi + \psi$. The value of $k = 1$ is used below, although the results are generalizable. For theoretical reasons, the domain of E is required to be a dense linear manifold. This allows us to consider solutions in the neighborhood of v by adding small perturbations, w, where w must satisfy the same boundary conditions as v. The second necessary assumption is that $E(v + \varepsilon w)$ is sufficiently differentiable such that ultimately its gradient may be found. This property permits a descent strategy to optimize E.

Consider that a minimum of the energy is obtained when infinitesimal changes in the solution v do not improve the energy. Our assumptions will allow us to take explicitly the derivative of $E(v + \varepsilon w)$ as $\varepsilon \to 0$. First, though, we must find the weak form of E, which comes from its first (Gateaux) variation. Substituting $\bar{w} = (v + \varepsilon w)$ for v, the first variation is

$$\frac{dE(\bar{w})}{d\varepsilon} = \int_\Omega \frac{d}{d\varepsilon} F(\mathbf{x}, \bar{w}, \bar{w}_x) d\Omega. \tag{11.10}$$

Applying the chain rule to the integrand,

$$\frac{\partial F(\mathbf{x}, \bar{w}, \bar{w}_x)}{\partial \varepsilon} = \frac{\partial F}{\partial \bar{w}} \frac{d\bar{w}}{d\varepsilon} + \frac{\partial F}{\partial \bar{w}_x} \frac{d\bar{w}_x}{d\varepsilon}. \tag{11.11}$$

Using the fact that we are looking for a stationary point of E provides the weak form of the variational energy,

$$\frac{dE(v)}{d\varepsilon} = \int_{\Omega} w \frac{\partial F}{\partial w} + w_x \frac{\partial F}{\partial w_x} d\Omega = 0. \tag{11.12}$$

Finally, the Euler-Lagrange form of the problem is found by performing integration by parts on the second term and applying boundary conditions, giving

$$\frac{dE(v)}{d\varepsilon} = \int_{\Omega} \left(\frac{\partial F}{\partial v} - \frac{d}{dx} \frac{\partial F}{\partial v_x} \right) w \, d\Omega, \tag{11.13}$$

where the integral is evaluated at $\varepsilon = 0$. Because w is arbitrary (within some restrictions), the gradient of E is then

$$\text{grad } E = \frac{\partial F}{\partial v} - \frac{d}{dx} \frac{\partial F}{\partial v_x} = 0. \tag{11.14}$$

A solution minimizing this equation will also provide a stationary point with respect to the global energy [53], by the variational principle. Similarly, the solution minimizing the global energy will satisfy the E-L equation. Technical details and proofs related to the variational formalism, including necessary conditions, existence, and uniqueness theorems, are summarized in, for example, [7].

For the optical flow problem, we then have

$$\frac{\partial \rho(C_{of})}{\partial v} - \frac{d}{dx} \frac{\partial |v_x|^2}{\partial v_x} = 0. \tag{11.15}$$

This simplifies to

$$I_x(I_x \cdot v + I_t) - v_{xx} = 0. \tag{11.16}$$

The first term minimizes the optical flow constraint, whereas the second, minimizing the Laplacian of the velocity field, came from the regularization term. This equation constitutes a local law, where we minimize a function, as opposed to the global law of the variational equation, requiring minimization of a functional. The next section will describe Thirion's demons algorithm which exploits this gradient to minimize E.

11.2.1   Demons Algorithm

The original demons algorithm was based on an heuristic argument that called for small "demons" to "push" the image around by its level sets until correspondence was achieved. Here, we show how the demons algorithm estimates a minimization of the Euler-Lagrange equations discussed above. Essentially, we view the method as trying to minimize each E-L term separately to yield an easy to implement gradient descent algorithm of the form, $V_{n+1} = V_n + \varepsilon \text{ grad } E$, where ε

is small and V is a solution. The method was first proposed by Thirion [70] as a fast mono-modality medical image registration algorithm. It has many of the advantages of finite difference techniques in general, in that it is easy to code and very fast. At the same time, implementations are restricted to uniform grids and may be difficult to debug.

Setting the first term of the optical flow E-L equation to zero defines a gradient constraint specified by Equation (11.4). A given point then yields a single linear equation in several unknowns, each corresponding to a different component of the velocity at the image point

$$I_x(I_x \cdot v + I_t) = 0. \tag{11.17}$$

In the case of two-dimensional images, for example, the solution for the velocity is a line in velocity space

$$v_u = \frac{-I_t I_x}{|I_x|^2} + \alpha I_x^{\perp}, \tag{11.18}$$

where α parameterizes the line and $I_x^{\perp}$ is the vector perpendicular to the spatial gradient of the image. To recover a unique estimate of the optical flow at a point, some methods, such as the demons technique, ignore the second term in Equation (11.18) and set the normal velocity as the registration solution. This choice is justified by the presence of the aperture problem in images, which suggests that variation along the image's level sets cannot be known.

The second term is similarly considered alone:

$$v_{xx} = 0. \tag{11.19}$$

The demons method estimates the second derivative of the solution with a Gaussian filter, such that

$$v_{xx} \approx G_\sigma \star v - v, \tag{11.20}$$

giving the solution $v = G_\sigma \star v$. The demons algorithm iteratively estimates the flow and smooths:

1. Initialize the solution as the identity, $\mathbf{V}_0 = \mathbf{Id}$.

2. Compute the velocity gradient $v_u = \frac{-I_t I_x}{|I_x|^2 + I_t^2}$ and set $\mathbf{V}_n = \mathbf{V}_n + v_u$. This is the descent in the optical flow direction.

3. Set $\mathbf{V}_{n+1} = G_\sigma \star (\mathbf{V}_n)$. This is the regularization update.

4. Go to 2.

Demons thus alternately updates the solution according to optical flow and then regularization constraints. Two heuristics are applied in the algorithm. The first is related to the fact that the optical flow field, v, holds only for very local displacements but, especially in medical imaging, large displacements need be estimated. The total field, V, is thus stored and regularized, resulting in an estimate of elastic behavior. This approach corresponds to the Lagrangian view of image registration, characterized by regularizing the field with respect to the origin. Note that in Step 3, the smoothing could also be applied to the velocity field alone, which estimates a fluid model. This corresponds to the Eulerian view, where only the incremental update field is regularized. The second heuristic is that the term I_t^2 is added to the denominator. This serves to stabilize the solution in the presence of large I_t in the numerator, but is not necessary for a solution. An example applying this algorithm to problems that fit its priors is below.

Atlas-Based Segmentation Example.

Volumetric and shape analysis of the brain parenchyma and its constituent tissues from MRI images is an important tool for studying diseases. Manual segmentation of the brain is a tedious task that requires substantial time and effort by trained personnel and in general can suffer from large inter-observer variability and poor reproducibility. Due to the lack of defined boundaries, automatic segmentation of the brain is a difficult problem that requires some *a priori* information and cannot be accomplished by relying solely on the image.

Atlas-based segmentation is one approach in solving the brain volume delineation problem where manual segmentation is done only once on an atlas image [21]. To segment a second (subject) image, the atlas is non-rigidly or deformably registered to the subject image. The output of the registration process is a displacement or deformation field which maps or aligns the atlas to the subject image. If the registration is accurate enough, the output deformation field can then be applied to an atlas label image or mask to automatically segment the subject image. This process is illustrated in Figures 11.5 to 11.7 and may be applied both for structural analysis [21] as well as neurosurgical localization [22].

In this example we will use two 3D proton density MR images of the brain as shown in Figure 11.5 where the top image will be used as the atlas and the bottom image as the subject to be segmented. The images have been pre-processed to align along the AC–PC (anterior commissure–posterior commissure) line, resampled to isotropic voxels and clipped with a brain mask to remove the skull and skin. For the atlas image, several subcortical and cerebellar structures were hand-segmented. Figure 11.6 shows the manual segmentations as an overlay on top of the atlas and subject images. It can be seen clearly that due to the complex differences between subjects, the atlas labels do not initially fit the subject image.

Effective atlas-based segmentation requires a robust non-rigid registration scheme. This example uses a multi-resolution implementation of the demons algorithm [69, 71], results of which are shown in Figure 11.7. The output

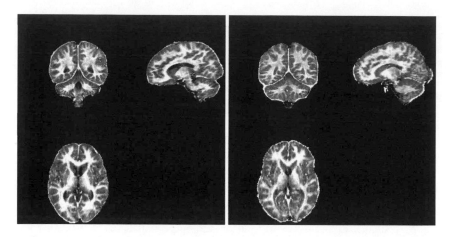

Figure 11.5. PD images of the brain used in the atlas-based segmentation example. The left image is used as the *atlas* image and right image as the *subject* image to be segmented. The images have been aligned along the AC-PC line, resampled to isotropic voxels, and clipped with a brain mask to remove the skull and skin. [Images and segmentations courtesy of Nancy C. Andreasen at The University of Iowa Psychiatry Department]

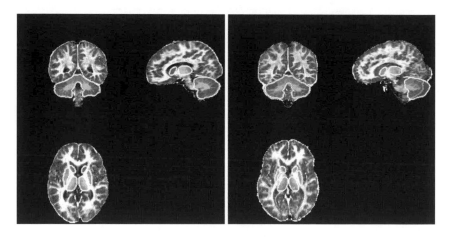

Figure 11.6. (See Plate XIV) For the atlas (left) image in Figure 11.5, several subcortial and cerebellar structures were hand segmented. The segmentations are shown as an overlay on top of the atlas image (left) and subject image (right).

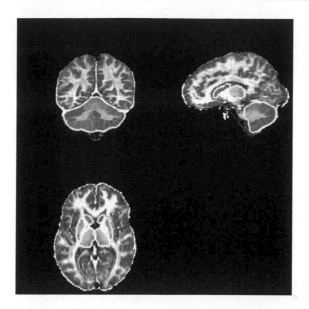

Figure 11.7. (See Plate XV) A multi-resolution demons algorithm was used to non-rigidly register the atlas image onto the subject image. The output deformation field was then used to warp the atlas image, producing the top image. The warp applied to the atlas labels produced the overlay in the bottom image, shown over the subject image.

deformation field was used to warp the atlas image to "match" the subject image (top image). The output field was also used to warp the atlas labels. The warped atlas labels are then used to segment the same structures of the subject. The automatically generated segmentations are show as an overlay on top of the subject in the bottom image.

11.2.2 Multimodality Extension.

Correspondence between different modalities is an equally important area of research to the mono-modality case. Rather than registering similarly appearing structures, more often these problems require the correspondence of functional data to structural data. Thus, the basic optical flow constraint may no longer be applied. Research in this area has focused on the development of information theoretic or statistical criteria which should be optimized during the registration.

The most commonly applied inter-modality similarity criterion is the mutual information [58]. Using the Euler-Lagrange equations from before, the goal, here, is to minimize

$$\frac{\partial \rho(C_{mi})}{\partial v} - v_{xx} = 0. \tag{11.21}$$

The mutual information (MI) criterion seeks a transformation that makes the joint histogram of the image pair sparse, in the sense that few bins contain many of the samples. The metric is defined with respect to the entropy and joint entropy of the samples,

$$\text{MI}(A,B) = H(A) + H(B) - H(A,B), \tag{11.22}$$

where $H(\cdot) = -\int \text{Pr}(\cdot) \ln \text{Pr}(\cdot) d\Omega$. For non-rigid image registration, the concern is to maximize this criterion globally, while computing derivatives which vary locally.

Mutual information derivatives are computed here so that MI's global statistical nature is combined with the need for locality of its gradients. The following scheme serves both needs such that the similarity term is given by,

$$C_{mi}(I,J,v) = w_1 \, \text{MI}_{global} + w_2 \, \text{MI}_{local}. \tag{11.23}$$

The derivative estimate for this function follows that given in [83], where Parzen windowing with Gaussian kernel functions is chosen for the estimation of the densities. The global term seeks to maximize the MI globally and involves taking samples of fixed image pixels, A_{global}, globally. The local term seeks to compensate for local non-stationarities lost in the global measure and takes its samples, A_{local}, only within a given radius. The union of these samples is taken to be $A = \{a_j\}$. The B samples, also drawn from the fixed image, estimate the entropy. The derivative is with respect to the local vector field v (using Viola's terminology),

$$\frac{\partial \, \text{MI}(I,J,v)}{\partial v} = \frac{1}{N_B} \sum_{b_i \in B} \sum_{a_j \in A} (b_i - a_j)^T [W_{ab}(b_i, a_j) \psi_{ab}^{-1} - \tag{11.24}$$

$$W_c(c_i, c_j) \psi_c^{-1}] \frac{d}{dv} (b_i - a_j). \tag{11.25}$$

The probability estimates are in W. The sample c_i is a pairing of corresponding pixels, $c_i = (a_i, q_i)$. The value q_i is found by applying v and finding the sample in the moving image corresponding to a_i.

An important aspect of this approach is that when estimating the local derivative, the B samples are taken only at the point under consideration. This is equivalent to using the local contribution to a global translation derivative as the local force. This may be seen by considering the case when the sum over the B samples is taken over every pixel to get the global translation derivative. Each local contribution is a single term in the sum over b_i. If each of these terms was placed in a vector field, F_{MI}, this would be the field that provides the local gradient $\nabla \Pi_{MI}$. Note that the computation is then linear in the number of A samples, making the approach relatively efficient.

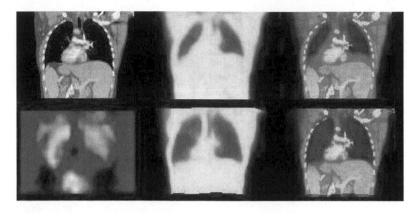

Figure 11.8. (See Plate XII) PET-CT image fusion. The original CT fixed image is at top left. The Jacobian of the transformation is at bottom left. The original PET transmission image is at top center. The warped PET transmission is at bottom center. The overlay of original images is at top right. The warped PET is fused with the CT at bottom right.

Positron emission tomography (PET) in conjunction with x-ray computed tomography (CT) aids in thoracic cancer screening [51]. The non-rigid deformations occur as the patient breathes during the long (minutes) PET scan, whereas the brief (seconds) CT scan is taken during a breath hold. The PET transmission scan typically shows the lungs in a relatively relaxed state in comparison to the CT, making it difficult to correlate clinically the anatomical and functional images. Non-rigid intra-subject registration is needed here to fuse the images, as shown in Figure 11.8, aiding in localization of neoplasm. A similar result with PET and MR images is shown in Figure 11.9.

11.3. Variational Framework for Computational Anatomy

Variational optimization uses the global formulation that is dual to the Euler-Lagrange equations. One of its main advantages over the E-L form isk that one explicitly constructs the global energies that one is minimizing as part of the optimization process. These energies generally have norms associated with them and affect a measure of formal distances between anatomies. This information is useful for applications in change detection that require spatial normalization and atlas construction. Furthermore, the variational view naturally suits physical models that give the (often diffeomorphic) anatomical deformations the one is trying to recover.

Another advantage is that the energies of the variational framework are closely connected to probabilities in Bayesian theory. The image registration problem may be formulated in Bayesian terms by taking the priors as regularization and

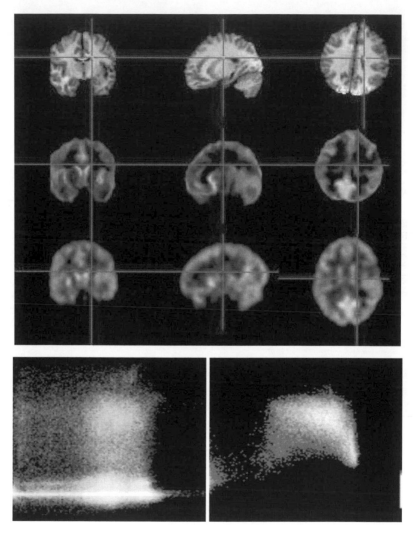

Figure 11.9. PET-MR inter-subject image fusion (top group of images). The PET activation image (bottow row) is transformed to the fixed MR image (top row). The result is in the middle row. The joint histogram is shown before (bottom left) and after the transformation (bottom right). Note the relatively large anatomical deformation of the PET image. This example is not usual clinical practice.

the likelihood as the data term,

$$\Pr(J \circ T^{-1} | I) = \Pr(I \circ T | J) \Pr(J \circ T^{-1}) / \Pr(I). \tag{11.26}$$

The transformation, T (e.g., $T = \mathbf{X} + \vec{u}$), maps between the atlas (fixed) image I and the observed (moving) image J. The variational energy associated with the Bayesian formula comes from the natural log when Gaussian assumptions are made for the probabilities,

$$\Pi(I, J, T) = \int_{\Omega} w_1 \ln \Pr(I \circ T | J) + w_2 \ln \Pr(I) +$$
$$w_3 \ln \Pr(J) + w_4 \ln \Pr(T) d\Omega, \tag{11.27}$$

where we define $\Pr(J \circ T^{-1}) = \Pr(J) \Pr(T^{-1})$ and $\Pr(T) = \Pr(T^{-1})$. This integral is taken over the $\mathbb{R}^d$ domain, Ω, with d giving the dimension. The Gaussian variances, w_i, have become the weights on the terms. Prior knowledge may be involved in each of the terms that result from this approach. The first may involve the brightness constancy assumption, as in optical flow [40]. The second and third derives from restricting the observed images or as an outlier term with robust statistical techniques [43]. The fourth restricts the permissible transformation space and also makes the problem well-posed.

The algorithms given below will optimize the Bayesian probabilities via a variational energy without priors on J or I,

$$\Pi(I, J, T) = w_1 \Pi_{\sim}(I, J, T) + w_2 \Pi_R(T) d\Omega. \tag{11.28}$$

The critical step in the development of variational algorithms is the specific definition of $\Pi_{\sim}$ and Π_R, as we saw in the case of optical flow. This is where one must convert prior knowledge about data, transformations, efficiency, and algorithmic capabilities into practice. Here, each term may be composed of a sum of similarity and regularization terms derived from statistical, continuum mechanical, or other models. In the cases that follow, the Π_R will always relate to norms of linear operators such that $\Pi_R = \|L\vec{u}\|^2$, though statistical priors (from, for example, principle components analysis of a database of deformations) may also be used [15, 46, 19, 60]. The finite element method systematically allows this flexibility by assigning a variational term to each prior.

Finite element methods as implemented in [41, 28, 87] solve the problems that follow. The method works on non-uniform domains, is very efficient and uses a systematic programming style. A sparse linear system ($\mathbf{Ku} = \mathbf{f}$) is always constructed, the solution of which minimizes the E-L equations for the problem. The construction of the linear system may derive from the variational form using the Rayleigh-Ritz method (when the problem's linear operator is positive-definite and bounded below) or the weak form using the Galerkin method [41]. The latter is the more general and therefore more common approach.

11.3.1 Priors for Computational Anatomy

Computational anatomy is one of the major areas in which non-rigid image registration plays a fundamental research role. The goal of the field, originated by British biologist D'Arcy Thompson [72], is to quantify shape differences in populations and subsets of populations via the transformations that map between individuals [34]. Quantitative understanding of what is normal and abnormal within given groups would enable one to incorporate an image-based change detection program as part of clinical practice. This is especially important for groups that have high risk for diseases such as breast or lung cancer. Neuroanatomical application of morphometry is also used to understand the progression of diseases such as schizophrenia and aids in localizing areas of specific anatomical compromise. Thus, it is important to define both basic measures on the anatomical space (such as the Jacobian) or alternative distance measures such as given by linear differential operators. Furthermore, these measures are important for insuring that the transformations are anatomically legitimate.

An important prior in computational anatomy is that the topology of anatomy within a given class should be preserved by the transformations mapping that class. This results in a diffeomorphism requirement, which the demons algorithm is known to violate often. The diffeomorphism group defined on domain Ω is formally stated as

$$G = \{\vec{g}\colon \Omega \to \Omega \mid \forall \vec{h} \in G : \vec{h} \text{ and } \vec{h}^{-1} \text{ differentiable}\}. \tag{11.29}$$

The identity transformation is then $\vec{h}^{-1}(\vec{h}(\mathbf{x},t),t) = \mathbf{Id}$. In image registration, we also enforce that the transformation be zero on the boundary of Ω. The diffeomorphism requirement may be measured by the transformation's Jacobian, $\mathcal{J}$, which is the determinant of its deformation gradient (noting that large-scale folds will not be detected by $\mathcal{J}$). The deformation gradient in d dimensions is a $d \times d$ matrix denoted

$$[\mathbf{D}_{ij}(\mathbf{T})] = [\frac{\partial u_i}{\partial x_j}] + \mathbf{I}, \tag{11.30}$$

with $\mathbf{I}$ the identity matrix. Its determinant measures the local dilatation or contraction due to the action of the vector field, $\mathcal{J}(\vec{g}) = |[\mathbf{D}_{ij}\vec{g}(\vec{x})]|$, where $|\cdot|$ is the determinant. Diffeomorphisms have the property that they are continuously parameterizable (e.g., in time) and have positive Jacobian, thus making them ideal for measuring anatomical variability of similar populations. Datasets of Jacobian images are often used to study shape differences in populations via the methods of SPM [2, 1], as illustrated in Figure 11.10.

A second measure of anatomical distance may be defined from a Euclidean and/or differential norm, L, as is used for regularization. The Euclidean distance will measure simply the size of the vector field. The differential norm may measure, for example, smoothness of the vector field or the amount of expansion or

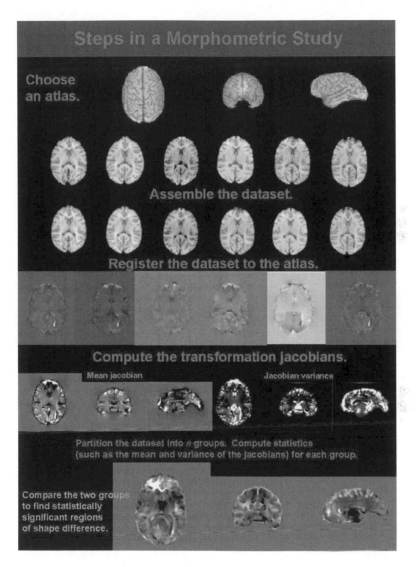

Figure 11.10. A morphometric study workflow is illustrated. The study in this example results in a t-field, shown at the bottom of the image, which can be used to localize regions with statistically significant shape difference between populations.

contraction needed. The Navier operator measures a weighted sum of both of these components of the transformation field,

$$L = \mu\nabla^2 + (\lambda + \mu)\vec{\nabla}(\vec{\nabla}\cdot). \tag{11.31}$$

The first term is the Laplacian, as was used in the optical flow case. The second term explicitly penalizes excess expansion or contraction and aids in keeping the transformation's Jacobian positive. Anatomical distances are computed by tracking the value of the operator's norm at different points during the progress of the registration. The Euclidean norm is often used in conjuction with Equation (11.31) in order to add numerical stability [42]. Many other operators are available [53].

11.3.2 Cauchy-Navier Regularization.

The regularization induced by the Cauchy-Navier operator may correspond to two different physical models. If it is applied in the Lagrangian frame, it models the behavior of a homogeneous linear elastic body. If it regulates only the flow or velocity field in; the Eulerian frame, the behavior is of a Newtonian fluid. Both of these viewpoints are valuable in image registration.

We now investigate the continuum mechanical view of this linear differential operator by looking at it with respect to the (slightly more general) classical linear elastic regularization. In this case, the image is considered a potentially inhomogeneous elastic body that is deforming under external forces. Consider the problem domain,

$$\Omega: \mathbb{R}^d \to \mathbb{R}^p. \tag{11.32}$$

The linear elastic regularizer is given by the generalized Hooke's law relating stresses to strains,

$$\sigma_{ij} = C_{ijkl}\varepsilon_{kl}, \tag{11.33}$$

where indicial notation is used. The elastic coefficients, C_{ijkl}, are components of a tensor describing the material properties of the body. The strains are given as $\varepsilon_{kl} = 1/2(u_{kl} + u_{lk})$ such that, in matrix form, $\sigma = \mathbf{D}\varepsilon$. External forces applied to a linear elastic body are denoted $\vec{f}$, giving a variational integral describing the potential energy of the configuration,

$$\Pi_R + \Pi_\sim = \int_\Omega (\sigma^T\varepsilon - \mathbf{f}\cdot\vec{u})d\Omega, \tag{11.34}$$

where σ is stress, ε is strain, and the surface tractions are ignored. The solution field is denoted $\vec{u}: \mathbb{R}^d \to \mathbb{R}^d$, whereas the external forces, $\vec{f}: \mathbb{R}^d \to \mathbb{R}^d$, will derive from a similarity constraint such as C_{of} or C_{mi}, as above. This integral is an example of the cost = deformation - similarity variational formulation. The solution of the integral in Equation (11.34) by the finite element method will provide $\vec{u}^*$, an optimal solution.

It is also worth noting that the linear elastic regularization term gives an energy inner product on a linear space [41] and is thus suitable as a computational anatomical metric. Furthermore, it may be shown, using a procedure similar to the one given previously for regularized optical flow, that the Euler-Lagrange equations for the elastostatic potential are of the form

$$\sigma_{ij,j} + f_i = 0. \tag{11.35}$$

Assuming an homogeneous material, the linear operator for these equations is in fact the Cauchy-Navier operator,

$$L\vec{u} = \mu\nabla^2\vec{u} + (\lambda + \mu)\vec{\nabla}(\vec{\nabla} \cdot \vec{u}). \tag{11.36}$$

Material properties are determined by the constants λ, μ. Integrating $|L\vec{u}|$ over the domain Ω gives the functional norm $\|L\vec{u}\|$. This equation governs Newtonian fluid regularization when $\vec{u}$ is replaced with v and the measurments are in the Eulerian frame.

11.3.3 Large and Small Deformation Registration from the Lagrangian Frame

The elastic regularization model, given above, may be used to develop both small and large deformation image registration algorithms.

Small Deformation Formulation

The elastostatic registration algorithm for finding $\vec{u}^*$ is one of the earliest applied to medical imaging [5]. Elastostatic registration requires one to estimate repeatedly optimal $\vec{u}^*$ as the similarity term driving the problem is highly non-linear. Note that treating the problem from the perspective of elastostatics requires that at each iteration the same total deformation energy is balanced with a total similarity energy. As the total deformation energy increases quadratically, and the similarity term is non-linear, it may become increasingly difficult to balance these terms and to escape local minima.

The optimization problem as formulated by Gee [27] finds the solution $\vec{u}^*$ as

$$\vec{u}^* = \operatorname{argmin} \vec{u} \left\{ \Pi_R(\vec{u}) + \int_\Omega \ln \operatorname{Pr}(I(\mathbf{X} + \vec{u}))|J(\mathbf{x}))d\Omega \right\}. \tag{11.37}$$

The Lagrangian frame tracks the deformation with respect to the original configuration such that, $\vec{u} = \mathbf{X} - \mathbf{x}$. Here, the deformation norm is defined as $\sqrt{\Pi_R}$.

The above equation contains no time-dependence. We reformulate the problem statement now using the time domain. The image forces, here, act explicitly in time on the continuum,

$$\vec{u}^{\star}(t_n) = \text{argmin } \vec{u} \ \{ \ \Pi_R(\mathbf{X},\vec{u}) + \int_{t=0}^{t=t_n} \Pi_{\sim}(t)dt \ \}. \tag{11.38}$$

The potential is always minimized with respect to the original undeformed configuration, $\mathbf{X}$, as the forces increase. In this case the energy norm is a function of time, $\|L\vec{u}\|(t)$. Noting that the force is given by the similarity gradient, $\vec{b}_{\sim} = \nabla\Pi_{\sim}$, it is typically approximated

$$\int_0^t \vec{b}_{\sim}(\mathbf{X}) \approx \sum_i \vec{b}_i(\mathbf{x}), \tag{11.39}$$

using a sum over discretized time points. The quadratic penalty is always applied to the total forces as measured from the original configuration. This is the purely Lagrangian reference frame and makes this formulation only valid for "small" deformation image registration. The definition of small here is relative, but may be considered as deformations which do not require long, curving paths. This is often sufficient for neuroanatomical image registration. Nonetheless, a large deformation formulation is also useful.

Large Deformation Formulation

The large deformation formulation for image registration uses the Lagrangian reference frame for a specified time interval. After the deformation reaches a given total size, as measured by its Jacobian or a deformation-based norm, the reference frame is reset to the current point. This practical check of the deformation field, along with a constraint on the size of the change in deformation field at each iteration, allows one to enforce a homeomorphic map. This methodology is called a transient quadratic (TQ) approach, an algorithm for which is given in Algorithm 1. A related method for boundary value problems is found in [57]. This extension to the traditional elastostatic algorithm allows time-control points $\{t_0, \cdots, t_{n+1}\}$ to be introduced into variational Equation (11.38). This creates a recursive problem,

$$\vec{u}^{\star}(t_{i+1}) = \text{argmin } \vec{u} \ \{ \ \Pi_R(\mathbf{x}(t_i),\vec{u}) +$$
$$\int_{t_i}^{t_{i+1}} \Pi_{\sim}(\mathbf{x}(t_i),\vec{u}) \ \}, \tag{11.40}$$

where the $\vec{u}$ is regularized with respect to the continuum configuration at time t_i, $\mathbf{x_i} = \mathbf{X} + \vec{u}^{\star}_{t_i}$, as it varies to time t_{i+1}. Between each control point Equation (11.38)

Algorithm 1 Transient Quadratic Registration.

The progress of the algorithm is parameterized by its iterations, t. The value of u and f are always taken to be estimates of the total solution as measured from $\mathbf{x}$, the configuration at the last time-control point.

1: Initialize $\vec{u}(t=0) = \vec{0}$, $\Pi(t=0) = \infty$.
2: **for all** Resolutions **do**
3:     **while** $t \in \{t_0, \cdots, t_n\}$ and $\Pi(t) - \Pi(t-1) > \varepsilon$ **do**
4:         $\vec{f_t}+ = \nabla\Pi_\sim(I, J, \vec{u_t})$.   Compute the gradient of the similarity.
5:         $\mathbf{K}_t = \nabla\Pi_R(I, J, \vec{u_t})$.   Compute the gradient of the regularization.
6:         $\vec{u}_{t+\delta t} = \mathbf{K}_t^{-1}\mathbf{f}_t$.   Solve the linear system.
7:         $\Pi(t+\delta t) \leftarrow \Pi_\sim(t+\delta t) + \Pi_R(t+\delta t)$
            Compute the total energy.
8:         $t \leftarrow t + \delta t$.   Increment the time step.
9:         if $\mathcal{J} < \varepsilon$ add a new time control-point and reset the coordinate system.
10:    **end while**
11: **end for**

---

is solved. Given this basis, the anatomical distance of the TQ model is defined as

$$D_{TQ}(t_n + t) = \sum_{i=0}^{n} \|L\vec{u}(t_i)\| + \sqrt{\Pi_R(\mathbf{x}(t_n))(t)}, \tag{11.41}$$

where L is the selected operator. A dataset is ordered by this metric in Figure 11.11. This may be viewed as a piecewise approximation to the length of a curved path.

The argument against traditional linear elastic registration is that the regularization penalties increase quadratically with the size of the deformation. This formulation overcomes that drawback by using the regularization for a specified time interval, after which a regridding strategy is employed as in [11]. Both the

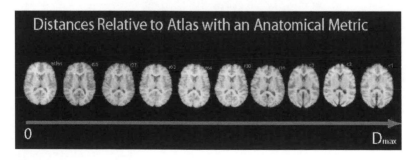

Figure 11.11. Anatomical distances of a human and chimp dataset with respect to the L metric as given by the TQ algorithm.

coordinate system and the field are reset to the identity. The past state of the algorithm is stored separately. The medium thus gains some plastic-like properties, maintaining the elastic regularization for a given time and then taking on a new shape until new stresses are applied. The total fields are stored for each time and then concatenated at the end of the algorithm. Local minima are avoided in these algorithms by using a multi-resolution strategy in the outer loop to propagate solutions from coarser to finer levels. The boundary conditions fix the border of the image. This algorithm is used below with the normalized cross-correlation similarity term for anatomical modeling and compared to the sum of squared differences criterion.

Quantifying Human Lung Motion

The lung can be described as an elastic body, whose elasticity originates from the network of fibers connecting the blood vessels, airways, and interstitial space. Pulmonary pathology directly affects the compliance of the lung. A key aspect of pulmonary physiology that can be measured by MRI is lung motion. Elastostatic registration is used here to compute a dense motion field over the lung parenchyma at each sampling time of a serial MR image sequence. The motion is estimated using the pulmonary vasculature and parenchyma as natural sources of spatial markers.

The changing lung volume during the respiratory cycle affects the average proton density within the pulmonary region resulting in intensity distortions, although relative (linear) brightness is maintained. The normalized cross correlation (NCC) measures the linear dependence between signals and may provide performance superior to sum of squared differences when (local) signal variations are present in intra-modality data [5]. The measure may also be applied to intermodality data [39]. An example of performance on brain data is in Figure 11.12. The result is superior to sum of squared differences in the right ventricle region where intensity inhomogeneity exists and similar elsewhere.

The cross-correlation [47] measures the correlation between signals over a region of interest,

$$CC = \langle I(\mathbf{x_I}), J(\mathbf{x_J}) \rangle, \tag{11.42}$$

where $\mathbf{x_I}$ and $\mathbf{x_J}$ are regions in correspondence. The normalized cross-correlation is the quotient of the cross-correlation divided by a normalization factor,

$$NCC = \frac{\langle I(\mathbf{x_I}), J(\mathbf{x_J}) \rangle}{(\langle I(\mathbf{x_I}), I(\mathbf{x_I}) \rangle \langle J(\mathbf{x_J}), J(\mathbf{x_J}) \rangle)^{\frac{1}{2}}}, \tag{11.43}$$

which removes dependence upon linear scaling (brightness). A shorthand notation is

$$NCC = \frac{\gamma_1}{\gamma_2}. \tag{11.44}$$

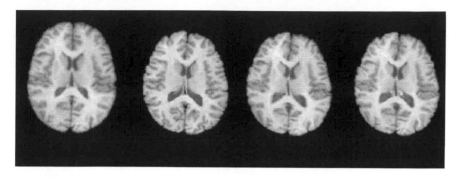

Figure 11.12. The moving image is at left and the fixed image left of center. The NCC result is right of center. The sum of squared difference result is at right.

The range is $[-1, 1]$, with 1 indicating perfect correlation.

The region $\mathbf{x_I}$ is given by $(\mathbf{X_I} + \vec{u}_I)$ when using the NCC in image registration. Its analytical derivative is used as the force term. Define $dI = \sum_i I(x_i)\nabla I(x_i)$ and $dJ = \sum_i J(y_i)\nabla I(x_i)$ where y_i is in correspondence with x_i. Then the derivative is expressed

$$\nabla NCC = \frac{1}{\gamma_2}(dI - dJ\gamma_1). \tag{11.45}$$

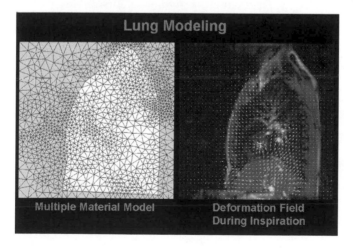

Figure 11.13. A customized finite element discretization for the lung registration problem. Physiologic material properties are used over the thorax and body and "air" material properties over the background. The anatomy-specific image registration results in an inspiration-modeling deformation field, overlayed at right.

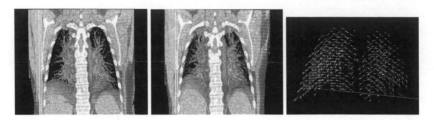

Figure 11.14. Estimation of pulmonary deformation using linear elasticity with the normalized cross-correlation similarity metric. High-resolution CT volumes at end-inspiration (top left) and end-expiration (top right) are registered using a multi-resolution strategy to obtain the volumetric deformation of the lungs seen with expiration (bottom). The two driving forces for passive expiration, diaphragmatic recoil and intercostal relaxation, are evident in this vector field.

Region sizes of 3^d or 5^d give a reasonable balance of performance and complexity.

This similarity measure is used to register sequences of inspiratory/expiratory pulmonary images. Figure 11.13 shows the results for one image pair in a breathing sequence. Note that the elastic recoil of the diaphragm, the main driving force for expiration, is captured by the registration. The lung and thorax are modeled with a physiologically elastic material; the compliance of the background is significantly higher to mimic air so that deformation within the background incurs negligible cost. A similar result using high-ressolution CT (HRCT) volumes is shown in Figure 11.14.

11.3.4   Diffeomorphic Flow-Based Image Registration

Compressible and incompressible fluids have the unique property of behavior that corresponds to specific diffeomorphism groups. The theoretical work by Miller et al. [56], focuses on finding geodesic paths across these diffeomorphism manifolds by extending the viscous formulation to optimize in time. Here, we investigate some aspects of this framework and give an algorithm for estimating these transformations using the transport equation with a compressible fluid model.

A flow-based formulation in terms of the transport ordinary differential equation finds $\vec{g}$ composed of continuous time incremental deformations or velocity fields, $\vec{v}$, each of which is diffeomorphic. Formally, this requires the solution to satisfy the following initial value flow equation, expressed in the Eulerian frame,

$$\begin{cases} \dfrac{D\vec{g}}{dt} & = \vec{v}(\vec{g}(\cdot,t),t), \\ \vec{g}(\cdot,0) & = \vec{g}_0. \end{cases} \tag{11.46}$$

The material derivative is denoted by the linear operator D/dt. The initial value, $\vec{g}_0$, is often the identity, **Id**. The velocity field, $\vec{v}$, is computed as the gradient of an energy functional as in Equation (11.28). This framework allows curved deformation paths due to the introduction of the non-linear material derivative.

The final solution is found by integrating the incremental solutions from time zero to time $t = 1$,

$$\vec{g}(1) = \int_{t=0}^{t=1} \vec{v}(\cdot, t) \, dt. \tag{11.47}$$

Let us consider what this correspondence formulation means. The basic notion is that the time-integrated deformation, $\vec{g}(t)$, should start at the initial value and, at time $t = 1$, satisfy a prior equivalence constraint. For the sum of squared differences case, this means $\vec{g}(t = 1)$ is the map between the moving and fixed domains and that it must satisfy the constraint given by the criterion, $C_{ssd} = (I(\vec{g}) - J)^2$.

The variational problem for domain matching under the transport equation, with this boundary condition, is then

$$\vec{g}^{\star} = \operatorname{argmin} \vec{v}(t) \left\{ \int_0^1 \|\vec{v}\|_L^2 + \Pi_{\sim} \, dt \right\}, \tag{11.48}$$

subject to Equation (11.46). A norm induced by the linear operator serves as the regularizer. The similarity equivalence generalizes the variational problem. This is a similar formulation to that given by Dupuis [25]. However, the looseness of the similarity definition belies Dupuis's minimizer existence proofs which used the sum of squared differences metric. A gradient descent algorithm may still estimate a geodesic path minimizing this energy at fixed time. The locally optimal infinitesimal velocity is computed,

$$\vec{v}(\cdot, t) = \operatorname{argmin} \vec{v} \in \mathcal{G} \left\{ w_1 \Pi_{\sim}(I, J, \vec{v} \circ \vec{g}) + w_2 \|\vec{v}\|_L^2 \right\}, \tag{11.49}$$

which gives a gradient in the space of diffeomorphic flows. The choice for L is often the Cauchy-Navier operator in Equation (11.31) plus a weighted Euclidean norm. The choice for $\sim$ as usual derives from assumptions about the intensity distributions, for example, that the (negative) mutual information is (minimal) maximal. The solution satisfies the (local) minimization of the integral Equation (11.48), with the distinction that this solution is generated from a flow, making the formulation appropriate for large deformations [35]. A result of this approach using the sum of squared intensity differences criterion is illustrated in Figure 11.15.

The diffeomorphic flow may be considered to induce a distance on the space of diffeomorphisms in analogy to curve length. The distance between a pair of

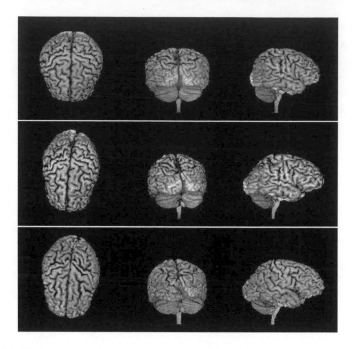

Figure 11.15. Non-rigid diffeomorphic registration with the moving image at top, the registered image at center, and the fixed image at bottom.

domains connected via a diffeomorphic flow is a shortest path defined, as in [35],

$$D(I,J) = \inf_{\vec{v} \in G} \{ \sqrt{\Pi_R(I,J)} \mid J \circ \vec{g}^{-1}(0) \sim J \wedge J \circ \vec{g}^{-1}(1) \sim I \}. \qquad (11.50)$$

This definition provides an appropriate notion of the curved geometric distance between two instances of anatomy. If one's velocity field is not smooth, or the domain configurations (I,J) are not reachable from each other, then the updates to $\vec{g}$ will need to be infinitesimally small to maintain diffeomorphism and the endpoint condition will never be met, resulting in infinite distance.

Algorithm 2, using the finite element method for diffeomorphic variational image registration, is stated below. The linear system associated with the minimization problem is constructed using the finite element method and written in the form $\mathbf{Kv} = \mathbf{f}$. The output of Algorithm 2 is the total Jacobian as well as the series of deformation fields, $\{\vec{g}(t_i)\}$. Note that regularization is applied only to the velocity field, $\vec{v}$, in Algorithm 2, so that large deformations are not unduly penalized. The boundary conditions fix the borders of the domain in order to enforce that the domain boundaries are homeomorphic.

Algorithm 2 Diffeomorphic Variational Image Registration.

The progress of the algorithm is parameterized by time. The time-integrated field
is $\vec{g}$. The incremental update field is $\vec{v}$.

1: Initialize $\vec{g}(t=0) = \vec{0}$, $\Pi(t=0) = 0$.

2: **for all** Resolutions **do**

3:    **while** $t \in \{t_0, \cdots, t_n\}$ and $\Pi(t) - \Pi(t-1) > \varepsilon$ **do**

4:       if $\mathcal{J}(g(t)) < \varepsilon_{\mathcal{J}}$, Enforce diffeomorphism.

5:       $\vec{\tilde{f}}_t = \nabla \Pi_\sim(I, J, \vec{g}_t)$.   Compute the gradient of the similarity.

6:       $\tilde{\mathbf{K}}_t = \nabla \Pi_R(I, J, \vec{g}_t)$.   Compute the gradient of the regularization.

7:       $\vec{v}_{t+\delta t} = \tilde{\mathbf{K}}_t^{-1} \vec{\tilde{f}}_t$.   Solve the linear system.

8:       $\vec{g}_{t+\delta t} = \vec{g}_t + \delta t (\vec{v}_{t+\delta t} \circ \vec{g}_t)$.

          Right compose the update field with the total field using the material
          derivative.

9:       $\Pi(t+\delta t) \leftarrow \Pi(t) + \int_\Omega \|L\vec{v}(\vec{x}, t)\|^2 \, d\Omega$.

          Add the incremental contribution to the total energy.

10:      if $\mathcal{J} < \varepsilon$ reset the coordinate system.

11:      $t \leftarrow t + \delta t$.   Increment the time step.

12:    **end while**

13: **end for**

11.3.5   Computational Anatomy Examples

An example of two registered images and the geodesic mid-point between them
is in Figure 11.16. The geodesic mid-point is defined as the point in time at which
the anatomical distance $D(t)$ is equivalent to half of its maximum value at $D(1)$.
Note that these distances should be independent of rigid transformation which
should be part of the initialization.

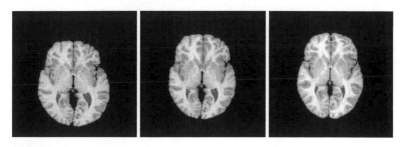

Figure 11.16. Illustration of mean geodesic L-norm distance. The original moving image
is first, its warp to the mean geodesic point is second. The final result is third.

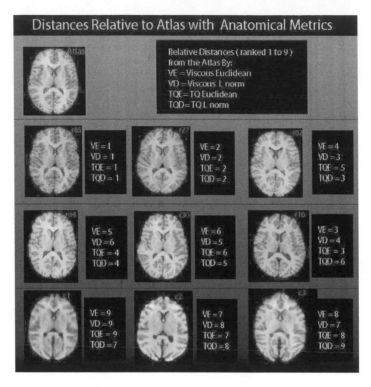

Figure 11.17. Summary of relative anatomical distances. The closest images are the same for all algorithms and for all images. As expected for all measures, the chimpanzee cortices are ranked as least similar to the atlas. These results reveal the ability of both algorithms to make intuitive measures of shape differences as well as the importance of using the differential-operator-induced distance.

Similarly, Figure 11.17 displays the relative ranking in shape similarity given by both the TQ and diffeomorphic fluid algorithms for Euclidean and differential energy-based distances. The distances are measured from an atlas MRI volume with cortex segmented to nine volumes (six female humans, 3 female chimpanzees) of segmented cortical images. One slice from each of the TQ results is shown in Figure 11.18 while the mean intensity image is shown in Figure 11.19 along with the original atlas. The registrations' Jacobians also show that the minimum Jacobians from the TQ algorithm are uniformly larger and the deviation from unity smaller. This is desirable for morphometry in the sense that topological errors are less likely. At the same time, the close values for post-registration intensity differences given by the TQ and viscous algorithm indicate that this advantage was gained without sacrificing the similarity optimization.

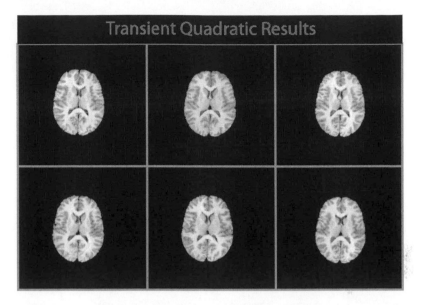

Figure 11.18. The set of TQ results for the human cortices. The same slice is shown for all images and should be compared with the atlas in figure 11.17. The images are ordered r16, r64, r85 (in the top row) and r27, r30, r62 in the bottom row.

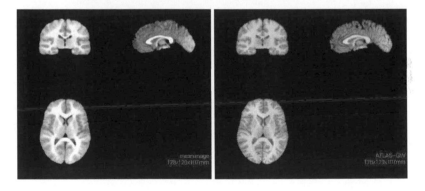

Figure 11.19. The mean intensity image from the TQ registration is at left. The atlas image is at right. The registrations were computed at $\frac{1}{4}$ resolution.

11.3.6 Diffeomorphic Landmark Matching

Human landmarking is an invaluable tool in gaining anatomically correct image registrations in cases where noise, lack of features, or pure anatomical complexity make automated methods unreliable [75]. Bookstein's point-based control revealed the power of this approach for studies of human and non-human shape variability [8]. A landmark is defined as a pair $(\mathbf{p}_i, \mathbf{q}_i)$ of corresponding sub-domains on Ω. Here, we illustrate a similarity term that uses constraints given by *a priori* corresponding points in anatomical space,

$$\Pi_{lm} = \sum_i \frac{1}{\sigma_i} |\vec{g}(\mathbf{p}_i) - \mathbf{q}_i|^2, \tag{11.51}$$

where σ_i is a variance. The diffeomorphic transformation, $\vec{g}$, is used here, as in [42]. Adding regularization results in a well-posed optimization problem,

$$\vec{g}^\star = \operatorname{argmin} \vec{v}(t) \left\{ \int_0^1 \|\vec{v}\|_L^2 + \Pi_{lm} \ dt \right\}. \tag{11.52}$$

In this case, the matching may be defined as inexact if σ is large and approaching exactness as $\sigma \to 0$. A synthetic example using this constraint is shown in Figure 11.20. Similarity terms for iterative closest-point-matching may also be formulated.

A second example using landmarks for biomechanical modeling is illustrated in Figure 11.21. Preoperative imaging is increasingly being used in surgical planning and guidance. However, neurosurgery poses an interesting challenge to this application. When a portion of the skull is removed at the beginning of a procedure, the brain shifts slightly in response to gravity and atmospheric pressures, inherent material properties of the brain tissue, loss of CSF during surgery, and other factors discussed in [64]. Even though the deformation is typically on the order of 1 cm, it is sufficient to warrant the registration of any preoperative scans to the new intraoperative brain position before planned interventions are executed. A simulation of this clinical situation is shown in Figure 11.21, in which a brain modeled as a linear elastic material is deformed by gravity. The deformation field is recovered by solving optimization problem Equation (11.52).

11.4   Review

This chapter has attempted to provide some of the theory, context, and applications of non-rigid medical image registration in its current state. The formulations given are meant to be general enough to maintain their relevance at least for the moment. Although examples were not given, it is noted that curves and surfaces, when parameterized, may be matched using variations of the volumetric formulations given here. Algorithms for solving the Monge-Kantorovich problem for medical image registration are also currently being investigated. Future work will certainly make the applications here look primitive, but hopefully the general spirit (of optimization in the face of ill-posedness) will persevere.

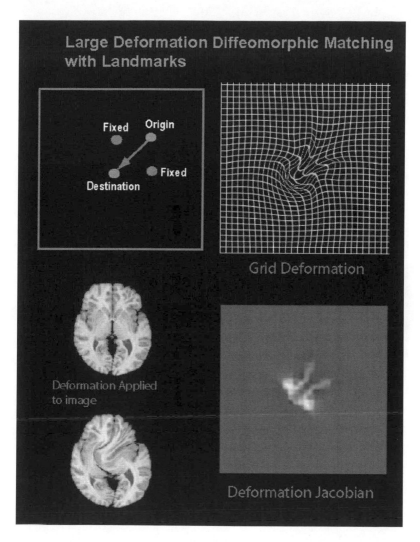

Figure 11.20. Large deformation diffeomorphic landmark matching is illustrated with an example in which one landmark is forced to pass between two others. This type of transformation would typically induce folding. The grid illustrates the smoothness of the transformation.

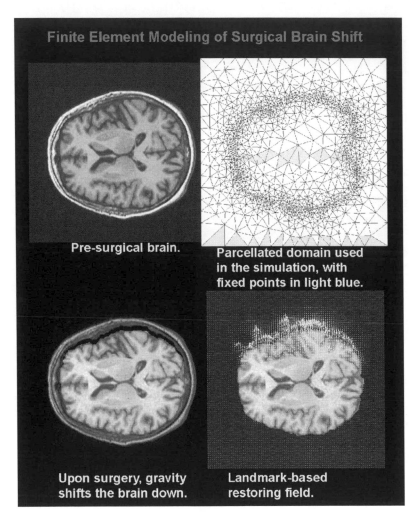

Figure 11.21. Surgical brain shift is simulated using a linear elastic finite element method and recovered with a set of curve landmarks along the brain surface. A uniform gravity load is initially applied to a patient brain to simulate surgical brain shift. The surgical table is situated beneath the patient's skull, while gravity is acting downward. At the same time, the exposed brain surface is landmarked with a curve (see color art). The intraoperative brain position contains the same anatomical landmarks, which are registered to their pre-operative position. The recovered deformation field generated from the interaction of the landmarks with the linear elastic deformation model is also illustrated.

References

[1] Ashburner, J., Hutton, C., Frackowiak, R., Price C., Johnsrude, I., and K. Friston. "Identifying global anatomical differences: Deformation-based morphometry," *Hum. Brain Mapp.* 6(1998): 348–357.

[2] Ashburner, J., Czernansky, J., Davatzikos, C., Fox, N., Frisoni, G., and P. Thompson. "Computer-assisted imaging to assess brain structure in healthy and diseased brains," *Lancet Neurology* 2(February 2003): 79–88.

[3] Avants, B. and J. C. Gee. "The shape operator for differential image analysis," in *Proceedings of Information Processing in Medical Imaging 2003*, Ambleside, UK, C. Taylor and J. Noble, eds. Springer-Verlag, LNCS2732. 101–113.

[4] Avants, B. and J. C. Gee. "Formulation and evaluation of variational curve matching with prior constraints," in *Biomedical Image Registration*, J. Gee, J. Maintz, and M. Vannier, eds., Heidelberg: Springer-Verlag, 2003. 21–30.

[5] Bajcsy, R. and C. Broit. "Matching of deformed images," in *6th Int. Conf. Pattern Recog.* (1982): 351–353.

[6] Batchelor, P., *et al.* "Measures of folding applied to the development of the human fetal brain," *IEEE Transactions on Medical Imaging* 21(8)(2002): 953–965.

[7] Blanchard, P. and E. Bruning, *Variational Methods in Mathematical Physics : A Unified Approach.* Berlin: Springer-Verlag, 1992.

[8] Bookstein, F. *Morphometric tools for landmark data: geometry and biology,* New York: Cambridge University Press, 1992.

[9] Bro-Nielsen, M. and C. Gramkow. "Fast fluid registration of medical images," in *Visualization in Biomedical Computing* (Proceedings of VBC'96, Hamburg, Germany 1996), K.H. Hohne, ed. Springer LNCS1131(1996): 267–276.

[10] Brown, L. "A survey of image registration techniques," *ACM Comput. Surveys* 24(1992): 325–376.

[11] Christensen, G., Rabbitt, R., and M. Miller. "Deformable templates using large deformation kinematics," *IEEE Transactions On Image Processing* 5(10)(October 1996): 1435–1447.

[12] Christensen, G., Miller, M., Vannier, M., and U. Grenander. "Individualizing neuroanatomical atlases using a massively parallel computer," *IEEE Computer* 29(1)(1996): 32–38.

[13] Christensen, G., Rabbitt, R., and M. Miller. "Deformable templates us-
    ing large deformation kinematics," *IEEE Transactions on Medical Imaging*
    5(10)(1996):1435–1447.

[14] Cooke, A., DeVita, C., Gee, J., Alsop, D., Detre, J., Chen, W., and
    M. Grossman. "Neural basis for sentence comprehension deficits in fron-
    totemporal dementia," *Brain and Language, in press*.

[15] Cootes, T., Taylor, C., and J. Graham. "Active shape models - their training
    and application," *Computer Vision and Image Understanding* 60(1995):
    38–59.

[16] Csernansky, J., *et al.* "Hippocampal morphometry in schizophrenia
    by high dimensional brain mapping," *Proc. Natl. Acad. Sci. (USA)*
    95(19)(1998): 11406–11411.

[17] Dale, A., Fischl, B., and M. Sereno. "Cortical surface-based analysis I:
    Segmentation and surface reconstruction," *Neuroimage* 9(2)(1999): 179–
    194.

[18] Dale, A., Fischl, B., and M. Sereno. "Cortical surface-based analysis II:
    Inflation, flattening, and a surface-based coordinate system," *Neuroimage*
    9(2)(1999): 195–207.

[19] Davies, R., Cootes, T., Waterton J., and C. Taylor. "An efficient method
    for constructing optimal statistical shape models," in *Medical Image
    Computing and Computer-Assisted Intervention* (Proceedings of MICCAI
    2001), W. Niessen and M. Viergever, eds. Springer, LNCS2208(October
    2001): 57–65.

[20] Davatzikos, C. and R. Bryan. "Using a deformable surface model to ob-
    tain a shape representation of the cortex," *IEEE Transactions on Medical
    Imaging* 15(6)(1996): 785–795.

[21] Dawant, B., Hartmann, S., Thirion, J., Maes, F., Vandermeulen, D., and
    P. Damaerel. "Automatic 3-d segmentation of internal structures of the head
    in MR images using a combination of similarity and free-form transfor-
    mations: Part I, methodology and validation on normal subjects," *IEEE
    Transactions on Medical Imaging* 18(10)(1990): 32–42.

[22] Dawant, B., Li, R., Cetinkaya, E., Kao, C., , Fitzpatrick, J., and P. Konrad.
    "Computerized atlas-guided positioning of deep brain simulators: An fea-
    sibility study," in *Workshop on Biomedical Image Registration*, Philadel-
    phia, 2003, (J. Gee and J. Maintz, eds.), 142–150.

[23] DeVita, C., Moore, P., Dennis, K., Gee, J., Lin, M., and M. Grossman. "Cognitive correlations of voxel-based morphometric analyses of gray matter atrophy in frontotemporal dementia," *Brain and Language* (2002).

[24] Dubb, A., Gur, R., Avants, B., and J. Gee. "Characterization of sexual dimorphism in the human corpus callosum using template deformation," *Neuroimage* 20(2003): 512–519.

[25] Dupuis, P, Grenander, U., and M. Miller. "Variational problems on flows of diffeomorphisms for image matching," *Quarterly of Applied Mathematics* 56(3)(1998) 587–600.

[26] Fletcher, T., Joshi, S., Lu, C., and S. Pizer. "Gaussian distributions on Lie groups and their application to statistical shape analysis," in *Proceedings of Information Processing in Medical Imaging 2003*, Ambleside, UK, C. Taylor and J. Noble, eds. Springer-Verlag, LNCS2732. 450–462,

[27] Gee, J. "On matching brain volumes," *Pattern Recognition* 32(1999): 99–111.

[28] Gee, J. and D. Haynor. "Numerical methods for high-dimensional warps," in *Brain Warping*, A. Toga, ed. San Diego: Academic Press, San Diego, 1999. 101–113.

[29] Gee, J., Reivich, M., and R. Bajcsy. "Elastically deforming a 3D atlas to match anatomical brain images," *Journal of Computer Assisted Tomography* 17(1993): 225–236.

[30] Gee, J., Haynor, D., Reivich, M., and R. Bajcsy. "Finite element approach to warping of brain images," in *Proceedings of SPIE: Medical Imaging* 2164(1994): 327–337.

[31] Gee, J. and R. K. Bajcsy. "Elastic matching: Continuum mechanical and probabilistic analysis," in *Brain Warping*, A. Toga, ed. San Diego: Academic Press, San Diego, 1999. 183–197.

[32] Gerig, G., Styner, M., Shenton, M., and J. Lieberman. "Shape versus size: Improved understanding of the morphology of brain structures," in *Medical Image Computing and Computer-Assisted Intervention* (Proceedings of MICCAI 2001), W. Niessen and M. Viergever, eds. Springer, LNCS2208(October 2001): 502–516.

[33] Good, C., Johnsrude, I., Ashburner, J., Henson, R., Friston, K., and R. Frackowiak. "A voxel-based morphometric study of ageing in 465 normal adult human brains," *NeuroImage* 14(2001): 21–36.

[34] Grenander, U. *General Pattern Theory*. New York: Oxford University Press, 1993.

[35] Grenander, U. and M. Miller. "Computational anatomy: An emerging discipline," *Quarterly of Applied Mathematics* 56(4)(1998): 617–694.

[36] Griffin, L. "The intrinsic geometry of the cerebral cortex," *Journal of Theoretical Biology* 166(3)(1994): 261–273.

[37] Gu, X., Wang, Y., Chan, T., Thompson, P., and S. Yau. "Genus zero surface conformal mapping and its application to brain surface mapping," in *Proceedings of Information Processing in Medical Imaging 2003*, Ambleside, UK, C. Taylor and J. Noble, eds. Springer-Verlag, LNCS2732. 172–184.

[38] Hadamard, J. *Lectures on the Cauchy problem in linear partial differential equations*. New York: Dover, 1952.

[39] Hermosillo, G., Chefd'Hotel, C., and O. Faugeras. "A variational approach to multi-modal image matching," *International Journal of Computer Vision* 50(3)(2002): 329–343.

[40] Horn, D. and B.G. Schunck. "Determining optical flow," *Artifical Intelligence* 17(1-3)(August 1981): 185–203.

[41] Hughes, T. *The Finite Element Method : Linear Static and Dynamic Finite Element Analysis*, first edition. Englewood Cliffs, NJ: Prentice-Hall, 1987.

[42] Joshi, S. and M. Miller. "Landmark matching via large deformation diffeomorphisms," *IEEE Transactions On Image Processing* 9(2000): 1357–1370.

[43] Kaneko, S., Satoh, Y., and S. Igarashi. "Using selective correlation coefficient for robust image registration," *Pattern Recognition* 36(5)(May 2003): 1165–1173.

[44] Kikinis, R., Shenton, M., *et al.* "A digital brain atlas for surgical planning, model-driven segmentation, and teaching," *IEEE Transactions on Visualization and Computer Graphics* 2(1996): 232–241.

[45] Lai, W., Rubin, D, and E. Krempl. *Introduction to Continuum Mechanics*. New York: Pergamon Press, 1978.

[46] Leventon, M., Grimson, E., and O. Faugeras. "Statistical shape influence in geodesic active contours," in *Proc. IEEE Conf. Comp. Vision and Patt. Recog.*, (2000) 4–11.

[47] Lewis, J. "Fast normalized cross-correlation," 1995.

[48] Mangin, J., Poupon, C., Clark, C., Le Bihan, D., and I. Bloch. "Distortion correction and robust tensor estimation for mr diffusion imaging," *Medical Image Analysis* 6(2002): 191–198.

[49] Mangin, J., *et al.* "Coordinate-based versus structural approaches to brain image analysis," *Artificial Intelligence in Medicine*, 2003, *in press.*

[50] Maintz, J. and M. A. Viergever. "A survey of medical image registration," *Medical Image Analysis* 2(1998):1–36.

[51] Mattes, D., Haynor, D., Vesselle, H., Lewellen, T., and W. Eubank. "PET-CT image registration in the chest using free-form deformations," *IEEE Transactions on Medical Imaging* 22(1)(January 2003):120–128.

[52] Meier, D. and E. Fisher. "Parameter space warping: shape-based correspondence between morphologically different objects," *IEEE Transactions on Medical Imaging* 21(1)(2002): 31–47.

[53] Mikhlin, S. *Mathematical Physics: An Advanced Course.* Amsterdam: North-Holland Publishing Company, 1970.

[54] Miller, M., Christensen, G., Amit, Y., and U. Grenander. "Mathematical textbook of deformable neuroanatomies," *Proc. Natl. Acad. Sci. (USA)* 90(24)(1993): 11944–11948.

[55] Miller, M., Joshi, S., and G. Christensen. "Large deformation fluid diffeomorphisms for landmark and image matching," in *Brain Mapping*, A. Toga, ed. San Diego: Academic Press, 1999. 115–131.

[56] Miller, M., Trouve, T., and L. Younes. "On the metrics and euler-lagrange equations of computational anatomy," *Annu. Rev. Biomed. Eng.*, vol. 4, pp. 375–405, 2002.

[57] Peckar, W., Schnorr, C., Rohr, K., Stiehl, H., and U. Spetzger. "Linear and incremental estimation of elastic deformations in medical registration using prescribed displacements," *Machine Graphics and Vision* 7(4)(1998): 807–829.

[58] Pluim, J., Maintz, J., and M. Viergever. "Mutual-information-based registration of medical images: a survey," *IEEE Transactions on Medical Imaging* 22(8)(2003): 986–1004.

[59] Rabbitt, R., Weiss, J., Christensen, G., and M. Miller. "Mapping of hyperelastic deformable templates using the finite element method," in *Vision Geometry IV*, R. A. Melter, A. Y. Wu, F. L. Bookstein, and W. D. Green, eds. SPIE, 1995. 252–265.

[60] Rueckert, D., Frangi, A., and J. Schnabel. "Automatic construction of 3-d statistical deformation models of the brain using nonrigid registration," *IEEE Transactions on Medical Imaging* 22(8)(2003): 1014–1025.

[61] Sebastian, T, Klein, P., Kimia, B., and J. Crisco. "Constructing 2D curve atlases," in *Mathematical Methods in Biomedical Image Analysis 2000* 70–77.

[62] Shen, D. and C. Davatzikos. "Hammer: Hierarchical attribute matching mechanism for elastic registration," *IEEE Transactions on Medical Imaging* 21(11)(November 2002): 1421–1439.

[63] Shenton, M., Gerig, G., McCarley, R., Szekely, G., and R. Kikinis, "Amygdala-hippocampus shape differences in schizophrenia: The application of 3D shape models to volumetric MR data," *Psychiatry Research Neuroimaging* 115(2002): 15–35.

[64] Skrinjar, O., Nabavi, A., and J. Duncan, "Model-driven brain shift compensation," *IEEE Transactions on Medical Imaging* 6(2002): 361–373.

[65] Sowell, E., Thompson, P., Holmes, C., Batth, R., Jernigan, T., and A. Toga. "Localizing age-related changes in brain structure between childhood and adolescence using statistical parametric mapping," *NeuroImage*, 6(1999): 587–597.

[66] Strang, G. *Introduction to Applied Mathemetics*. Wellesley: Wellesley-Cambridge Press, 1986.

[67] Styner, M. and G. Gerig. "Medial models incorporating object variability for 3D shape analysis," in *Proceedings of Information Processing in Medical Imaging 2001*, 502–516.

[68] Talairach, J. and P. Tournoux. *Coplanar stereotaxic axis of the human brain*. New York: Thieme, 1988.

[69] Thirion, J. "Fast non-rigid matching of 3D medical image," Tech. Rep., Research Report RR-2547, Epidure Project, INRIA Sophia, May 1995.

[70] Thirion, J. "Non-rigid matching using demons," *IEEE Computer Vision and Pattern Recognition* (1996): 245–251.

[71] Thirion, J. "Image matching as a diffusion process: an analogy with Maxwell's demons," *Medical Image Analysis* 2(3)(1998): 243–260.

[72] Thompson, D. *On Growth and Form*. Cambridge: Cambridge University Press, 1917.

[73] Thompson, P. and A. Toga. "A surface-based technique for warping 3-dimensional images of the brain," *IEEE Transactions on Medical Imaging* 15(4)(1996): 402–417.

[74] Thompson, P., Schwartz, C., and A. Toga. "High-resolution random mesh algorithms for creating a probabilistic 3D surface atlas of the human brain," *Neuroimage* 6(1996): 19–34.

[75] Thompson, P. and A. Toga. "Anatomically-driven strategies for high-dimensional brain image warping and pathology detection," in *Brain Warping*, A. Toga, ed. San Diego: Academic Press, San Diego, 1999. 311–336.

[76] Thompson, P., *et al.* "Cortical variability and asymmetry in normal aging and alzheimer's disease," *Cerebral Cortex* 8(1998): 492–509.

[77] Thompson, P., Mega, M., and A. Toga. "Elastic image registration and pathology detection," in *Handbook of Medical Image Processing*, I. Bankman, R. Rangayyan, A. C. Evans, R. P. Woods, E. Fishman, and H. K. Huang, eds. San Diego: Academic Press, 1999.

[78] Thompson, P., Mega, M, Narr, K, Sowell, E., Blanton, R., and A. Toga. "Brain image analysis and atlas construction," in *Handbook of Medical Imaging, Vol.2: Medical Image Processing and Analysis*, (M. Sonka and J. M. Fitzpatrick, eds.), SPIE Press, July 2000. 1061–1129.

[79] Thompson, P., Mega, M., and A. Toga. *Brain Mapping: The Disorders*, chapter Disease-specific brain atlases, London: Academic Press, 2000. 131–177.

[80] Tikhonov, A. and V. Arsenin. *Solutions of Ill-Posed Problems.* Washington, DC: Winston, 1977.

[81] Toga, A. *Brain Warping*, San Diego: Academic Press, 1999.

[82] Van Essen, D., Drury, H., Joshi, S., and M. Miller. "Functional and structural mapping of human cerebral cortex: solutions are in the surfaces," *Proc. Nat. Acad. Sci. USA* 95(1998): 788–795.

[83] Viola, P. *Alignment by Maximizaition of Mutual Information*, Ph.D. thesis, MIT, 1995.

[84] Wang, Y. and L. Staib. "Shape-based 3D surface correspondence using geodesics and local geometry," *Computer Vision and Pattern Recognition* 200(II): 644–651.

[85] Wolpert, D. and W. Macready. "No free lunch theorems for optimization," *IEEE Transactions on Evolutionary Computation* 1(1)(April 1997):67–82.

[86] Woods, R., Grafton, S., Holmes, C., Cherry, S., and J. Mazziotta. "Automated image registration: I General methods and intra-subject intra-modality validation," *Journal of Computer Assisted Tomography* 22(1998): 141–154.

[87] Zienkiewicz, O. *The Finite Element Method in Engineering Science.* New York: McGraw-Hill, 1971.

Part Four

Hybrid Methods - Mixed Approaches
to Segmentation

Hybrid Segmentation Methods

Celina Imielinska
Yinpeng Jin
Elsa Angelini
Columbia University

Dimitris Metaxas
Ting Chen
Rutgers University

Jayaram K Udupa
Ying Zhuge
University of Pennsylvania

12.1   Introduction

Image segmentation is the process of identifying and delineating objects in images. It is one of the most crucial among all computerized operations done on images acquired by using an image acquisition device. Any image visualization, manipulation, and analysis tasks require directly or indirectly image segmentation. In spite of several decades of research [82, 93], this still largely remains an open problem. It is the most challenging among all operations done on images such as interpolation, filtering, and registration. Since these latter operations require object knowledge in one way or another, they all depend to some extent on image segmentation.

351

Methods for performing segmentations vary widely depending on the specific application, imaging modality, body region, and other factors. There is currently no single segmentation method that can yield acceptable results for every medical image. Combining several segmentation techniques together to form a hybrid framework can sometimes significantly improve the segmentation performance and robustness comparing to each individual component. The hybrid segmentation approach integrates boundary-based and region-based segmentation methods that amplify the strength but reduce the weakness of both approaches. However, most previous work still requires significant initialization to avoid local minima. Furthermore, most of the earlier approaches use prior models for their region-based statistics, which we would rather avoid to increase usefulness in situations where a comprehensive set of priors may not be available. Although this area of research is in its infancy, several promising strategies have been reported. These pioneering methods include: utilizing the results of region-based approaches to assist in boundary finding [14, 15, 16, 17, 22, 23, 87, 100], and combining fuzzy connectedness and deformable boundary approaches [47, 58, 59, 60, 96].

We propose a *hybrid segmentation engine* that consists of component modules, for automated segmentation of radiological patient and the Visible Human data. We integrate boundary-based and region-based segmentation methods to exploit the strength of each method hopefully to cover the weakness of the other method. This powerful and promising approach combines fuzzy connectedness (FC) [122, 123], Voronoi diagram classification method [8, 52], Gibbs prior models (GP)[40], and deformable models (DM)[71], comprise components of the engine. Each of the component modules in the engine represents a stand-alone segmentation method, but the possibility and advantages of combining the modules in a cooperative fashion, irrespective of the class they represent, may improve performance. We can derive a large number of hybrid segmentation methods by integrating different subsets of the modules, and tailor them to serve the best a specific medical imaging application. Although this area of research is in its infancy, the preliminary results are very encouraging. In this chapter we describe one instance of such a hybrid method that combines fuzzy connectedness, Voronoi diagram classification and deformable models (FC-VD-DM), and Gibbs prior with deformable models (GP-DM) and test them on a number of clinical datasets and the Visible Human Project data [116].

The following is the organization of the chapter. In the next section, we survey image segmentation methods and present the rationale for proposing the hybrid segmentation engine. Subsequently, in the following three sections, we describe the hybrid segmentation engine and the hybrid methods derived from it, respectively. The final section contains experimental results that include segmentations of radiological and Visible Human Project datasets, together with (limited) evaluation of segmentation using quantification of accuracy for true delineation [124].

12.2 Review of Segmentation Methods

For brevity, throughout this paper, we shall refer to any multidimensional vector-valued image as a *scene*, and represent any scene **C** by a pair **C** $= (C, f)$, where C is a rectangular multidimensional array of volume elements, referred to as *voxels*, and f is a function that assigns a vector to each voxel in C whose components represent imaged property values called *scene intensities*. Scene segmentation may be thought of as consisting of two related processes: *recognition* and *delineation*. *Recognition* is the high-level process of determining roughly the whereabouts of an object of interest in the scene. *Delineation* is the low-level process of determining the precise spatial extent and point-by-point composition (material membership percentage) of the object in the scene. Humans are more qualitative and less quantitative. Computers are more quantitative and less qualitative. Incorporation of high-level expert human knowledge algorithmically into the computer has remained a challenge. Most of the drawbacks of current segmentation methods may thus be attributed to the latter weakness of computers in the recognition process. We envisage, therefore, that the assistance of humans, knowledgeable in the application domain, will remain essential in any practical image segmentation method. The challenge and goal for image scientists are to develop methods that minimize the degree of this required help as much as possible. In the following subsections, we shall review the strategies currently available for delineation.

Approaches to delineation are studied far more extensively than those for recognition in image segmentation. In fact, as indicated in the last paragraph, commonly delineation is itself considered to be the entire segmentation process. At the outset, three classes of approaches to delineation may be identified *boundary-based, region-based*, and *hybrid* as described below. Boundary-based approaches focus on delineating the interface between the object and the surrounding co-objects in the scene. Region-based approaches concentrate on delineating the region occupied by the object in the scene. Hybrid approaches attempt to combine the strengths of both boundary-based and region-based approaches. In all three groups, it is possible to use hard (crisp) or fuzzy (in the sense of fuzzy subset theory) strategies to address geometric, shape-related, and topological concepts. In the rest of this section, we shall briefly review these three groups of approaches to delineation.

12.2.1 Boundary-Based Approaches

Mouse-controlled *manual boundary tracing* [121] is the simplest and the most readily available among all delineation methods. Unlike other methods, manual tracing does not require an initial developmental phase (which often can be very time-consuming) wherein the method is adapted and fine-tuned to each new application. Major drawbacks of manual tracing, however, are (1) the drudgery and time involved in tracing, and (2) its poor precision, which may vary considerably depending on the fuzziness/sharpness of the boundaries, window level setting used for scene display, brightness/contrast of the monitor, and on the actual size of the object/boundary. First attempts toward automating boundary tracing took *optimum boundary detection* approaches [64, 75, 76, 95], which pose boundary delineation as an optimization problem, that is, to pick that among all possible boundaries that can be drawn in the scene which optimizes a properly chosen objective function.

Inadequacies of these methods, especially globally optimal boundaries often differing substantially from the real boundaries, led to the so-called *active contour* or *deformable boundary* methods [26, 27, 41, 61, 65, 68, 69, 71, 78, 129], initiated by the ideas first presented in [61]. In these approaches, an initial boundary is specified somewhat close to the desired boundary (either explicitly by a human operator, or in an implicit fashion such as considering the segmented boundary in the previous slice as the initial boundary for the current slice in a slice-by-slice approach). A mechanical system is then set up to deform the boundary through external forces originating from the scene (such as from scene gradients), through internal forces exerted by the stiffness properties assumed for the boundaries, and external forces specified through user input. The final segmented boundary is taken to be that boundary for which an energy functional is minimized. An advantage of these methods is that, even when boundary information is missing in the scene in some parts of the object boundary, such gaps are filled in simply because of the closure and connectedness properties of the boundary as it is deformed. However, there is no guarantee that the resulting boundary would match well the real boundary of the object. In an attempt to avoid the post-delineation correction required by these methods, a different family of user-steered delineation methods called *live-wire* have emerged [36, 37, 38, 77, 111]. In these techniques, recognition by a human operator and delineation by the computer take place cooperatively, synergistically, and with a certain degree of continuity in a tightly coupled manner. By changing this degree, they allow within a single framework various degrees of automation and human interaction. These methods have been shown to be more reproducible and about 3-10 times faster than manual tracing [38] in certain applications and can be quickly (within minutes) adapted to a new application without requiring developmental time [37, 77].

Active shape and appearance methods [31, 30, 28, 29, 35, 45, 72] have emerged recently (and have become popular) in an attempt to overcome some of the inadequacies of the deformable boundary methods. These methods bring in constraints explicitly based on the shape of the boundary as well as the intensity patterns (appearance) in the vicinity of the boundary. The main premise is that, by creating statistical shape and appearance models (in some normalized fashion) for the objects to be segmented in the particular application, and by matching these models through smooth deformation to the information presented in a given scene, the object can be segmented.

Another class of boundary-based delineation techniques called *level-set methods* [67, 80, 112, 113] have emerged also for overcoming the inadequacies of the deformable boundary methods. They have several advantages compared to deformable models. They can handle changing topology naturally, and can deal with local voxel level deformations. In this approach, the boundary of an object is implicitly represented as the zero-level set of a time dependent function called the *level set* function. The manner in which this function evolves over time can be controlled by several entities including the geometric properties of the evolving boundary, scene gradients, and prior boundary shape information. The zero-level set of this function yields the segmentation. Level set methods have been actively pursued in scene segmentation and other applications [57, 112, 113]. For more on level sets and their mathematics, see Chapter 8, Isosurfaces and Level Sets.

12.2.2 Region-Based Approaches

The simplest among region-based methods is intensity thresholding [98]. Many methods to automatically find the thresholds in some optimum fashion have also been devised [1, 63, 81, 105, 108]. Region-growing methods [46, 94, 96, 132] evolved in an attempt to overcome the drawbacks of thresholding. The basic idea in region-growing techniques is to start from some specified seed voxels and subsequently add voxels in the vicinity of the growing region to the growing region if voxel intensity-based properties satisfy some conditions. If these conditions change adaptively with the growing region, then the repeatability of segmentations with respect to different seed points cannot be guaranteed.

Clustering or *feature-space partitioning* methods [32, 34, 120] are among the popular region-based delineation techniques. This is particularly true in brain MR image analysis [9, 12, 24, 42, 50, 73, 90, 91, 92]. The basic premise of these methods is that object regions are manifest as clusters in an appropriate feature space, and therefore, to segment the object regions, we need to identify and delineate the clusters in the scatter plot defined in the feature space. The commonly-used clustering methods in medical imaging are k-nearest neighbor [32], c-means [9], and fuzzy c-means [9] techniques. One drawback of clustering techniques as employed commonly is the requirement of multiple (two or more) features associated with every voxel. In MR image analysis, these are the imaged properties (such as T2-weighted, PD, and T1-weighted values). This also implies that the multiple acquisitions should be in registration or should be registered post hoc. Neural network techniques have also been used [44, 89, 99, 108, 132] for the classification of voxels into tissue classes based on the voxels' feature values. *Graph-based approaches* [102, 103, 104, 105, 106, 107, 122, 125, 126, 128, 133] to region delineation pose delineation as a graph problem and present a solution via graph search algorithms. Two actively pursued classes of methods in this group are *graph-cut* [10, 11, 125, 126, 128] and *fuzzy connectedness* [33, 48, 102, 103, 104, 105, 106, 107, 133, 122]. In graph-cut formulation, the scene is represented as a graph with voxels as its nodes and adjacency defining edges. Costs are assigned to edges based on scene intensity properties. A minimum cost cut then generates a segmentation. Such a cut is determined using various types of graph-cut algorithms. In fuzzy connectedness, affinity between two nearby voxels defines their local hanging-togetherness in the same object. Affinity is determined based on the distance between voxels as well as on the similarity of their intensity-based features. Fuzzy connectedness is a global fuzzy relation that assigns to every pair of voxels a strength of connectedness which is the strength of the strongest among all paths between the two voxels in the pair. The strength of a path is simply the smallest affinity of pairwise voxels along the path. Fuzzy connectedness captures the notion of a Gestalt even in the presence of noise, blurring, slow background variation, and natural heterogeneity. The objects are segmented via dynamic programming. The framework has been extended to relative, scale-based, iterative, and vectorial fuzzy connectedness [33, 48, 103, 104, 105, 106, 107, 122, 133], and to a variety of image segmentation applications.

The *finite mixture (FM)* model is a commonly used method for statistical scene segmentation because of its simple mathematical formula and the piecewise constant nature of ideal tissue images. However, being a histogram-based model, the FM model has an intrinsic limitation: no spatial information is taken into account. Using *Markov random field (MRF)*, the spatial information in an image can be encoded through contextual constraints of neighboring voxels [62, 21]. More importantly, *MRF-based* approaches can be combined with other techniques, (e.g., bias field correction) to form an effective approach for tissue segmentation.

Another important class of methods referred to as *watershed* is commonly used for region delineation [110, 107, 101, 88, 86, 7, 6]. In this approach, the n-dimensional scene is considered as a surface in an $(n+1)$-dimensional space. In the 2D scene, the region occupied by an object in the scene is considered to be the set of all those voxels which get flooded under certain conditions when the water level is gradually raised from a starting basin in the surface that corresponds to a set of voxels (seed region) in the object in the scene.

12.2.3   Hybrid Approaches

Each of the boundary-based and region-based approaches has its own strengths and weaknesses. Boundary-based methods are more sensitive to noise than region-based methods. Also, the latter are less affected when high frequency information is missing or compromised in the scene. Further, when the shape of the boundary is extremely complex, model-dependent boundary-based methods run into difficulties. On the other hand, boundary-based methods are better suited for incorporating prior object shape information into delineation as demonstrated by deformable boundary and active shape and appearance methods. They are also less affected by changes in gray-level distributions such as those caused by background intensity inhomogeneity [4] and scene-to-scene gray-scale variation [79] (both are problems encountered in MR image analysis), although region-based methods such as fuzzy connectedness [33, 48, 103, 104, 105, 106, 107, 122, 123, 133] have been shown to be resistant to these variations. The premise of hybrid methods is to exploit the strength of each method hopefully to cover the weakness of the other method. Several promising strategies have been reported. These methods include: using the results of region-based approaches to assist in boundary finding [14, 15, 16, 17, 22, 23, 87, 100], combining fuzzy connectedness and deformable boundary approaches [96, 57, 60, 59, 58, 47], and combining fuzzy connectedness with the Voronoi-diagram classification method [55, 54, 56, 3].

12.3   Hybrid Segmentation Engine

The hybrid segmentation engine, that has been implemented in the Insight Toolkit (www.itk.org) is an open collection of region- and boundary-based segmentation methods with its component modules as described in Figure 12.1, namely: fuzzy connectedness (FC), Voronoi diagram classification (VD), Gibbs prior models (GP-MRF), and deformable models (DM). We are capable of generating a large number of hybrid segmentation methods derived from the four modules that can be tailored to a specific medical imaging application and evaluated under a proposed framework for evaluation of segmentation [124]. We show examples, in Figure 12.2, of hybrid methods that we can generate. The modules are implemented as ITK filters. We note that under the FC and DM modules, several different filters have been implemented as explained in the next section. Some results using these methods have already been published [133, 106, 56, 54, 57, 55, 3, 20, 122, 123], and they have been successfully used for segmentation of multichannel images in several applications. This method uses the fact that medical images are inherently fuzzy (inhomogeneous), and we will describe the method in more detail below. In particular, the fuzzy connectedness module is implemented under a number of the ITK filters for: simple fuzzy connectedness [122, 123], vectorial scale-based fuzzy connectedness [133], vectorial relative fuzzy connectedness [107], and vectorial iterative relative fuzzy connectedness [106]. The boundary-based deformable model module [71] has been implemented in ITK as: 2D and 3D balloon force filters, marching cubes methods to construct deformable meshes close to the object surface, derived from a binary mask of a prior. The region-based Markov random field segmentation module that is driven by high order Gibbs prior models has been implemented in the Insight toolkit to handle single and multi-channel data [21]. Similarly, the Voronoi diagram classification

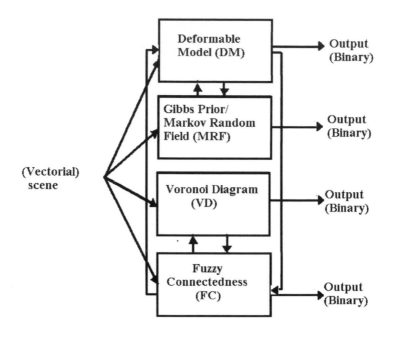

Figure 12.1. The ITK-Hybrid Segmentation Engine.

has been made available in ITK to handle both single channel and three-channel (RGB and multi-channel radiological) data. In this chapter, we limit our focus to a description of two classes of hybrid methods only, one that integrates fuzzy connectedness (FC), Voronoi diagram classification (VD), and deformable models (DM), and second that integrates Gibbs prior (GP) with deformable models.

12.4   Hybrid Segmentation: Integration of FC, VD, and DM

We present a hybrid segmentation method which requires minimal manual initialization, that integrates fuzzy connectedness (FC), Voronoi diagram classification (VD), and deformable models (DM). We will start with the fuzzy connectedness algorithm to generate a region with a sample of tissue that we plan to segment. From the sample region, we generate automatically homogeneity statistics for the

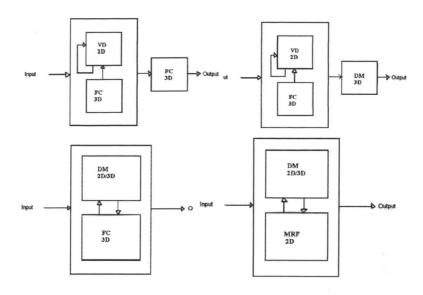

Figure 12.2. Examples of hybrid segmentation methods derived from the ITK-Hybrid Segmentation Engine.

VD classification that will return, in a number of iterations, an estimation of the boundary. Below, we describe briefly the component FC, VD, and DM methods.

12.4.1 Fuzzy Connectedness Algorithm (FC)

The simple fuzzy connectedness method, introduced in [123], uses the fact that medical images are inherently fuzzy. We define *affinity* between two *elements* in an image (i.e., pixels, voxels, or spels) via a degree of *adjacency* and the *similarity* of their intensity values. The closer the elements are and more similar their intensities are, the greater is the affinity between them. There are two important characteristics of a medical image. First, it has graded composition coming from material, blurring, noise, and background variation. Second, the image elements that constitute an anatomical object hang together in a certain way. Both these properties, *graded composition* and *hanging-togetherness* are fuzzy properties. The aim of fuzzy connectedness is to capture the global hanging-togetherness using image-based local fuzzy affinity.

Let us define a scene over a digital space (Z^n, α) as a pair $\Omega = (C, f)$, where C is an n-dimensional array of spels (space elements) and $f : C \rightarrow [0, 1]$. Fuzzy affinity, κ, is any reflexive and symmetric fuzzy relation in C, that is:

$$\kappa = \{((c,d),\mu_\kappa(c,d))|c,d \in C\},$$

$$\mu_\kappa : C \times C \to [0,1],$$

$$\mu_\kappa(c,c) = 1, \quad \text{for all} \quad c \in C, \tag{12.1}$$

$$\mu_\kappa(c,d) = \mu_\kappa(d,c), \quad \text{for all} \quad c,d \in C.$$

The general form of μ_κ can be written as follows: for all $c,d \in C$,

$$\mu_\kappa(c,d) = g(\mu_\alpha(c,d),\mu_\psi(c,d),\mu_\phi(c,d),c,d), \tag{12.2}$$

where: $\mu_\alpha(c,d)$ represents the degree of adjacency of c and d, $\mu_\psi(c,d)$ represents the degree of intensity homogeneity of c and d; $\mu_\phi(c,d)$ represents the degree of similarity of the intensity features of c and d to expected object features. Fuzzy κ-connectedness, K, is a fuzzy relation in C, where $\mu_K(c,d)$ is the strength of the strongest path between c and d, and the strength of a path is the smallest affinity along the path. To define the notion of a fuzzy connected component, we need the following hard binary relation, K_Θ, based on the fuzzy relation, K. Let $\Omega = (C,f)$ be a membership scene over a fuzzy digital space (Z^n,α), and let κ be a fuzzy spel affinity in Ω. We define a (hard) binary relation K_Θ in C as

$$\mu_{K_\Theta} = \begin{cases} 1 & \text{if } f\,\mu_K(c,d) \geq \Theta \in [0,1] \\ 0 & \text{otherwise.} \end{cases} \tag{12.3}$$

Let O_Θ be an equivalence class ([62] Chapter 10) of the relation K_Θ in C. A *fuzzy κ-component* Γ_Θ *of C of strength* Θ is a fuzzy subset of C defined by the membership function

$$\mu_{\Gamma_\Theta} = \begin{cases} f(c) & \text{if } f\,c \in O_\Theta, \\ 0 & \text{otherwise.} \end{cases} \tag{12.4}$$

The equivalence class $O_\Theta \subset C$, such that for any $c,d \in C$, $\mu_\kappa(c,d) \geq \Theta$, $\Theta \in [0,1]$, and for any $e \in C - O_\Theta$, $\mu_\kappa(c,d) < \Theta$. We use the notation $[o]_\Theta$ to denote the equivalence class of K_Θ that contains o for any $o \in C$. The *fuzzy κ- component of C that contains o* denoted $\Gamma_\Theta(o)$, is a fuzzy subset of C whose membership function is

$$\mu_{\Gamma_\Theta(o)} = \begin{cases} f(c) & \text{if } f\,c \in [O]_\Theta, \\ 0 & \text{otherwise.} \end{cases} \tag{12.5}$$

A *fuzzy $\kappa\Theta$-object of Ω* is a fuzzy κ-component of Ω of strength Θ. For any spel $o \in C$, a fuzzy $\kappa\Theta$-object of Ω that contains o is a fuzzy κ-component of Ω of strength Θ that contains o. Given κ, o, Θ and Ω, a fuzzy $\kappa\Theta$-object of Ω of strength $\Theta \in [0,1]$ containing o, for any $O \in C$, can be computed via dynamic programming [123].

12.4.2   Voronoi Diagram Classification (VD)

This algorithm, which is described in detail in [52], is based on repeatedly dividing an image into regions using VD and classifying the Voronoi regions based on a selected homogeneity classifier for the segmented anatomical tissue. We will use the algorithm as a component in the hybrid method where the classifiers for different tissue types will be generated automatically from the region segmented by the fuzzy connectedness method. VD and Decanal triangulation (DT) play a central role in the algorithm. In 2D, the VD for a set V of points is a partition of the Euclidean plane into Voronoi regions of points closer to one point of V than to any other seed point [97]. For any $p_i \in V, V = \{p_1,, p_n\}, p_i \in V, V = p_1, ..., p_n,$

$$VD(p_i) = \{x \in R^2 | d(x, p_i) \le d(x, p_j), \forall j \neq i, 1 \le j \le n. \qquad (12.6)$$

Similarly, we define the VD in 3D:

$$VD(p_i) = \{x \in R^3 | d(x, p_i) \le d(x, p_j), \forall j \neq i, 1 \le j \le n. \qquad (12.7)$$

Two Voronoi regions are adjacent if they share a Voronoi edge. The DT of V is a dual graph of the Voronoi diagram of V, obtained by joining two points whose Voronoi regions are adjacent.

12.4.3   Deformable Model (DM)

Three-dimensional deformable models used in [119, 70, 117, 127] are defined as models whose geometric forms (usually are deformable curves and surfaces) deform under the influence of internal and external forces. There are two major classes of 3D deformable models: explicit deformable models ([119, 70, 118, 117]) and implicit deformable models ([9, 16, 131, 130]). Although these two classes of models are different at the implementation level, the underlying principles are similar.

Explicit deformable models construct deformable surfaces explicitly using global and local parameters during the segmentation process. One can control the deformation process by changing the values of parameters or using user-defined external forces. Explicit deformable models are easy to represent and faster in implementation. However, the adaptation of model topology is difficult.

Implicit deformable models in [67] represent the curves and surfaces implicitly as a level set [112] of higher-dimensional scalar function. Implicit deformable models can have topological changes during the model deformation. However, the description of implicit deformable models is more complicated and the deformation process takes more time.

There are two different formulations for explicit deformable models: the energy-minimization formulation [61] and the dynamic-force formulation [26, 27]. The energy-minimization formulation has the advantage that its solution satisfies a minimum principle; while the dynamic force formulation provides the flexibility of applying different types of external forces onto the deformable model. In dynamic force formulation models, external forces push the model to the features of interest in the image and the model stops when these external forces equilibrate or vanish. The external forces can be potential forces, non-potential forces such as balloon forces, and the combination of both. Potential forces are derived from image information, e.g., the gradient information. They can attract the deformable model to the boundary of the object or other features of interest in the image. The balloon force was proposed by Cohen [26]. In [26] the deformable model starts as a small circle (2D) or sphere (3D) inside the object. The balloon force can help to push the model closer to the boundary by adding a constant inflation force inside the deformable model. Therefore the balloon-force-driven deformable model will expand like a balloon being filled with gas. However, for objects with complex surface structures, the deformable model may leak out of the boundary if it uses the balloon force exclusively. In medical imaging applications, we usually use the combination of both forces to achieve a better deformation performance.

In our framework, we use the dynamic force formulation of the explicitly parameterized deformable model. The deformable model has a simple description, easy to be interacted with, and can segment the object of interest with high efficiency. We use other modules in the hybrid framework to lead it out of local minima during the segmentation process.

Traditional deformable models such as snakes and balloons fail to segment 3D objects with complex surface structures. In [71] we use a super-quadric ellipsoid model which is capable of performing global and local deformation using tapering, bending, and twisting parameters. Nevertheless, it is difficult for this model to fit to objects with deep concavities on their surfaces, and the final segmentation result may be over-smoothed. Instead of using extensive user-defined constraints and external forces on the model surface, we try to initialize the geometry of the deformable model close enough to boundary features so that the gradient-derived forces can directly lead the model to the object surface. This gives us the motivation for integrating marching cubes methods [66, 74] into our framework.

We use the marching cubes method to create a surface mesh based on the 3D binary mask created by a prior model. We define a voxel in a 3D binary image volume as a cube created from eight pixels, four each from a slice. We give an index to the thirteen potential locations of nodes on the deformable model surface in this cube and define thirteen possible plane incidences onto which elements of the deformable model can lie.

Given that the binary mask is close enough to the surface of the object, the mesh created by the marching cubes method should be in the effective range of the gradient force. One can use this mesh as the initial deformable surface and apply gradient derived forces onto it from the very beginning of the fitting process. By doing so, the deformable model can fit well into concavities and convexities on the object surface. Plus, we skip the balloon fitting process so that the deformation process is shortened.

There are several other advantages in using the marching cubes method for 3D deformable mesh construction. The deformable surface created by the marching cubes method is close to the object surface so that we can assume that the global displacement is small enough to be neglected during the deformation. Therefore we only need to consider the local displacement during the deformable model fitting process. The deformable mesh created by the marching cubes method is composed of sub-pixel triangular elements. Therefore the final segmentation result can also achieve sub-pixel accuracy.

To integrate fuzzy connectedness, Voronoi diagram classification, and deformable models into a hybrid segmentation method, we must fit the result of the Voronoi diagram classification into the deformable model. For a 3D segmentation task, marching cubes is used to create an initial 3D deformable model surface, from a prior 3D binary mask. A mesh surface is constructed that is close enough to the actual surface of the object [21]. From the mesh, a 3D deformable model is created and a combination of the edge force and the gradient-based force is applied to the deformable model to improve the resulting segmentation (For more details see [21]).

12.4.4   The Hybrid Method: Integration of FC, VD, and DM.

The algorithm integrates two methods: fuzzy connectedness and a VD-based algorithm. We will outline the algorithm first, and explain the component steps later. The fuzzy connectedness algorithm is used to segment a fragment of the target tissue. From the sample, a set of statistics is generated automatically, in RGB and HVC color spaces, to define the homogeneity operator. The homogeneity operator will be used as a multi-channel classifier for the VD-based algorithm. As we mentioned, we will use, in the future, the deformable model, to determine the final (3D) smooth boundary of the segmented region. Below, we outline the hybrid method:

Step 1. We run the fuzzy connectedness algorithm to segment a sample of the target tissue, and generate statistics, average and variance, in three color channels, in two color spaces, RGB and HVC.

Step 2. Run the VD-based algorithm using multiple color channels, until it converges:

  (a) For each Voronoi region, classify it as interior/exterior/boundary region using multi-channel homogeneity operator;

  (b) Compute DT and display segments which connect boundary regions;

(c) Add seeds to Voronoi edges of Voronoi boundary regions;

(d) Go To Step 2(a) until the algorithm converges to a stable state or until the user chooses to quit.

Step 3. (optional) Use the deformable model to determine the final (3D) boundary and reset the homogeneity operator.

Implementation of Step 1. To initialize the fuzzy connectedness algorithm and establish the mean and standard deviation of voxel values and their gradient magnitudes, the user collects the pixels within the region of interest, by clicking on the image and selecting at each time a square region with 5×5 pixels. Then an initial seed voxel is selected to compute the fuzzy connectedness, using a dynamic programming approach [122, 124]. We determine the strength of the fuzzy connectedness Θ, $\Theta \in [0, 1]$, by letting the user select its threshold value interactively, such that the initially segmented sample of the target tissue resembles the shape of the underlying image. For a binary image with a roughly segmented sample of a tissue, we generate the strongest three channels in two color spaces, RGB and HVC [43], for average and variance, respectively. First, we define for the binary image, the smallest enclosing rectangle, a region of interest (ROI), in which we identify the segmented image and its background. Within the ROI, we calculate the mean and variance in each of the six color channels (R,G,B,H,V,C) for the object and its background, respectively. Then three channels with the largest relative difference in mean value and in variance value between the object and its background are selected, respectively. The homogeneity operator for the VD-based algorithm uses the expected mean/variance values of the object together with tolerance values, computed for each selected channel, for classifying the internal and external region **Implementation of Step 2.** We build an initial VD by generating some number of random seed points (Voronoi points) and then run QuickHull [5] to calculate the VD. Once the initial VD has been generated, the program visits each region to accumulate classification statistics and makes a determination as to the identity of the region. For each Voronoi region, the mean/variance value for the pre-selected channels are computed. If they are similar, then it is marked as internal, otherwise external. Those external regions that have at least one internal neighbor are marked as boundary. Each boundary region is divided for next iterations until the total number of pixels within it is less than a chosen number.

12.4.5 Hybrid Segmentation: Integration of Gibbs Prior Model and DM.

The integration of Gibbs prior with deformable models has been described in Chapter 7, Markov Random Fields.

12.5 Evaluation of Segmentation

A complete description of the evaluation framework for segmentation can be found in [124]. We will outline the main points of this approach and demonstrate it using this framework in a limited fashion (namely for assessment of accuracy of delineation) in our preliminary studies. For evaluating segmentation methods, three factors need to be considered for both recognition and delineation: *precision* (reproducibility), *accuracy* (agreement with truth, validity), and *efficiency* (time taken). To assess precision, we need to choose a figure of merit, repeat segmentation considering all sources of variation, and determine variations in figure of merit via statistical analysis. It is usually impossible to establish true segmentation. Hence, to assess accuracy, we need to choose a surrogate of true segmentation and proceed as for precision. In determining accuracy, it may be important to consider different "landmark" areas of the structure to be segmented depending on the application. To assess efficiency, both the computational and the user time required for algorithm and operator training and for algorithm execution should be measured and analyzed. Precision, accuracy, and efficiency are interdependent. It is difficult to improve one factor without affecting others. Segmentation methods must be compared based on all three factors. The weight given to each factor depends on application. Any method of evaluation of segmentation algorithms has to specify at the outset the application domain under consideration. We consider the application domain to be determined by the following three entities: A: An application or task (e.g., volume estimation of tumors); B: A body region (e.g., brain); P: An imaging protocol (e.g., FLAIR MR imaging with a particular set of parameters). An evaluation description of a particular algorithm α for a given application domain $< A, B, P >$ that signals high performance for α may tell nothing at all about α for a different application domain $< A', B', P' >$. For example, a particular algorithm may have high performances in determining the volume of a tumor in the brain on an MR image, but have low performance in segmenting a cancerous mass from a mammography scan of a breast. Therefore, evaluation must be performed for each application domain separately. The following additional notations are needed for our description.

Object: A physical object of interest in B for which images are acquired; example: brain tumor.

Scene: A 3D volume image, denoted by $C = (C, f)$, where C is a rectangular array of voxels, and $f(c)$ denotes the *scene intensity* of any voxel $c \in C$. C may be a vectorial scene, meaning that $f(c)$ may be a vector whose components represent several imaged properties. C is referred to as a binary scene if the range of $f(c)$ is $\{0, 1\}$.

S: A set of scenes acquired for the same given application domain: $< A, B, P >$.

In this section, we will evaluate segmentation methods using assessment of accuracy of true delineation by comparing true delineation of the scenes with the one obtained by a segmentation algorithm. Thus, we will not address the other two factors, precision and efficiency, but we have to note that these factors are extremely important. A complete evaluation of segmentation methods under all three factors must be done under precisely-defined experiments, where the acquired data and collection of true delineation and true recognition is performed following strict protocols. This ultimate segmentation evaluation study is beyond the scope of this chapter and was beyond the resources of the currently funded research; comprehensive segmentation evaluation remains a leading unexplored research area in medical image processing. For patient images, since it is impossible to establish absolute true segmentation, some surrogate of truth is needed. We will use manual delineation where the object boundaries are traced or regions are painted manually by experts.

Let S_{td} be the set of scenes containing true delineations for the scenes in S. For any scene $C \in S$, let C_o^M be the scene representing the fuzzy object defined by an object o of B in C obtained by using method M, and let $C_{td} \in S_{td}$ be the corresponding scene of true delineation, all under the application domain $< A, B, P >$. The following measures are defined to characterize the accuracy of method M under $< A, B, P >$ for delineation.

False Negative Volume Fraction:

$$FNVF_M^d(o) = \frac{|C_{td} - C_o^M|}{C_{td}} \qquad (12.8)$$

False Positive Volume Fraction:

$$FPVF_M^d(o) = \frac{|C_o^M - C_{td}|}{|C_{td}|} \qquad (12.9)$$

True Positive Volume Fraction:

$$TPVF_M^d(o) = \frac{|C_o^M \cap C_{td}|}{|C_{td}|} \qquad (12.10)$$

The meaning of these measures is illustrated in Figure 12.3 for the binary case. They are all expressed as a fraction of the volume of true delineation. $FNVF_M^d$ indicates the fraction of tissue defined in C_{td} that was missed by method M in delineation. $FPVF_M^d$ denotes the amount of tissue falsely identified by method M as a fraction of the total amount of tissue in C_{td}. $TPVF_M^d$ describes the fraction of the total amount of tissue in C_{td} with which the fuzzy object C_o^M overlaps. Note that the three measures are independent, that is, none of them can be derived from the other two. True negative volume fraction has no meaning in this context since it would depend on the rectangular cuboidal region defining the scene domain C.

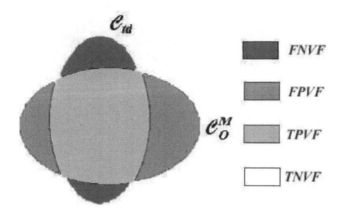

Figure 12.3. A geometric illustration of the three precision factors for delineation.

We will use this limited version of the evaluation framework in assessing the accuracy of segmentation compared with true delineation in some examples presented below.

12.6 Results

We have demonstrated the capabilities of the method in segmentation of various organs and tissues and our results are promising. The hybrid methods that are derived from the hybrid segmentation engine have been shown effective in delineating even complex heterogeneous anatomical tissue. We present results that involve both the Visible Human Project data and radiological patient data that have been segmented under a number of different projects. Some of the results were evaluated for accuracy of delineation using quantified measures from our proposed framework for evaluation of segmentation.

12.6.1 Segmentation of Brain Tumor from MRI data.

We have recently processed the radiological imaging data from a patient who underwent an orbito-zygomatic craniotomy for the removal of an anterior skull base meningioma. The following example shows this patient's MRI data that have been segmented with FC segmentation, Figure 12.4, using two different parameters for the strength of FC segmentation. In Figure 12.5, we overlay the segmented 3D tumor on intraoperative digital photographs obtained through the operating microscope during critical portions of the procedure. These are preliminary results for

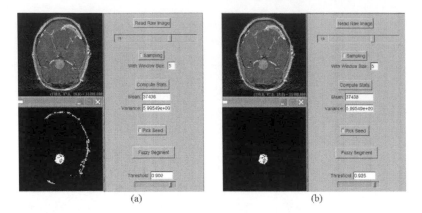

(a)   (b)

Figure 12.4. Fuzzy segmentation of a brain tumor from a patient MRI data, with FC threshold (a) 0.9 and (b) 0.925.

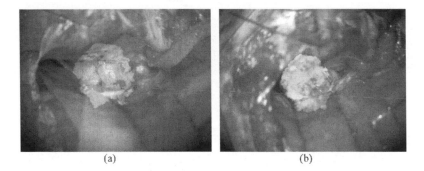

(a)   (b)

Figure 12.5. (See Plate XIII) 3D model of the tumor overlaid over two snapshots from a surgery for the removal of an anterior skull base meningioma: (a) before and (b) after tumor removal. [Courtesy of Dr. J. Bruce and Dr. A. D'Ambrosio]

a project, being developed at Columbia University Department of Neurological Surgery, to apply augmented reality for microscope-guided skull-base surgery.

12.6.2   Segmentation of Visible Human Project Data

We have developed atlases [51] from the Visible Human Project [116] data sets that are systematically incorporated into the anatomy curriculum at Columbia University College of Physicians and Surgeons. We have tested our segmentation methods on the color Visible Human Project data and compared the results with the hand segmentations delineated by an anatomy expert. Figure 12.6 shows the segmentation of the temporalis muscle using FC/VD segmentation. We il-

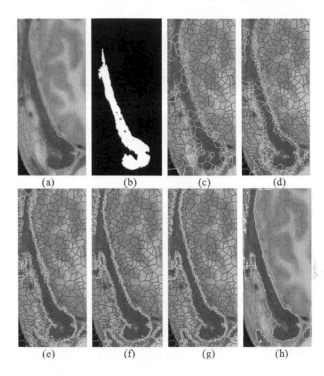

Figure 12.6. (See Plate XVI) Automated segmentation of temporalis muscle: (a) color Visible Human Male slice; (b) a fuzzy connected component; (c)–(g) iterations of the VD-based algorithm; (h) an outline of the boundary.

lustrate the steps of the FC/VD method, where Figure 12.6(b) shows the result of the temporalis muscle segmented with FC method, and from this sample we derive an homogeneity operator to classify the Voronoi region in the VD classification. In Figures 12.6(c)–(g), we show iterations of the VD classification, with the brightly colored boundary Voronoi regions that converge into the final boundary of the segmented region. Finally, in Figure 12.6(h) we outline the boundary with a subgraph of the Decanal triangulation, a dual graph to Voronoi diagram.

This process is easily applied to other types of tissue. In Figure 12.7, we show FC/VD segmentation used to delineate the highly heterogeneous visceral adipose tissue in the abdomen region. For smooth structures, the FC/VD segmentation process can be augmented with deformable models (FC/VD/DM). In Figure 12.8, we segmented left kidney from the Visible Human Project male dataset set using: FC, Figure 12.8(b); FC/VD followed by a selection of a connected component, Figure 12.8(c); FC/VD/DM, Figure 12.8(d); and in Figure 12.9 we show corresponding 3D models generated with the 3D Vesalius Visualizer.

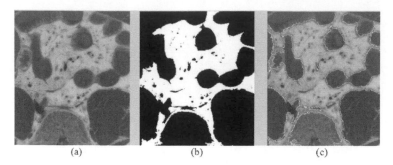

Figure 12.7. (See Plate XVII) FC-VD Segmentation of Visible Human Male adipose tissue in the abdomen region: (a) input data; (b) FC; (b) VD final boundary.

We compared the results for FC, FC/VD, and FC/VD/DM with hand segmentation of the left kidney using an accuracy assessment for true delineation. We present the quantified accuracy errors in Table 12.1. The result shows that the VD classification improves the FC segmentation, and DM calibrates the final segmentation. It also shows that we can make a fair comparison of accuracy of successive segmentation steps using quantifiable parameters, while it is impossible to assess the same error by mere observation.

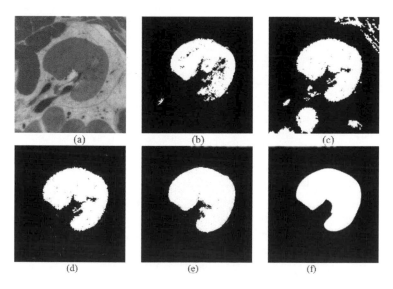

Figure 12.8. (See Plate XVIII) Segmentation of the Visible Human Male left kidney: (a) input data; (b) FC; (c) VD; (d) VD-CC; (e) DM; (f) hand segmentation.

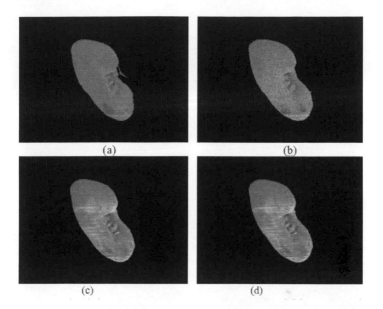

Figure 12.9. (See Plate XIX) 3D segmented and visualized with *Vesalius*^{TM} Visualizer, the Visible Human Male left kidney. 3D models of: (a) FC; (b)VD-CC; (c) DM; (d) hand segmentation.

	FC	VD–CC	DM
Area Difference %	17.2	5.6	0.91
FNVF	0.193	0.084	0.053
FPVF	0.021	0.028	0.062
TPVF	0.807	0.916	0.947

Table 12.1. Visible Human Project male dataset: kidney. Quantification of accuracy for true delineation.

12.6.3 3D Rotational Angiography.

In Figure 12.10 and Figure 12.11, we present segmentation of 3D rotational angiogram data that is processed by fuzzy connectedness (FC) and fitted with deformable model and the corresponding 3D model.

12.6.4 Quantification of Adipose Tissue from Whole Body MRI Scans

One of our collaborative projects, with Dr. Heymsfield, the Director of the St. Luke's Roosevelt Obesity Research Center, involves quantification of adipose tis-

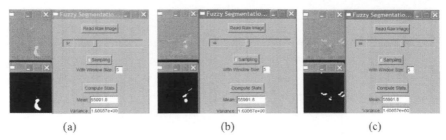

(a)   (b)   (c)

Figure 12.10. 3D rotational angiogram data segmented with fuzzy connectedness algorithm: (a)–(c) three segmented slices in the volume.

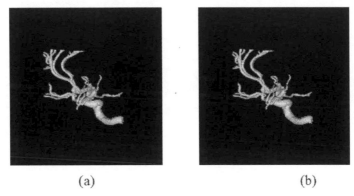

(a)   (b)

Figure 12.11. 3D rotational angiogram, segmented with (a) FC and (b) finished with DM segmentation.

sue from whole body MRI scans. We are in the process of building an integrated system for acquisition, imaging, reconstruction, and quantification of body composition. The existing approach at the Center uses manual segmentation of the adipose and other tissue from the MRI scans, a very time consuming and laborious process. The outcome of manual segmentations and corresponding 3D visualizations are presented in Figure 12.12(a) and (b). We used FC/VD hybrid segmentation on a abdominal dataset that consists of six slices in the volume. The preliminary results are promising. We used sixteen existing manual segmentations that have been collected over time from a number of experts working for the Center. They each were asked to segment the same MRI scan of an abdomen and to repeat the same segmentation three times in three month-intervals. We used these manual segmentations as a source for deriving true delineation of adipose tissue. Figures 12.13 and 12.14 show intra-operator and inter-operator variability for manual segmentation, respectively. We note that even though the protocol for manual segmentation was well specified, we cannot say that they were fully controlled. Despite the fact that this is not a perfect set of manual delineations, we

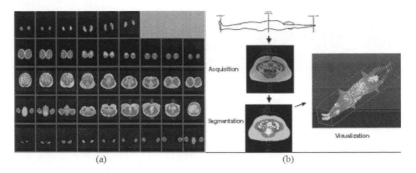

(a)   (b)

Figure 12.12. (See Plate III) Obesity Research Center: integrated system for quantification of adipose tissue from whole body MRI scans. [Courtesy of Dr. Heymsfield]

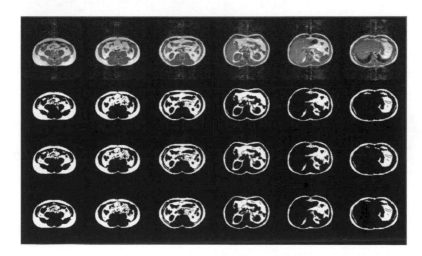

Figure 12.13. Quantification of adipose tissue from whole body MRI T1 weighted scans—6 slices from the abdomen. Evaluation of accuracy of segmentation— ground truth (delineation)— intra-operator variability.

used it to derived true delineation to conduct quantification of accuracy. We used the evaluation framework reported in [124]. Figure 12.15 shows the input slice, the fuzzy map, and the FC segmented region with user selected fuzzy connectedness value. Figure 12.16 shows visceral and subcutaneous tissue using FC/VD segmentation, followed by a basic smoothing of the segmented region.

We present accuracy quantification using parameters in equations comparing the outcome of FC/VD segmentation with the true delineation. Figure 12.17 shows three of six slices from our input data set. The true delineation was es-

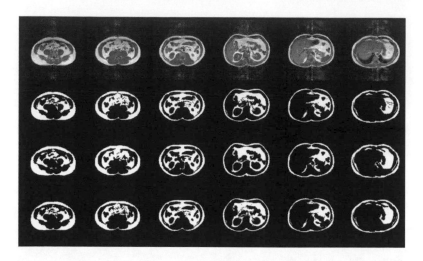

Figure 12.14. Quantification of adipose tissue from whole body MRI T1 weighted scans—6 slices from the abdomen. Evaluation of accuracy of segmentation ground truth (delineation) inter-operator variability.

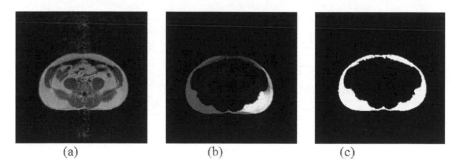

(a)   (b)   (c)

Figure 12.15. Fuzzy connectedness segmentation: (a) input image; (b) fuzzy scenemap; (c) result: segmented object with fuzzy connectedness level value (a user-defined threshold) of 0.025.

tablished by averaging sixteen manual delineations, combining the corresponding binary masks into a fuzzy object (with pixel values between 0 and 1) (for details, see [124]). The results of the segmentation of the data with FC/VD were compared with the true delineation. The three factors (FNVP, FPVF, TPVF) for measuring the accuracy of true delineation for individual slices in the data (the inter-slice distance in the data set was too large to treat as a contiguous volume) are computed. We observe in Table 12.2 that the simple measurement of area difference does not provide accurate evaluation in terms of overall performance,

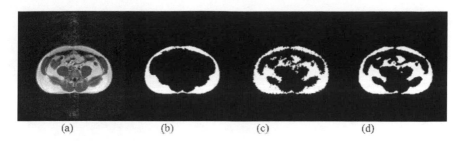

Figure 12.16. Quantification of adipose tissue from whole body MRI scans: (a) input data: MRI T1 image; (b) training: (simple FC); (c) segmented image: VD classification; (d) smoothed segmented image: visceral and subcutaneous tissue.

while parameters (FNVP, FPVF, TPVF) give better characterization of the accuracy measurement.

In general, our segmentation system gives 8% 9% accuracy in these slices, comparing to the best possible accuracy of 5%–7% from Table 12.2. As for the efficiency measurement, an experienced human operator usually takes about 15– 20 minutes to finish a delineation session on adipose tissue for a seven-slice MRI T1 dataset. While using our semi-automatic system, only a few mouse-clicks are needed, and the segmentation for 2D slice images is usually finished in real time (less than one second).

Ground Truth	Area Difference (%)	FNVF (%)	FPVF (%)	TPVF (%)
Figure (b)	1.3	8.2	9.4	91.8
Figure (f)	0.2	7.7	7.9	92.3
Figure (j)	3.9	8.1	11.9	91.9

Table 12.2. Accuracy measurements of the FC/VD segmentation shown in Figure 12.17 (c), (g), and (k).

12.7 Conclusions

We have proposed a hybrid segmentation engine with its component modules that are boundary-based and region-based segmentation methods implemented in ITK. From the engine, we can derive a number of hybrid methods that integrate boundary- and region- based approach to segmentation into powerful new tools that exploit the strength of each method and hopefully cover the weaknesses of the other method. In particular we presented an instance of a hybrid method that combines the fuzzy connectedness, Voronoi diagram classification, and deformable models and tested it on the Visible Human Project and radiological patient data.

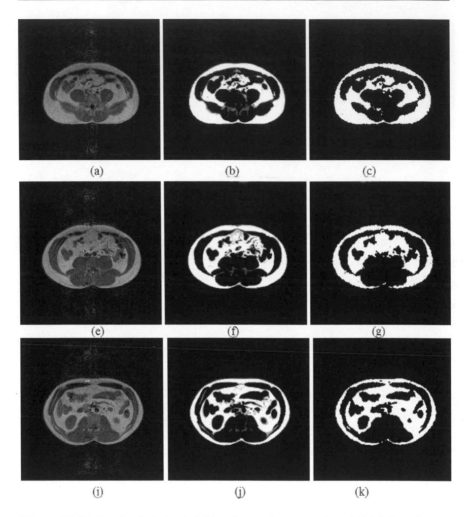

Figure 12.17. Results depicting hybrid and manual segmentation: (a)(e)(i) input images (MRI T1); (b)(f)(j): ground truth images; (c)(g)(k): FC/VD hybrid segmentation.

In some examples, like the quantification of adipose tissue from whole body MRI scans, we have quantified accuracy error for true delineation using three factors: FNVP, FPVF, TPVF. Under the framework, we have demonstrated that the results from the hybrid segmentation methods are close to those obtained from manual delineations (surrogates of ground truth). The preliminary results are very encouraging, and hybrid segmentation methods have a potential to become important tools in medical image processing.

References

[1] Abutaleb, A. "Automatic thresholding of gray-level pictures using two-dimensional entropy," *Computer Vision, Graphics and Image Processing* 47(1989): 22–32.

[2] Adalsteinson, D. and J. A. Sethian. "A fast level set method for propogating interfaces," *Journal of Computational Physics* 118(1995): 269–277.

[3] Angelini, E., Imielinska, C., Jin, Y., and A. Laine. "Improving statistics for hybrid segmentation of high-resolution multichannel images," in Proceedings of SPIE: Medical Imaging 4684(2002): 401–411.

[4] Axel, L., Costantini, J., and J. Listerud. "Intensity corrections in surface-coil MR imaging," *American Journal of Roentgenology* 148(1987): 418–420.

[5] Barber, C., Dobkin, D., and Huhdanpaa H. "The quickhull algorithm," in *Convex Hull* The Geometry Center, University of Minnesota, 1995.

[6] Beucher, S. and C. Lantuejoul. "Use of watersheds in contour detection," in *Proceedings of the International Workshop on Image Processing*, Real-Time Edge and Motion Detection/Estimation (1979) 17–21.

[7] Beucher, S. "The watershed transformation applied to image segmentation," *10th Pfefferkorn Conference on Signal and Image Processing in Microscopy and Microanalysis* (1992): 299–314.

[8] Bertin, E., Parazza, F., and J. Chassery. "Segmentation and measurement based on 3D VoronoiDiagram: application to confocal microscopy," *Computerized Medical Imaging and Graphics*, 17(1993): 175–182.

[9] Bezdek, J., Hall, L., and L. Clarke, "Review of MR image segmentation techniques using pattern recognition," *Medical Physics* 20(1993): 1033–1048.

[10] Boykov, Y. and M. Jolly. "Interactive graph cuts for optimal boundary & region segmentation of objects in N-D images," in *Proceedings of ICCV*, Part I, 105–112, 2001.

[11] Boykov, Y., Veksler, O. and R. Zabih. "Fast approximate energy minimization via graph cuts," *IEEE Transactions on Pattern Analysis and Machine Intelligence* 23(2001): 1222–1239.

[12] Brandt, M., Bohan, T., Kramer, L. and J. Fletcher. "Estimation of CSF, white and gray matter volumes in hydrocephalic children using fuzzy

clustering of MR images," *Computerized Medical Imaging and Graphics* 18(1994): 25–34.

[13] Carlbom, I., Terzopoulos, D., and K. Harris. "Computer-assisted registration, segmentation, and 3D reconstruction from images of neuronal tissue sections," *IEEE Transactions on Medical Imaging* 13(2)(1995): 351–362.

[14] Chakraborty, A., Staib, L. and J. Duncan. "Deformable boundary finding influenced by region homogeneity," in *Proceedings of Computer Vision and Pattern Recognition (CVPR'94)* (Seattle, WA, June, 1994):624–627.

[15] Chakraborty, A. and J. Duncan. "Integration of boundary finding and region-based segmentation using game theory," in *Proceedings of Information Processing in Medical Imaging 1995*, Y. Bizais, *et al.*, editor, Dordrecht, The Netherlands: Kluwer Academic Publishers, 1995. 189–201.

[16] Chakraborty, A., Staib, L. and J. Duncan. "Deformable boundary finding in medical images by integrating gradient and region information," *IEEE Transactions on Medical Imaging* 15(1996): 859–870.

[17] Chakraborty, A., Staib, L. and J. Duncan. "An integrated approach to surface finding for medical images," in *Proceedings of Mathematical Methods in Biomedical Image Analysis*, (1996): 253–262.

[18] Chan, T. and L. Vese. "Active contours without edges," *IEEE Transactions on Image Processing* 10(2)(2001): 266–277.

[19] Chang, Y. and X. Li. "Adaptive image region-growing," *IEEE Transactions on Image Processing* 3(1994): 868–872.

[20] Chen, T. and D. Metaxas. "Image segmentation based on the integration of Markov random fields and deformable models," in *Medical Image Computing and Computer-Assisted Intervention* (Proceedings of MICCAI 2000), S. Delp, A. DiGioia, and B. Jaramaz, eds. Springer, LNCS1935(October 2000): 256–265.

[21] Chen, T. *Integration of Gibbs Prior Models, Marching Cubes Methods and Deformable Models in Medical Image Segmentation*. Ph.D. dissertation (University of Pennsylvania, May 2003).

[22] Chu, C. and J. Aggarwal. "The integration of image segmentation maps using region and edge information," *IEEE Transactions on Pattern Analysis and Machine Intelligence* 15(1993): 1241–1252.

[23] Chu, S. and A. Yuille. "Region competition: Unifying snakes, region growing and Bayes/MDL for multi-band image segmentation," *IEEE Transactions on Pattern Analysis and Machine Intelligence* 18(1996): 884–900.

[24] Clark, M., Hall, L., Goldgof, D., Clarke, L., Silbiger, M. and C. Li. "MRI segmentation using fuzzy clustering techniques: Integrating knowledge," *IEEE Engineering in Medicine and Biology Magazine* 13(1994): 730–742.

[25] Clark, M., Hall, L., Goldgof, D., Clarke, L., Velthuizen, R., Murtagh, F. and M. Silbiger. "Automatic tumor segmentation using knowledge-based techniques," *IEEE Transaction on Medical Imaging* 17(1998): 187–201.

[26] Cohen, L. "On active contour models and balloons," *Computer Vision, Graphics, and Image Processing: Image Understanding* 53(2)(1991): 211–218.

[27] Cohen, L. and I. Cohen. "Finite element methods for active contour models and balloons for 2D and 3D images," *IEEE Transactions on Pattern Analysis and Machine Intelligence* 15(11)(1993): 1131–1147.

[28] Cootes, T., Hill, A., Taylor, C. and J. Haslam. "The use of active shape models for locating structures in medical images," *Image and Vision Computing* 12(1994): 355–366.

[29] Cootes, T., Taylor, C., Cooper, D., and J. Graham. "Active shape models–their training and application," *Computer Vision and Image Understanding* 61(1995): 38–59.

[30] Cootes, T., Edwards, G., and C. Taylor. "Comparing active shape models with active appearance models," in *Proceedings of the British Machine Vision Conference*, T. Pridmore and D. Elliman, eds. 1(1999): 173–182.

[31] Cootes, T., Edwards, G., and C. Taylor. "Active appearance models," *IEEE Transactions on Pattern Analysis and Machine Intelligence* 23(2001): 681–685.

[32] Cover, T. and P. Hart. "Nearest neighbor pattern classification," *IEEE Transaction on Information Theory*, 13(1967): 21–27.

[33] Cutrona, J. and N. Bonnet. "Two methods for semi-automatic image segmentation based on fuzzy connectedness and watersheds," in *France-Iberic Microscopy Congress* (Barcelona 2001): 23–24.

[34] Duda, R. and P. Hart. *Pattern Classification and Scene Analysis*. New York: John Wiley and Sons, 1973.

[35] Duta, N. and M. Sonka. "Segmentation and interpretation of MR brain images: An improved active shape model," *IEEE Transactions on Medical Imaging* 17(1998): 1049–1062.

[36] Falcao, A. and J. Udupa. "Segmentation of 3D objects using live wire," in *Proceedings of SPIE* 3034(1997): 228–239.

[37] Falcao, A., Udupa, J., Samarasekera, S., Sharma, S., Hirsch, R. and R. Lotufo. "User-steered image segmentation paradigms: Live wire and live lane," *Graphical Models and Image Processing* 60(1998): 233–260.

[38] Falcao, A., Udupa, J. and F. Miyazawa. "An ultra-fast user-steered image segmentation paradigm: Live wire-on-the-fly," *IEEE Transactions on Medical Imaging* 19(2000): 55–62.

[39] Fritsch, D., Pizer, S., Yu, L., Johnson, V., and E. Chaney. "Segmentation of medical image objects using deformable shape loci," in *Proceedings of Information Processing in Medical Imaging 1997* Berlin: Springer (1997): 127–140.

[40] Geman, S. and D. Geman. "Stochastic relaxation, Gibbs distributions, and the Bayesian restoration of images," *IEEE Transactions Pattern Analysis and Machine Intelligence* 6(1984):721–741.

[41] Geiger, D., Gupta, A., Costa, L. and J. Vlontzos. "Dynamic programming for detecting, tracking, and matching deformable contours," *IEEE Transactions on Pattern Analysis and Machine Intelligence* 17(1995): 294–302.

[42] Goldszal, A., Davatzikos, C., Pham, D., Yan, M., Bryan, R. and S. Resnick. "An image-processing system for qualitative and quantitative volumetric analysis of brain images," *Journal of Computer Assisted Tomography* 22(1988): 827–837.

[43] Gong,Y. and M. Sakauchi. "Detection of regions matching specified chromatic features," *Computer Vision and Image Understanding* 61(2)(1995): 263–269.

[44] Hall, L., Bensaid, A., Clarke, L., Velthuizen, R., Silbiger, M. and J. Bezdek, "A comparison of neural network and fuzzy clustering techniques in segmenting magnetic resonance images of the brain," *IEEE Transactionn o Neural Networks* 3(1992): 672–682.

[45] Hamarneh, G. and T. Gustavsson. 'Combining snakes and active shape models for segmenting the human left ventricle in echocardiographic images," *IEEE Computers in Cardiology* 27(2000) :115–118.

[46] Haralick, R. and L. Shapiro. "Image segmentation techniques," *Computer Vision, Graphics, and Image Processing*, 29(1985): 100–132.

[47] Herlin, I., Nguyen, C., and C. Graffigne. "A deformable region model using stochastic processes applied to echocardiographic Images," in *Proceedings of Computer Vision and Pattern Recognition (CVPR'92)* (Urbana, IL, June, 1992) 534–539.

[48] Herman, G. and B. Carvalho. "Multiseeded segmentation using fuzzy connectedness," *IEEE Transactions on Pattern Analysis and Machine Intelligence* 23(2001): 460–474.

[49] Herndon, R., Lancaster, J., Toga, A., and P. Fox. "Quantification of white matter and gray matter volumes from T1 parametric images using fuzzy classifiers," *Journal of Magnetic Resonance Imaging* 6(1996): 425–435.

[50] Hillman, G., Chang, C., Ying, H. and J. Yen. "Automatic system for brain MRI analysis using a novel combination of fuzzy rule-based and automatic clustering techniques," in imiitProceedings of SPIE, 16–25, 1995.

[51] Imielinska, C. and P. Molholt. "The Vesalius Project–could it be a revolution in medical education?," *Communications of the ACM, in press.*

[52] Imielinska, C., Downes, M, and W. Yuan. "Semi-automated color segmentation of anatomical tissue," *Journal of Computerized Medical Imaging and Graphics* 24(April 2000): 173–180.

[53] Imielinska, C., Laino-Pepper, L., Thumann, R., and R. Villamil. "Technical challenges of 3D visualization of large color data sets," in *Proceedings of the Second Visible Human Project Conference* (Oct. 1998, Bethesda, MD).

[54] Imielinska, C., Metaxas, D., Udupa, J., Jin, Y., and T. Chen. "Hybrid segmentation of the Visible Human data," in *Proceedings of the Third Visible Human Project Conference* (Oct. 2000, Bethesda, MD).

[55] Imielinska, C., Metaxas, D., Udupa, J., Jin, Y., and T. Chen. "Hybrid segmentation of anatomical data," in *Medical Image Computing and Computer-Assisted Intervention* (Proceedings of MICCAI 2001), W. Niessen and M. Viergever, eds. Springer, LNCS2208(October 2001): 1048–1057.

[56] Jin, J., Imielinska, C., Laine, A., Udupa, J., Shen, W., and S. Heymsfield. "Segmentation and evaluation of adipose tissue from whole body MRI scans," in *Medical Image Computing and Computer-Assisted Intervention* (Proceedings of MICCAI 2003), R. Ellis, and T. Peters, eds. Springer, LNCS2878(October 2003): 635–642.

[57] Jin, J., Imielinska, C., and A. Laine. "A homogeneity-based speed term for level-set shape detection," in *Proceedings of SPIE: Medical Imaging* 4684(2002).

[58] Jones, T. *Image-Based Ventricular Blood Flow Analysis*, Doctoral Dissertation (University of Pennsylvania, 1998).

[59] Jones, T. and D. Metaxas. "Automated 3D segmentation using deformable models and fuzzy affinity," in *Proceedings of Information Processing in Medical Imaging 1997* Berlin: Springer (1997): 113–126.

[60] Jones, T. and D. Metaxas. "Image segmentation based on the integration of pixel affinity and deformable models," *Proceedings of IEEE CVPe*, (Santa Barbara, CA, June 1998).

[61] Kass, M., Witkin, A., and D. Terzopoulos. "Snakes: Active contour models," *International Journal of Computer Vision* 1(4)(1987): 321–331.

[62] Kaufmann, A. *Introduction to the Theory of Fuzzy Subsets, Vol. I.* New York: Academic Press, 1975.

[63] Li, C. and C. Lee. "Minimum entropy thresholding," *Pattern Recognition* 26(1993): 617–625.

[64] Liu, H. "Two- and Three-dimensional boundary detection," *Computer Graphic Processing* 6(1977): 123–134.

[65] Lobreg, S. and M. Viergever. "A discrete dynamic contour model," *IEEE Transactions on Medical Imaging* 14(1995): 12–24.

[66] Lorensen, W. and H. Cline. "Marching cubes: a high resolution 3D surface construction algorithm," *Computer Graphics* 21(4)(1987): 163–168,

[67] Malladi, R., Sethian, J., and B. Vemuri. "Shape modeling with front propogation: A level set approach," *IEEE Transactions on Pattern Analysis and Machine Intelligence* 17(2)(1995): 158–175.

[68] McInerney, T. and D. Terzopoulos. "A dynamic finite element surface model for segmentation and tracking in multidimensional medical images with application to cardiac 4D image analysis" *Computerized Medical Imaging and Graphics* 19(1)(1995): 69–83.

[69] McInerney, T. and D. Terzopoulos. "Deformable models in medical image analysis: A survey," *Medical Image Analysis* 1(1996): 91–108.

[70] McInerney, T. and D. Terzopoulos. "A dynamic finite element surface model for segmentation and tracking in multidimensional medical images with application to cardiac 4D image analysis" *Computerized Medical Imaging and Graphics* 19(1)(1995): 69–83.

[71] Metaxas, D. *Physics-Based Deformable Models: Application to Computer Vision, Graphics, and Medical Imaging*. Norwell, MA: Kluwer Academic Publishers, 1996.

[72] Mitchell, S., Lelieveldt, B., Geest, R., Reiber, J., and M. Sonka. "Multistage hybrid active appearance model matching: Segmentation of left and right ventricles in cardiac MR images," *IEEE Transactions on Medical Imaging* 20(2001): 415–423.

[73] Mohamed, F., Vinitski, S., Faro, S., Gonzalez C., Mack, J., and T. Iwanaga. "Optimization of tissue segmentation of brain MR images based on multispectral 3D feature map," *Magnetic Resonance Imaging* 17(1999): 403–409.

[74] Montani, C., Scateni, S., and R. Scopigno, "Discretized marching cubes," in *Proceedings of IEEE Visualization 1994* IEEE Computer Society Press (October 1994): 281–287.

[75] Montanari, U. "On the optimal detection of curves in noisy pictures," *Communications of the ACM* 14(1971): 335–345.

[76] Morgenthaler and A. Rosenfeld. "Multidimensional edge detection by hypersurface fitting," *IEEE Transactions on Pattern Analysis and Machine Intelligence* 3(1981):482–486.

[77] Mortensen, E. and W. Barrett. "Intelligent scissors for image composition," in *Proceedings of Computer Graphics (SIGGRAPH'95)* (1995): 191–198.

[78] Neuenschwander, W., Fua, P., Szekely, G., and O. Kubler. "Initializing snakes," in *Proceedings of IEEE CVPR* (1994): 658–663.

[79] Nyul, L. and J. Udupa. "On standardizing the MR image intensity scale," *Magnetic Resonance in Medicine* 42(1999): 1072–1081.

[80] Osher, S. and J. Sethian. "Fronts propogating with curvature-dependent speed: Algorithms based on Hamilton-Jacobi formulations," *Journal of Computational Physics* 79(1988): 12–49.

[81] Otsu, N. "A threshold selection method from gray-level histograms," *IEEE Transaction on Systems, Man, and Cybernetics* 9(1979): 62–66.

[82] Pal, N. and S. Pal. "A review of image segmentation techniques," *Pattern Recognition* 26(1993): 1277–1294.

[83] Paragios, N. and R. Deriche. "Geodesic active contours and level sets for the detection and tracking of moving objects," *IEEE Transactions on Pattern Analysis and Machine Intelligence* 22(2000): 266–280.

[84] Paragios, N. and R. Deriche. "Geodesic active region: a new framework to deal with frame partition problems in computer vision," *Journal of Visual Communication and Image Representation* 13(2002): 249–268.

[85] Paragios, N. and R. Deriche. "Geodesic active region and level set methods for supervised texture segmentation," *International Journal of Computer Vision* 46(2002): 223–247.

[86] Park, J. and J. Keller, "Snakes on the Watershed," *IEEE Transactions on Pattern Analysis and Machine Intelligence* 23(2001): 1201–1205.

[87] Pavlidis, T. and Y. Liow. "Integrating region growing and edge detection," *IEEE Transactions on Pattern Analysis and Machine Intelligence* 12(1990): 225–233.

[88] Pei, X. and M. Gabbouj. "Robust image contour detection by watershed transformation," in *Proceedings of IEEE Workshop on Nonlinear Signal and Image Processing* (1997): 315–319.

[89] Penedo, M., Carreira, M., Mosquera, A., and D. Cabello, "Computer-aided diagnosis: a neural-network-based approach to lung nodule detection," *IEEE Transactions on Medical Imaging*, 17(1998): 872–880.

[90] Pham, D. and J. Prince. "Adaptive fuzzy segmentation of magnetic resonance images," *IEEE Transactions on Medical Imaging*, 18(1999): 737–752.

[91] Pham, D. "Spatial models for fuzzy clustering," *Computer Vision and Image Understanding* 84(2001): 285–297.

[92] Pham, D., Prince, J., Dagher, A., and C. Xu. "An automated technique for statistical characterization of brain tissue in magnetic resonance imaging," *International Journal on Pattern Recognition and Artificial Intelligence* 11(1997):1189–1211.

[93] Pham, D., Xu, C., and J. Prince. "Current methods in medical image segmentation," *Annual Review Biomedical Engineering* 2(2000): 315–337.

[94] Pong, T., Shapiro, L., Watson, L., and R. Haralick. "Experiments in segmentation using facet model region grower," *Computer Vision, Graphics, and Image Processing* 25(1984): 1–23.

[95] Pope, D., Parker, D., Gustafson, D., and P. Clayton. "Dynamic search algorithms in left ventricular border recognition and analysis of coronary arteries," *IEEE Proceedings of Computers in Cardiology* 9(1984): 71–75.

[96] Poon, C. and M. Braun. "Image segmentation by a deformable contour model incorporating region analysis," *Physics in Medicine and Biology* 42(1997): 1833–1841.

[97] Preparata, F. and M. Shamos. *Computational Geometry*. New York: Springer, 1985.

[98] Prewitt, J. and M. Mendelsohn. "The analysis of cell images," *Annals of the New York Academy of Science* 128(1966): 1035–1053.

[99] Reddick, W., Glass, J., Cook, E., Elkin, T., and R. Deaton. "Automated segmentation and classification of multispectral magnetic resonance images of brain using artificial neural networks," *IEEE Transactions on Medical Imaging,* 16(1997): 911–918.

[100] Ronfard, R. "Region-based strategies for active contour models," *International Journal of Computer Vision* 13(1994): 229–251.

[101] Saarinenl, K. "Color image segmentation by a watershed algorithm region adjacency graph processing," in *Proceedings of IEEE ICIP* (1994): 1021–1025.

[102] Saha, P., Udupa, J., and D. Odhner. "Scale-based fuzzy connected image segmentation: theory, algorithms, and validation," *Computer Vision and Image Understanding* 77(2000): 145–174.

[103] Saha, P. and J. Udupa. "Relative fuzzy connectedness among multiple objects: Theory, algorithms, and applications in image segmentation," *Computer Vision and Image Understanding,* 82(2001): 42–56.

[104] Saha, P. and J. Udupa. "Fuzzy connected object delineation: axiomatic path strength definition and the case of multiple seeds," *Computer Vision and Image Understanding* 83(2001): 275–295.

[105] Saha, P. and J. Udupa. "Optimum image thresholding via class uncertainty and region homogeneity," *IEEE Transactions on Pattern Analysis and Machine Intelligence* 23(2001): 689–706.

[106] Saha, P. and J. Udupa. "Iterative relative fuzzy connectedness and object definition: theory, algorithms, and applications in image segmentation," in *Proceedings of IEEE Workshop on Mathematical Methods in Biomedical Image Analysis*, (Hilton Head, South Carolina, 2000): 28–35.

[107] Saha, P, Udupa, J., and R. Lotufo. "Relative fuzzy connectedness and object definition: theory, algorithms, and applications in image segmentation," *IEEE Transactions on Pattern Analysis and Machine Analysis* 24(2002): 1485–1500.

[108] Sahiner, B., Chan, H., Petrick, N., Wei, D., Helvie, M., Adler, D. and M. Goodsitt. "Classification of mass and normal breast tissue: a convolution neural network classifier with spatial domain and texture images," *IEEE Transactions on Medical Imaging*, 15(1996): 598–610.

[109] Sahoo, P., Soltani, S. and A. Wong. "A survey of thresholding techniques," *Computer Vision, Graphics, and Image Processing* 41(1988): 233–260.

[110] Salembier, P. and M. Pardas. "Hierarchical morphological segmentation for image sequence coding," *IEEE Transactions on Image Processing* 3(1994): 639–651.

[111] Schenk, A, Prause, G. and H. Peitgen. "Local cost computation for efficient segmentation of 3D objects with live wire," in *Proceedings of SPIE: Medical Imaging* in *Proceedings of SPIE* 4322:1357–1364, 2001.

[112] Sethian, J. *Level Set Methods: Evolving Interfaces in Geometry, Fluid Mechanics, Computer Vision, and Material Sciences*. Cambridge: Cambridge University Press, 1996.

[113] Sethian, J. *Level Set Methods and Fast Marching Methods Evolving Interfaces in Computational Geometry, Fluid Mechanics, Computer Vision, and Materials Science*. Cambridge: Cambridge University Press, 1999.

[114] Shafarenko, L., Petrou, M., and J. Kittler. "Automatic watershed segmentation of randomly textured color images," *IEEE Transactions on Image Processing*, 6(1997): 1530–1544.

[115] Siddiqi, K., Lauziere, Y., Tannenbaum, A. and W. Zucker. "Area and length minimizing flows for shape segmentation," *IEEE Transactions on Image Processing* 7(1998):433–443.

[116] Spitzer, V., Ackerman, M., Scherzinger, A., and D. Whitlock, D. "The Visible human male: A technical report," *Journal of the American Medical Informatics Association* 3(2)(1996): 118–130.

[117] Staib, L. and J. Duncan. "Deformable fourier models for surface finding in 3D images," in *Visualization In Biomedical Computing 1992* (Proceedings of VBC92, Chapel Hill, NC 1992), R. A. Robb, ed. SPIE 1808(1992): 90–104.

[118] Staib, L. and J. Duncan. "Boundary finding with parametrically deformable models," *IEEE Transaction on Pattern Analysis and Machine Intelligence* 14(11)(1992): 1061–1075.

[119] Terzopoulos, D. "Regularization of Inverse Visual Problems Involving Discontinuities," *IEEE Transaction on Pattern Analysis and Machine Intelligence* 8(4)(1986): 413–424.

[120] Trivedi, M. and J. Bezdek. "Low-level segmentation of aerial images with fuzzy clustering," *IEEE Transaction on Systems, Man, and Cybernetics* 16(1986): 589–598.

[121] Udupa, J. "Interactive segmentation and boundary surface formation for 3D digital images," *Computer Graphics and Image Processing* 18(1982): 213–235.

[122] Udupa, J. and S. Samarasekera. "Fuzzy connectedness and object definition," in *Proceedings of SPIE: Medical Imaging* 2431(1995): 2–11.

[123] Udupa, J. and S. Samarasekera. "Fuzzy connectedness and object definition: Theory, algorithms, and applications in image segmentation," *Graphical Models and Image Processing* 58(1996): 246–261.

[124] Udupa, J., *et al.* "Methodology for evaluating image segmentation algorithms," in *Proceedings of SPIE: Medical Imaging* 4684(2002): 266–277.

[125] Urquhart, R. "Graph theoretical clustering based on limited neighborhood sets," *Pattern Recognition* 15(1982): 173–187.

[126] Veksler, O. "Image segmentation by nested cut," in *Proceedings of IEEE CVPR 2000* 1(2000): 339–344.

[127] Whitaker, R. "Volumetric deformable models: Active blobs," in *Visualization In Biomedical Computing 1994* (R. A. Robb, ed.), (Mayo Clinic, Rochester, Minnesota) SPIE (1994): 122–134.

[128] Wu, Z. and R. Leahy. "An optimal graph theoretic approach to data clustering: theory and its application to image segmentation," *IEEE Transactions on Pattern Analysis and Machine Intelligence*, 15(1993): 1101–1113.

[129] Xu, C. and J. Prince. "Snakes, shapes, and gradient vector flow," *IEEE Transactions on Image Processing* 7(1998): 359–369.

[130] Xu, C. and J. Prince. "Generalized Gradient Vector Flow External Forces for Active Contours," *Signal Processing* 71(2)(December 1998): 131–139.

[131] Yezzi, A., Kichenassamy, S., Kumar, A., Olver, P., and A. Tannenbaum. "A geometric snake model for segmentation of medical imagery," *IEEE Transactions on Medical Imaging* 16(2)(April 1997): 199–209.

[132] Zhu, Y. and Z. Yan. "Computerized tumor boundary detection using a Hopfield neural network," *IEEE Transactions on Medical Imaging* 16(1997): 55–67.

[133] Zhuge, Y. and J. Udupa. "Vectorial scale-based fuzzy connected image segmentation," in *Proceedings of SPIE: Medical Imaging* 4684(2002): 1476–1487.

[134] Zucker, S. "Region growing: childhood and adolescence," *Computer Graphics Image Processing* 5(1976): 382–399.

Index

active contours, 209, 220, 354
active shape and appearance, 354
adaptive thresholding, 22
affine transform, 264
    Jacobian, 267
    optimizer update, 265
    parameter scaling, 266
aliasing, 37, 39
angiography
    CT, 141, 164
    digital subtraction, 25
    MR, 141
    MRI, 163
anisotropic
    diffusion, 105, 212
    scaling transform, 256
atlas-based segmentation, 317

B-spline, 289
balloon force, 230
Bayesian theory, 55, 114, 181, 185, 321
binomial kernel, 30
boundary conditions, 105
box filter, 41
breast cancer, 167
brightness, 22

Cauchy-Navier, 326, 327, 333
central differences, 21, 30, 110, 194, 200, 280
central limit theorem, 30
change detection
    as registration application, 239

clustering, 355
color map, 24
computational anatomy, 324
computerized tomography, 21
conductance, 106, 107
continuous functions, 19
contrast, 22
contrast enhancement, 35
convolution, 25, 66, 104
    convolution theorem, 38
    for differentiation, 28, 194, 226
    properties of, 27
CT, 5, 7, 21, 22, 44, 50, 124, 141, 164, 240, 295, 321
    angiography, 164
    colonoscopy, 170
CTA, 164
curvature, 112, 113, 126, 194, 211
    flow, 113, 114, 205
    from discrete grids, 202
    Jacobian, 33
    Laplacian, 112
    mean, 195, 202, 205, 212, 308
    total, 195, 211, 212

deformable model, 352, 358
deformable registration, 302, 307
    demons algorithm, 315, 317
deformable surfaces, 195
delta function, 39
demons
    algorithm, 315
    registration, 307

Demons algorithm, 315, 316, 324
difference of Gaussians, 33, 34
    DOG, 33, 34
        difference of offset Gaussians, 31
        DoOG, 31
differential geometry, 194
differentiation, by convolution, 28, 194,
        226
diffusion equation, 104
discrete Fourier series, 45
discrete Fourier transform, 44
discrete functions, 19
discrete grid, 244

elastic, 317, 326, 327, 329, 332, 336,
        338
elementary quaternions, 274
energy functionals, 114
energy minimization
    MRF, 186
Euler-Lagrange, 314, 315, 321, 323
Eulerian frame, 311, 317, 326, 327, 332
expectation-maximization, 78

features, 47
    feature space, 48
FEM-based registration, 307
FFT, 45
FGMM, 59
filtered back projection, 44
filtering, 4
    linear, 26, 27, 38, 104
    nonlinear, 103, 105
    unsharp masking, 34
finite difference registration, 307
finite differences, 110, 200
finite element, 126, 311, 314, 323, 327,
        331, 334, 340
fluid, 311, 317, 326, 332, 336
Fourier
    discrete Fourier transform, 44
    fast Fourier transform, 45
    Fourier series, 44
    Fourier series, discrete, 45
    Fourier transform, 104
    Fourier transform, n-D, 42

Fourier transform, projection, 44
Fourier transform, rotation of, 43
rotational invariance of, 43
functions, continuous vs. discrete, 19
fuzzy connectednes, 131
fuzzy connectedness, 127, 132, 139, 352,
        358

gain field correction, 50
Gaussian, 26, 39, 41, 42, 226, 289
    classifiers, 63
    difference of, 34
    distribution, 48, 50
    extruded distribution, 50
    kernel, 39, 41, 104, 320
    mixture models, 78
Gaussian blurring, 26, 103, 194
Gibbs prior, 182, 233, 352
gradient, 30, 194, 225
    descent, 115
    magnitude, 30
    vector flow, 226
graph-based segmentation, 355
grid mapping, 244

Hadamard, 309
hard connectedness, 134, 138, 141
Hessian, 79, 195
histogram equalization, 24
Hotelling matrix, 61
Hounsfield units, 9, 21
hybrid segmentation, 351

identity transform, 251
image interpolation, 278
    B-spline, 279, 281, 284
    in registration problems, 242
    linear, 278, 281, 284, 295, 302
    nearest neighbor, 278, 281
image processing, 19
    medical, 4
impulse, 39
intensity-based registration, 241
inter-subject registration, 240

interpolation
    B-spline, 279, 280
    image, 242, 278
    linear, 39, 41, 278
    nearest-neighbor, 39–41, 278
isosurface, 194

Jacobian, 109, 248, 324, 326, 334, 336
    matrix, 31, 249

k-means, 59, 60, 74, 80, 81, 87
kernel, 25
    binomial, 30
    box, 41
    Gaussian, 41
    pyramid, 41
kNN, 59, 64, 69, 124, 355

Lagrangian frame, 310, 317, 326, 328
landmark, 338
landmark-based registration, 240
Laplacian, 31, 33, 112
    zero crossing, 33
level set, 126, 194, 355
    surface processing, 210
level sets, 221
linear operator, 19, 20, 104, 313, 323,
        324, 326
low pass, 104

magnetic resonance imaging, 5, 10
mammography, 167
marching cubes, 226, 230, 357
Markov random fields, 125, 181, 228,
        356, 357
maximum likelihood estimation, 79
mean curvature, 195, 202, 205
mixture modeling, 59
MLP, 59
morphometry, 324, 336
MRA, 163
MRF, 125, 181, 228, 356, 357
    cliques, 184
    Gibbs priors, 183
    posterior energy functions, 185
MRI, 5, 10, 49, 50, 93, 124, 138, 141,
        162, 163, 168, 188, 232, 240,
        281, 290, 291, 295, 317, 321,
        336, 355
    angiography, 163
multi-modality registration, 240, 295
multi-resolution registration, 295
multiple sclerosis, 161, 162
mutual information, 241, 288, 295, 302,
        319, 320
    density estimation, 289
    entropy, 288
    Parzen windowing, 289
    probability density function (pdf),
        288

narrow-band method, 202
Navier, 326
NMR, 5
noise reduction, 25, 26
non-rigid registration, 307–309, 312, 321,
        334, 338
nonlinear operator, 103
normal, 194
normalized cross correlation, 330
nuclear medicine, 13
numerical methods, 199
Nyquist
    Nyquist criterion, 40
    Nyquist frequency, 40, 44

operators, linear, 20
optical flow, 312, 313, 315, 316, 319,
        323, 326, 327
optimization
    in registration problems, 242, 295

partial differential equation, 34, 103
Parzen windowing, 59, 66, 68, 87, 124,
        289, 320
PDE, 34, 103
penalty, 114
perceptron, 59
PET, 15, 240, 321
pyramid filter, 41

quaternion composition
    matrix notation, 277

quaternions, 267
    elementary quaternions, 274
    matrix representation, 277
    scalar, 268
    tensor, 268
    versor, 268

radiation treatment planning
    as registration application, 240
Radon transform, 44
region-based segmentation, 355
registration, 4, 181, 239, 307
    affine, 241
    applications of, 239
    concepts, 240
    coordinates origin, 254
    criteria, 240
    deformable, 241, 302, 307
    demons, 307
    FEM-based, 307
    finite difference, 307
    image center, 254
    image interpolation, 242
    information theory, 288
    intensity-based, 241
    inter-subject, 240
    interpolators, 278
    landmark-based, 240
    metrics, 279
    multi-modality, 240, 295
    multi-resolution, 295
    non-rigid, 307
    of serial images, 239
    optimizers, 294, 295
    parameter optimization, 242
    progressive, 248
    rigid, 241
    segmentation-based, 240
    spatial transformation, 241, 243
    transforms, 243
    validation, 301
    variational, 307
regularization, 28, 310, 312, 315, 316,
        321, 324, 327, 329, 334, 338
rigid transform, 264
    in 3D, 273

rotation in 2D, 257
rotation optimization
    exponential factor, 260
rotation transform, 257

sampling, 44
sampling artifact, 39
scale space, 42
scaling optimization
    exponential factor, 260
scaling transform, 253
scaling transform optimization, 255
scatterplots (scattergrams), 53
segmentation, 4, 121, 317
    atlas based, 317
    level sets, 209
segmentation-based registration, 240
separability, 42
Shannon sampling theorem, 40
shape matrix, 195, 211
similarity metrics
    difference density, 284
    mutual information, 288, 295, 302
    mean squares, 280
    normalized correlation, 281
    registration framework and, 279
similarity transform, 264
    in 3D, 276
    matrix representation, 277
sinc function, 41
smoothing, 26
snakes, 209, 220
sparse-field method, 204
spatial transformation
    in registration problems, 241
SPECT, 13, 15, 240
spherical linear interpolation, 274
statistical pattern recognition, 47
surface, 308
    curvature, 194
    normal, 194

tangent plane, 195
theorem
    central limit, 30

convolution, 38
    Shannon sampling, 40
thresholding, 21
    adaptive, 22
Tikhonov, 310, 313, 314
total curvature, 195
transforms
    Jacobian, 248, 281, 284, 286
    registration framework and, 243
    rotation, 257
    taxonomy, 251
translation transform, 252

ultrasound, 15
unsharp masking, 34
up-wind method, 200

variational calculus, 310–312, 314, 315,
    321, 323, 326, 328, 333, 334
variational partial differential equations,
    114
variational registration, 307
versor
    exponentiation, 272
    numerical representation, 274
    optimization, 271
versor composition
    non-commutative, 270
Voronoi diagram classification, 352, 358

windowing
    intensity, 22
    Parzen, 59, 66, 68, 87, 320

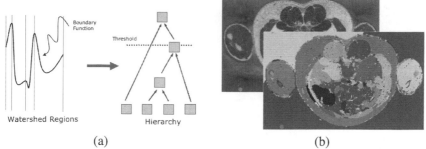

(a)   (b)

Plate I. (From Figure 5.1) Watershed segmentation: (a) a 1D height field (image) and its watersheds with the derived hierarchy; (b) a 2D color data slice from the NLM Visible Human Project and a derived watershed segmentation. [Example courtesy of Ross Whitaker and Josh Cates, Univ. of Utah]

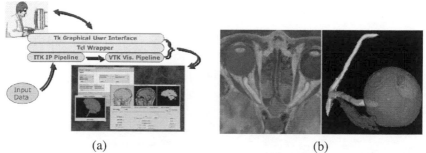

(a)   (b)

Plate II. (From Figure 5.2) An ITK watershed segmentation application with visualization provided by VTK and with the user interface provided by Tcl/TK. This example shows how ITK can be be integrated with multiple packages in a single application: (a) data flow and system architecture for the segmentation user application; (b) anatomic structures (the rectus muscle, the optic nerve, and the eye) outlined in white (shown in 2D but extending in the third dimension) and the resulting visualization.) [Example courtesy of Ross Whitaker and Josh Cates, Univ. of Utah]

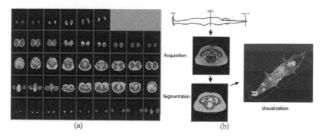

Plate III. (From Figure 12.12) Obesity Research Center: integrated system for quantification of adipose tissue from whole body MRI scans. [Courtesy of Dr. Heymsfield]

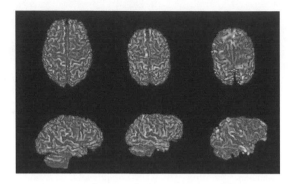

Plate IV. (From Figure 11.1) The gray-white interfaces of a human (left) and a chimp (far right) are registered by their mean curvature. The result is shown in the center column.

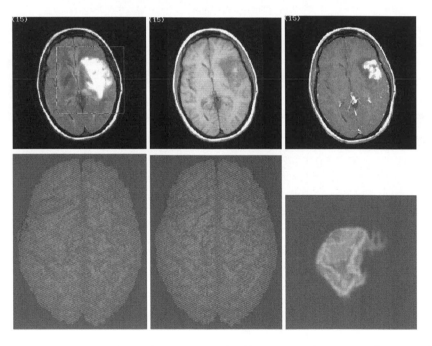

Plate V. (From Figure 6.8) Brain tumor MRI FLAIR sequence. Top row: a slice each of a FLAIR scene, a T1-weighted scene, and a T1 with Gadolinium scene. Bottom row: a surface rendition of brain plus edema segmented from the FLAIR scene; brain plus edema and enhancing aspect of tumor segmented from the two T1 scenes; edema plus enhancing tumor volume rendered.

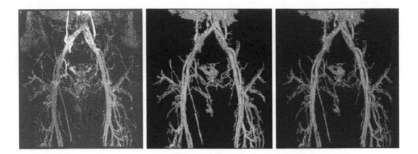

Plate VI. (From Figure 6.6) A MIP rendition of an MRA scene acquired with the blood-pool contrast agent MS325 (left), the entire vascular tree segmented via generalized fuzzy connectedness and volume rendered (middle), and the arteries and veins separated by using iterative relative fuzzy connectedness (right.)

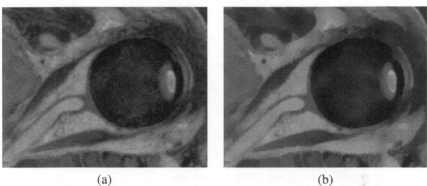

(a)   (b)

Plate VII. (From Figure 4.4) The result of vector valued anisotropic diffusion on a color image: (a) original; (b) filtering with dissimilarity defined as difference of the first two principal components of the Jacobian, with $t = 2.4$ and $K = 0.7$.

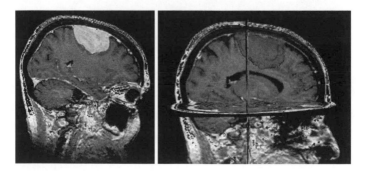

Plate VIII. (From Figure 8.12) Segmentation of MRI data using level sets: (a) a 2D slice from an MRI image; (b) 3D rendering of the level-set tumor segmentation.

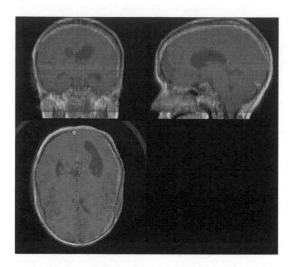

Plate IX. (From Figure 10.34) Registration start with the center of the CT image initially aligned with the center of the MR-T1 image. The CT image is shown as a color overlay on top of a gray-scale MR-T1 image. The yellow color corresponds with the bright intensity regions of the CT image.

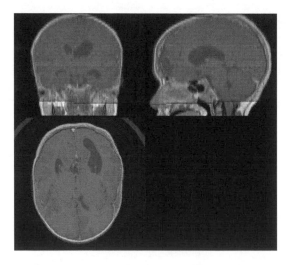

Plate X. (From Figure 10.35) Results of the multi-resolution registration process described in Section 10.4.1. A color overlay on top of a gray-scale MR-T1 image; the yellow color corresponds with the bright intensity regions of the CT image.

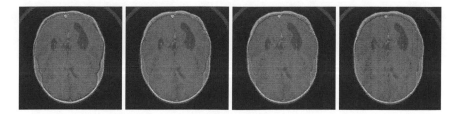

Plate XI. (From Figure 10.36) Results of the multi-resolution registration process in Section 10.4.1 after registration at the first (coarsest) level (left), the second level (middle-left), the third level (middle-right) and the fourth (finest) level (right). The CT and MR-T1 images are shown as a checkerboard where each block alternately displays data from each dataset.

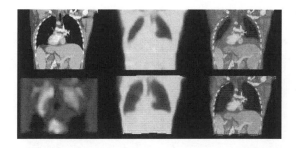

Plate XII. (From Figure 11.8) PET-CT image fusion. The original CT fixed image is at top left. The Jacobian of the transformation is at bottom left. The original PET transmission image is at top center. The warped PET transmission is at bottom center. The overlay of original images is at top right. The warped PET is fused with the CT at bottom right.

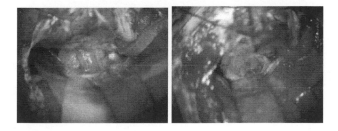

Plate XIII. (From Figure 12.5) 3D model of the tumor overlaid over two snapshots from a surgery for the removal of an anterior skull base meningioma: (a) before and (b) after tumor removal. [Courtesy of Dr. J. Bruce and Dr. A. D'Ambrosio]

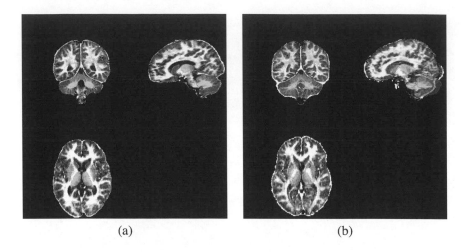

(a)   (b)

Plate XIV. (From Figure 11.6) For the atlas image (a), several subcortial and cerebellar structures were hand segmented. The segmentations are shown as an overlay on top of the atlas image (left) and subject image (b).

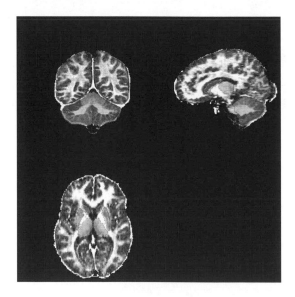

Plate XV. (From Figure 11.7) A multi-resolution demons algorithm was used to non-rigidly register the atlas image onto the subject image. The output deformation field was then used to warp the atlas image, producing the top image. The warp applied to the atlas labels produced the overlay in the bottom image, shown over the subject image.

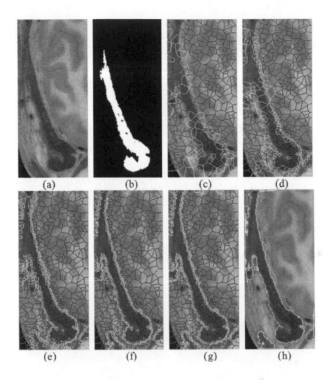

Plate XVI. (From Figure 12.6) Automated segmentation of temporalis muscle: (a) color Visible Human Male slice; (b) a fuzzy connected component; (c)-(g) iterations of the VD-based algorithm; (h) an outline of the boundary.

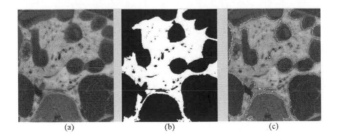

Plate XVII. (From Figure 12.7) FC-VD segmentation of Visible Human Male adipose tissue in the abdomen region: (a) input data, (b) FC; (c) VD final boundary.

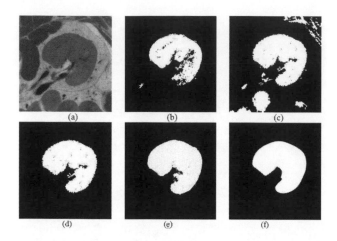

Plate XVIII. (From Figure 12.8) Segmentation of the Visible Human Male left kidney: (a) input data; (b) FC; (c) VD; (d) VD-CC; (e) DM; (f) hand segmentation.

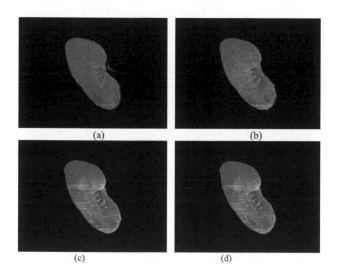

Plate XIX. (From Figure 12.9) 3D segmented and visualized with *Vesalius^{TM}* Visualizer, the Visible Human Male left kidney. 3D models of: (a) FC; (b)VD-CC; (c) DM; (d) hand segmentation.

Printed and bound by CPI Group (UK) Ltd, Croydon, CR0 4YY

29/10/2024

01780546-0001